T0329118

Working with Dynamic Crop Models

ACKNOWLEDGMENTS

The authors gratefully acknowledge their home institutions:

- INRA (Institut National de la Recherche Agronomique)
- The University of Florida, Agricultural and Biological Engineering Department
- ACTA, head of the network of French Technical Institutes for Agriculture

The following projects provided an invaluable framework and support for discussing and applying methods for working with dynamic crop models:

- Réseau Mixte Technologique Modélisation et Agriculture, funded by a grant from the French Ministry for Agriculture and Fisheries
- The project 'Associate a level of error in predictions of models for agronomy and livestock' (2010–2013), funded by a grant from the French Ministry for Agriculture and Fisheries
- AgMIP, the Agricultural Model Intercomparison and Improvement Project
- The FACCE-JPI project MACSUR (Modelling European Agriculture with Climate Change for Food Security)

In addition, we owe thanks to:

- Luc Champolivier, CETIOM, for providing field water content unpublished data
- Sylvain Toulet, intern at INRA in 2012, Juliette Adrian, intern at ACTA in 2013, and Lucie Michel, intern at ACTA-INRA, for their contributions to the examples
- Senthold Asseng of the University of Florida, for working with the authors and making use of the new material in this book in the graduate course 'Simulation of Agricultural and Biological Systems'
- All the students on the courses we have given on crop modeling, who have enriched our thinking with their questions and remarks.

Cover image
Field of lavender crops. Copyright ITEIPMAI (Institut technique interprofessionnel des plantes à parfum médicinales et aromatiques), www.iteipmai.fr.

Working with Dynamic Crop Models

Methods, Tools, and Examples for Agriculture and Environment

Second Edition

Daniel Wallach

AGIR Agrosystèmes et développement territorial
INRA
Toulouse, France

David Makowski

UMR agronomie
INRA AgroParisTech
Thiverval-Grignon, France

James W. Jones

Agricultural and Biological Engineering Department
University of Florida
Gainesville, Florida, USA

François Brun

ACTA - Réseau des Instituts des filières animales et végétales
Toulouse, France

AMSTERDAM • BOSTON • HEIDELBERG • LONDON
NEW YORK • OXFORD • PARIS • SAN DIEGO
SAN FRANCISCO • SINGAPORE • SYDNEY • TOKYO
Academic Press is an imprint of Elsevier

Academic Press is an imprint of Elsevier
32 Jamestown Road, London NW1 7BY, UK
225 Wyman Street, Waltham, MA 02451, USA
525 B Street, Suite 1800, San Diego, CA 92101-4495, USA

Notice
No responsibility is assumed by the publisher for any injury and/or damage to persons
or property as a matter of products liability, negligence or otherwise, or from any use
or operation of any methods, products, instructions or ideas contained in the material
herein. Because of rapid advances in the medical sciences, in particular, independent
verification of diagnoses and drug dosages should be made

British Library Cataloguing-in-Publication Data
A catalogue record for this book is available from the British Library

Library of Congress Cataloging-in-Publication Data
A catalog record for this book is available from the Library of Congress

ISBN : 978-0-12-397008-4

For information on all Academic Press publications
visit our website at elsevierdirect.com

Typeset by MPS Limited, Chennai, India
www.adi-mps.com

Printed and bound in United States of America

13 14 15 16   10 9 8 7 6 5 4 3 2 1

  Working together
to grow libraries in
developing countries

www.elsevier.com • www.bookaid.org

Contents

Section 2
Methods

Preface

This second edition covers, for the most part, the same subjects as the first, but other than that it is a very different book. Such methods for working with system models as uncertainty and sensitivity analysis, parameter estimation (model calibration), and model evaluation and data assimilation are all still presented. One major difference is that this edition includes details of how to use R, a very widely used free programming language and platform, to implement all of the methods that are presented. A second major difference is in the examples that are used throughout the book. At the time of the first edition, there were relatively few examples of the application of the proposed techniques to dynamic system models in agronomy. The situation has evolved, and this allows us to show throughout how the methods are applied in real cases. In addition, every chapter has been extensively rewritten, and a substantial amount of new material has been added. Finally, introductory chapters have been added, so that this book can now be used as a self-contained introduction to modeling and modeling methods.

This book is meant as both a textbook for students and a resource for researchers who want to learn the methods for working with dynamic system models. The examples are from agronomy (models for crop growth and water balance), environment (models for soil carbon and soil nitrogen), ecology (population models), and plant pathology (disease dynamics) but the principles and methods apply to dynamic system models in general.

The very wide range of possible applications of agronomic dynamic system models has been discussed and illustrated in numerous journal papers and books. Nonetheless, there is a certain sense of disappointment and impatience that real-world applications have been nowhere near as widespread as imagined. We suggest that one major cause lies in the lack of information, methods, and tools for working with these models.

Most of the modeling efforts in the past have been devoted to better understanding of the processes and interactions involved in the complex systems treated in agro-ecology. This has led to major advances in our knowledge of these systems and our capacity to represent that knowledge in the form of mathematical equations. This aspect of modeling is far from over. As just a few examples we can mention studies of the mechanisms of drought resistance in plants, the effect of increased CO_2 concentration on plant processes, the relationship between QTLs and plant behavior in the field, and the dissemination of pollen and thus of genetic modification between fields.

However, additional research aimed at understanding processes is not suf-ficient to produce models that are useful for solving real-world problems. The difficulty is that for systems as complex as those studied in agronomy and in environmental sciences, models are necessarily only rough approxi-mations to the real world. This has two implications. First, it becomes very important to reduce prediction error as far as possible. Better understanding of processes is of course a major path for improving predictions, but parame-ter estimation is another very important path since the quality of model pre-dictions depends crucially on both the functional forms of the equations and on the values of the parameters. Injecting additional real-time information (data assimilation) can also lead to a major improvement in prediction. Parameter estimation and data assimilation are covered in detail in this book. Secondly, it is essential to obtain realistic estimates of the error or uncer-tainty in model predictions. In fact, predictions are not very useful if one knows that the errors are important but one has no estimate of that error. Uncertainty analysis, sensitivity analysis, and model evaluation, all concerned with quantifying the uncertainty in model predictions, are also covered in detail in this book.

Even if one accepts that better methodology could play a major role in leading to more useful models, why a book specifically devoted to method-ology for dynamic system models? After all, all of the methods are covered elsewhere. The answer is that it seems important to bring together in one place state-of-the-art methods that are useful for these models, and to illus-trate them with examples from agro-ecology. A second motivation is that it is not sufficient to have a basic understanding of these methods; it is neces-sary to be able to apply them in practice. Previously, this required pro-gramming the methods oneself, a difficult undertaking. This is no longer the case, thanks to the R statistical software. This software is free, and there are packages that implement all of the methods presented in this book. This book presents in detail the R implementation of the methods in question.

The organization of this book is as follows: The first section covers pre-liminary material. Its first chapter explains the principles of dynamic sys-tem models. Several examples of dynamic system models are presented in detail, to illustrate concepts for developing such models and the types of models in question. The other chapters concern statistics (Chapter 2), the R language (Chapter 3), and simulation (Chapter 4). Together these chapters provide the background for understanding the rest of this book. This limits the prerequisites for this book to very basic notions of statistics and programming.

Section 2 concerns methods for working with dynamic system models. The order of the chapters follows approximately the order in which the meth-ods would be used in an overall modeling project. Chapter 5 covers uncer-tainty analysis and sensitivity analysis. These methods allow one to explore the behavior of the model, get an idea of the uncertainties in prediction, and quantify different contributions to uncertainty. Chapter 6 concerns parameter estimation with classical statistical methods, also sometimes referred to as model calibration. The emphasis is on how statistical principles can be applied to complex system models. Chapter 7 concerns Bayesian parameter estimation. Currently, this method of parameter estimation is quite rare for dynamic system models, but it is reasonable to suppose that in the future it will attain much more importance, as it has in statistics in general. In this approach one uses prior information about the parameters, which is very well adapted to dynamic system models where there is information based on studies of the individual processes. Chapter 8 concerns data assimilation, which applies to the case where one has real-time observations that can be used to improve model predictions for the specific situation that is observed. Chapter 9 concerns the problem of model evaluation. A major emphasis is on the distinction between apparent error and prediction error, and on estimating prediction error. However, the chapter also covers other aspects of model evaluation, such as evaluating how well a model classifies outcomes, or how to evaluate probabilistic predictions. Chapters 5–9 describe and explain the methods in question, using examples of applications to models in agronomy, and present detailed explanations of how the methods can be implemented with R. The last chapter, Chapter 10, applies methods from the previous chapters to a simple dynamic system model, to show how the methods can be combined in an overall analysis.

There is an R package (called ZeBook) that accompanies this book. Like all R packages, this package can be freely obtained from any of the R download sites. It is most easily obtained by using the packages tab within the R software. The package contains many of the R programs used to create the examples in the book. In each case the model is coded as a function called by a main program, so that one could easily adapt these programs to apply the methods in question to one's own model. The package also contains code and data for use with the exercises, to minimize the work not directly related to the main point of the exercises. The package contents are self-documented and are also described in Appendix 2 in this book. The package contents will evolve over time, to include additional models and data.

The book can be used for a one semester postgraduate course. The chapters can then be presented in the order in which they appear in this book. This corresponds to approximately one week per chapter, which implies a choice in each chapter of the major points. The self-learner has more leeway. The basic material in Section 1 is needed for the rest of the book, but of course one can skip some of those chapters if one is already familiar with the material. In Section 2 one can study the chapters in any order, since each chapter is self-contained.

In writing this book, D. Wallach took major responsibility for the chapters on statistical notions, frequentist parameter estimation, and model evaluation, D. Makowski was in charge of the chapters on uncertainty and sensitivity analysis, Bayesian parameter estimation, and data assimilation, Jim Jones handled the chapters on basics of system models and simulation, and François Brun handled the chapters on the R language and the case study, in addition to helping everywhere with the R programs and with preparing the R package.

Daniel Wallach
David Makowski
James Jones
François Brun

Section 1

Basics

## Chapter 1

Basics of Agricultural System Models

1. INTRODUCTION

Agricultural systems are complex combinations of various components. These components contain a number of interacting biological, physical, and chemical processes that are manipulated by human managers to produce the most basic of human needs—food, fiber, and energy. Whereas the intensity of management varies considerably, agricultural production systems are affected by a number of uncontrolled factors in their environments, being exposed to natural cycles of weather, soil conditions, and pests and diseases. In comparison to physical and chemical systems, agricultural systems are much more difficult to manage and control because of the living system components that respond to their physical, chemical, and biological environmental conditions in highly nonlinear, time-varying ways that are frequently difficult to understand. These interactions and nonlinearities must be considered when attempts are made to understand agricultural systems' responses to their environments and management. Dynamic system models are increasingly being used to describe agricultural systems to help scientists incorporate their understanding of the interactions among components for use in predicting performance of agricultural systems for better achieving goals of farmers (food production, profits) and of society (environmental quality, sustainability).

In this chapter, we present concepts of system models, with examples of various agricultural system models. We will also present methods for developing system models. Many agricultural system models already exist, and in some cases, there are multiple models of the same agricultural system. Examples include multiple models of cropping systems (Rosenzweig et al., 2013; Rotter et al., 2012); there are over 27 wheat crop production system models (Asseng et al., 2013). Thus, someone interested in analyzing a particular system could select from existing models and not have to worry about how the model was developed. However, it is also very important for model users to understand the models if they adopt an existing one. An understanding of model development methods will allow them to attain a deeper

Working with Dynamic Crop Models. DOI: http://dx.doi.org/10.1016/B978-0-12-397008-4.00001-0

understanding of a particular model's capabilities and limitations than they would otherwise have. Each model is a simplification of the real system, with assumptions that may or may not be acceptable for a particular application. It will help model users understand the assumptions and relationships used in specific models and enable them to better judge the suitability of a model for their purpose, thereby reducing the potential for using an existing model for purposes for which it was not designed or evaluated. It will also help model users determine if modifications are needed to an existing model, and if so, it will give them a basis for evaluating how this can be done.

System models can be viewed in two different, complementary ways. First, a model can be treated as a system of differential or difference equations that describe the dynamics of the system. Second, the model can be treated as a set of static equations that describe how responses of interest at particular times depend on explanatory variables. We present and discuss these two viewpoints in this chapter. As we shall see, the different methods described in this book may call for one or the other of these viewpoints.

In this chapter, we start off in Section 2 by presenting general systems concepts and definitions that are needed in modeling systems. Then in Section 3, we go through the process of developing a model of a system with two simple interacting components that will help give students an intuitive understanding of the system modeling process. In Section 4, we will present other forms frequently used to present agricultural systems models in various types of applications. Example system models will be presented in Section 5 for several important agricultural system components, demonstrating some of the key features and relationships used in model development.

2. SYSTEM MODELS

A system is a set of components and their interrelationships that are grouped together by a person or a group of persons for the purposes of studying some part of the real world. Usually, a group of persons work together to define the system that they intend to analyze, and in many cases individuals are from different disciplines because of the scope of the system they intend to study. The selection of the components to include in a particular system depends on the objectives of analysts and on their understanding and perspectives of the real world. A system may have only a few components that interact or it may have many components that interact with each other and that may be affected by factors that are not included in the system. Conceptually, a system may consist of only one component; however, in this book we focus on the more common situation that exists in agricultural systems where the complexities of interactions among components are required to understand the performance of the system being studied.

2.1 Systems Approach

A systems approach starts with the definition of the system to be studied and includes development of a system model and the use of that model to analyze the system. Thus, it is important that all participants in a study have the same understanding of the objectives of the analyses that will be performed and of the system and its components. The systems approach provides an effective framework for interdisciplinary research, and thus it is important that the analyses include all disciplines at the start of a project and that an explicit definition of the system, its components, the interactions among system components, and the interactions of those components with other factors are agreed upon and communicated.

2.2 System Environment and Boundary

A system has an environment as well as components and interactions among them. The boundary of a system separates the system components from its environment. A system environment may include anything in the real world except the components being studied. We usually describe a system environment as factors that affect the behavior of components in the system but are not affected by them. For example, many agricultural systems are affected by weather conditions, but regardless of how the system components behave, they do not affect weather. A cropping system (Figure 1.1) may include soil and a crop which interact with each other and with the environment. The environment may have several variables that define the characteristics of the environment that directly influence some of the system component processes.

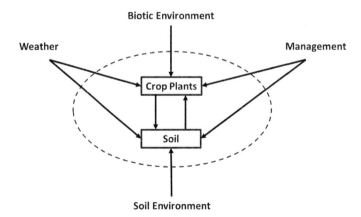

FIGURE 1.1   Schematic diagram of a cropping system model showing interactions between soil and crop components in the system and influences of explanatory variables from the environment. In this system, it is assumed that weather, management, pests in the biotic environment and soil variables in the soil environment (e.g., below 1 m in depth) influence the system but are not affected by system components.

The soil and crop system is usually defined as a homogenous field or representative area in the field that is exposed to weather throughout the course of a season. The boundary of this system would be an imaginary box immediately surrounding this representative uniform area in the field (e.g., with dimensions of 1 ha by 1 ha or of 1 m by 1 m in area), and a lower boundary in the soil at a 1 m depth. The environment would be everything in the soil and atmosphere outside of this imaginary box that would affect the soil and crop behavior in the system. Note that each system component may have multiple variables that describe the conditions of system components at any particular time. For example, the soil component may include soil water content, mineral nitrogen content, and soil organic matter that change from day to day. The crop component may include above ground biomass, leaf area, and depth of roots in the soil.

This example also illustrates another important point; assumptions have to be made when choosing a system's components and its environment. Here it is assumed that the temperature and humidity in the canopy are equal to the values of these variables in the air mass above the canopy. In reality, the soil and crop conditions affect canopy temperature and humidity through the transfer of heat and water vapor into the air above the crop, and thus the system does affect the immediate environment of the crop. This interaction may be important in many situations such that the system would include the canopy temperature and humidity. In the example, the assumption is that those effects are small relative to the influences of external factors, such as that of the external air mass. This assumption is made in most, but not all, existing cropping system models.

This example of a system and its environment can also be used to illustrate the implications of incorporating additional components into a system. If canopy air temperature and humidity are added to the soil and crop components, the system will include another component, the canopy air environment. This will cause the model to become more complex in that explicit mathematical relationships would be needed to model the dynamics of canopy air temperature and humidity in addition to the soil and crop conditions. The environment for this expanded system would still include weather conditions above the crop canopy.

2.3 System Model and Simulation

2.3.1 System Model

A system model is a mathematical representation of the system, including all of the interrelationships among components and effects of the environment on those components. The model description includes all of the equations of processes that cause components to change, all of the environmental variables that influence components in the system, all important properties of the system's components, and all of the assumptions that were made to develop a particular model. It is common to find two models of the same system that are very different due to choices made by those developing each model. For example, the 27 wheat models used in the Asseng et al. (2013) study differ in the mathematical formulations used to model the same overall components and in the assumptions that were made to develop each model. This is true even though most of these wheat crop models have the same components and respond to the same environmental factors. The differences in the models are due to differences in assumptions among models about what controls the development and growth processes in the crop and how soil water and nitrogen change over time. These differences lead to differences in mathematical functions used to represent system dynamics and even to the variables used to describe the status of the crop and soil conditions at any point in time.

2.3.2 Simulation

Simulation refers to the numerical solution of the system model to produce values of the variables that represent the system components over time. Although in some literature, a model is referred to as a "simulation model," this terminology hides an important and distinctive difference between a system model and its solution. Computer code should be developed based on the mathematical description and explicit assumptions. Otherwise, the assumptions and relationships may be hidden in the computer code and not easily understood or communicated to each modeling team member and to the outside world. Simulation is discussed in more detail in Chapter 4.

2.3.3 General Form of a Dynamic System Model

The general form of a dynamic system model in continuous time is:

$$\frac{d(U_1(t))}{dt} = g_1(U(t), X(t), \theta)$$

$$\vdots \tag{1}$$

$$\frac{d(U_S(t))}{dt} = g_S(U(t), X(t), \theta)$$

Where t is time, $U(t) = [U_1(t), \ldots, U_S(t)]^T$ is the vector of state variables (defined below) at time t, $X(t)$ is the vector of environmental variables at time t (sometimes referred to as exogenous, forcing, or driving variables), θ is the vector of system component properties that are included in the model to compute rates of change, and $g_1, g_2, \ldots, g_S$ are mathematical relationships that describe the relationships among state variables, parameters, and environmental variables. The state variables $U(t)$ could include, for example, leaf area index (leaf area per unit soil area), crop biomass, root depth, soil water content in each of several soil layers, etc. The explanatory variables $X(t)$ typically include climate variables (such as daily maximum and minimum temperatures) and management variables (such as irrigation dates and amounts). As discussed in section 2.5 below, in the rest of this book we will use $X(t)$ and θ

to refer to explanatory variables and parameters, respectively, which is not quite the same definition as above.

The model of Equation (1) is a system of first order differential equations that describe rates of changes in each of the state variables of the system. It is dynamic in the sense that the solution (simulation) of the system of equations describes how the state variables evolve over time. It describes a system in the sense that there are several state variables that interact.

Continuous time dynamic systems are sometimes modeled using discrete time steps; in this case, the models are mathematically represented as a set of difference equations. We will later show how this is useful in simulation of the dynamic system model and the interpretation of dynamic models as response functions. By writing the left-hand side of Equation (1) as a discrete approximation of the derivative, it is straightforward to develop the general form of a dynamic system model in discrete time:

$$U_1(t + \Delta t) = U_1(t) + g_1[U(t), X(t), \theta]\Delta t$$
$$\vdots$$

$$(2)$$

$$U_S(t + \Delta t) = U_S(t) + g_S[U(t), X(t), \theta]\Delta t$$

There is a very important concept underlying the use of Equations (1) and (2) to represent dynamic systems. Note that numerical methods are almost always needed to solve for behavior of system model variables over time, and Equation (2) is a highly useful and simple approach for solving dynamic system models. However, if the intent is to closely approximate the exact mathematical solution of a dynamic system model (i.e., if it could be solved analytically), one needs to select the time step Δt such that the numerical errors associated with approximating continuous time by discrete time steps are acceptable. Operationally when solving the model, one can evaluate the effects of choosing different values of Δt on important model results. On the other hand, model developers may select the Δt when developing the model because of the level of understanding of processes in the system and on the availability of environmental variables. In this case, all of the model equations and explanatory variables have to be developed specifically for the chosen time step, and this may require different approximations to the processes than those used when solving a model using continuous time. The mathematical form of such discrete time models may be represented as in Equation (2), and such models were referred to as 'functional models' by Addiscot and Wagenet (1985). An example is modeling soil water flow vs. soil depth and over time using the Richards equation (continuous time model) vs. using a so-called tipping bucket approach (with a daily time step). This will be discussed more in Chapter 4.

In some agricultural system models, such as many dynamic crop models, Δt is one day. One reason for this choice of time steps in many crop models is that highly important weather data may be available only at a daily time step (e.g., daily rainfall). In this case, the Δt on the left side becomes 1 and it disappears on the right-hand side of the equation. Note that Equation (2) is an approximation to the continuous model form. The choice of difference equations to model dynamic systems may mean that the model developer must use functional relationships that approximate the underlying physical, chemical, or biological processes involved. Thus, the development of functions (g_i) that compute changes in the state variables need to take into account the time step used in discrete time models.

2.4 State Variables $U(t)$

State variables are quantities that describe the conditions of system components at any particular time. These state variables change with time in dynamic system models as components interact with each other and with the environment. A system may have only a few or many state variables. State variables play a central role in dynamic system models. The collection of state variables determines what is included in the system under study. A fundamental choice is involved here. For example, if it is decided to include soil mineral nitrogen within the system being studied, then soil mineral nitrogen will be a state variable and the model will include an equation to describe the evolution over time of this variable. If soil mineral nitrogen is not included as a state variable, it could still be included as an explanatory variable, i.e., its effect on plant growth and development could still be considered. However, in this case the values of soil mineral nitrogen over time would have to be supplied to the model; they would not be calculated within the model. The limits of the system being modeled are different in the two cases.

The choice of state variables is also fundamental for a second reason. It is assumed that the state variables at time t give a description of the system that is sufficient for calculating the future trajectory of the system. For example, if only root depth is included among the state variables and not variables describing root geometry, the implicit assumption is that the evolution of the system can be calculated on the basis of just root depth.

Furthermore, past values of root depth are not needed. Whatever effect they had is assumed to be taken into account once one knows all the state variables at time t. Given a dynamic model in the form of Equation (1) or (2), it is quite easy to identify the state variables. A state variable is a variable that appears both on the left side of an equation, so that the value is calculated by the model, and on the right side, since the values of the state variables determine the future trajectory of the system.

2.5 Explanatory Variables and Parameters

We adopt definitions for parameters and explanatory variables that are consistent with the various applications used throughout the rest of this book.

Parameters are quantities that are unknown and are not measured directly. They must be estimated using observations of system behavior. Examples are the photoperiod sensitivity of a crop cultivar or the relative rate of decomposition of soil organic matter. These quantities are properties of the components of a system model. There may be other characteristics that can be measured, such as the soil depth or the carbon content in soil organic matter. Those are referred to as explanatory variables. In order to simulate dynamic system models, the values of all state variables must be known at time $t = 0$ when the simulation begins. Some or all of these state variable initial values could possibly be measured and thus not be considered as parameters. However, initial conditions of state variables can also be considered as parameters if they are not measured but are rather estimated from past data.

Explanatory variables include the variables that are measured directly or known in some other way for each situation where the model is applied. They are variable in the sense that they may vary between situations. They include all environmental and management variables as well as system component properties that are known or measured.

Parameters and constants are used in the mathematical relationships of the system model, quantifying how component state variables change. Constants are quantities that remain the same regardless of the conditions and assumptions made about model components. Examples of constants include the molecular weight of glucose, the gravitational constant, and the number of minutes in a day.

Environmental variables help explain the influence of the environment on the system and are assumed not to be influenced at all by the system. Explanatory variables (including environmental variables) and parameters can be recognized by the fact that they appear only on the right-hand side of Equations (1) and (2). They enter into the calculation of the system dynamics but are not themselves calculated by the model. Another difference between explanatory variables and parameters is that explanatory variables are measured or observed for each situation where the model is applied. Thus for example daily maximum and minimum temperatures are measured values that influence crop growth and soil water processes in a system but are assumed not to be influenced by the status of the crop and soil water. Parameters, on the other hand, are estimated from a set of past data. Generally, an explanatory variable can differ depending on the situation, while a parameter is by definition constant across all situations of interest.

The explanatory variables likewise imply a basic decision about what is important in determining the dynamics of the system. In the chapter on model evaluation, we will discuss in detail how the choice of explanatory variables affects model predictive quality. Briefly, adding additional explanatory variables have two opposite effects. On the one hand, added explanatory variables permit one to explain more of the variability in the system, and thus offer the possibility of improved predictions. On other hand, the additional explanatory variables normally require additional equations and may lead to a need for more parameters that need to be estimated, which can lead to additional errors and thus less accurate predictions (see Chapter 9).

3. DEVELOPING DYNAMIC SYSTEM MODELS

3.1 Methods

We present one method that is widely used in developing models of dynamic systems that was first used in modeling industrial management systems (Forrester, 1961, 1971). This approach has been used to schematically represent various agricultural, biological, and ecological systems and to develop the mathematical relationships to represent interactions among system components (e.g., Dent and Blackie, 1979; Odum, 1973; Penning de Vries and van Laar, 1982; Jones and Luyten, 1998). A number of software packages have been developed to help users develop these diagrams of dynamic systems and to develop a set of first order differential equations for the system (e.g., Stella, http://www.isee-systems.com/softwares/Education/StellaSoftware.aspx; and MATLAB, http://www.mathworks.com/products/matlab/). The basic idea of this approach is to define the components of a system, define the state variables that are to be used to represent each of the components in the system, and then identify the processes that cause the state variables to change. This approach is particularly effective in conceptualizing what components to include and how they interact during model development and it provides an easily understood picture for documenting the model structure in publications. One key feature of these diagrams is that first order differential equations like those in Equation (1) can be developed from the diagram itself. This feature will be shown below.

A system component may have more than one state variable to represent its status at any particular time, such as variables of leaf, stem, and grain biomass of a crop component. In Forrester diagrams, each state variable is represented by a box or a compartment in the Forrester terminology (Figure 1.2). Generally, these compartments can represent any state variable that changes over time in response to flows into and/or out of the compartments, including mass, volume, energy, numbers of entities, dollars, etc. Compartments may be used to represent a state variable that is assumed to be homogenous over some spatial dimension, such as the average soil water content in a soil 1 m in depth and 1 km^2 in area. If model developers want to represent spatial variability in their model, then their overall spatial domain can be divided into multiple compartments within which state variables are assumed to be homogenous within the smaller spatial units, and processes to represent the transport among spatial units of the variable that is 'stored' in the compartment. We will show an example of such a system model later in this chapter.

A conceptual model represented by Forrester diagrams shows a network of system compartments representing all relevant components in the system. The compartments are connected by lines that represent the flow of

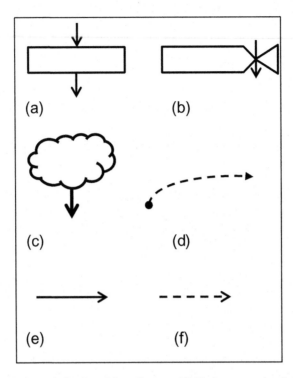

FIGURE 1.2 Basic symbols adapted from Forrester (1961) that are used to build conceptual system models: (a) the 'level' of the material that is stored in the compartment, which is a state variable of a system component; (b) a rate of transport or transformation of the stored material; (c) a source of material that enters the system from the environment or a sink that represents flow of material from the system into the environment (with the arrow pointing to the cloud symbol); (d) an auxiliary variable affecting some compartment or flow; (e) pathway for material flow; and (f) pathway for information flow, showing influences among components.

quantities among compartments and between the system environment and the components. Dashed lines in the diagrams represent information flow or influence of some auxiliary variable (such as a parameter). Figure 1.2 shows an abbreviated set of symbols from Forrester (1961); these are the most commonly used symbols and the ones needed to represent the overall structure of a system model. Another way to understand this approach is that it usually represents a mass, energy, or other material balance, with flows of some quantity among compartments.

The 'level' represents the state variable at a point in time and is shown by a rectangle. Conceptually, these levels should be selected based on what is stored and flows among compartments, such as mass or volume of water or biomass. This is important when developing the overall system of equations for the model. For example, the water volume in a soil component of a particular dimension is what is stored and what flows into and from the component. However, one may be interested in the concentration of water in the soil volume, which can be computed by dividing the soil water volume by the total volume of the soil component. Note that water volume (e.g., mm^3 of water per unit area) is the variable that is transported into and out of the homogenous unit area compartment and is stored, not the concentration (the volumetric soil water content, mm^3[water]/mm^3[soil volume]). First, developing the Forrester diagrams using variables that are stored in the compartments as state variables helps one develop the right system equations and also check the equations for dimensional consistency.

The 'rate' symbol always has a solid arrow representing the pathway for material flow, or rate of accumulation of some variable, with the arrow showing direction of flow. Flow can occur in both directions, so the solid lines can have arrowheads on each end. These rate symbols represent processes that cause the level or state variable in a compartment to change. Thus, there is always one or more equations that are required to represent each flow. Auxiliary variables are shown by circles with dashed lines pointing to the rate or level symbols showing influence of explanatory variables, parameters, and constants. For example, temperature of the environment may affect the biomass accumulation rate in a crop model. Dashed lines without the circle are used to show causal influences on rates or levels.

In practice, system modelers usually develop an initial diagram showing all of the compartments that represent state variables and show rates of transformation or transport among compartments without showing auxiliary variables because they may not be known yet. After a compartment model is constructed, even though it may not show auxiliary variables, one can write the overall mathematical model representing the system (Equation (1)) using the structure of the Forrester diagram and component descriptions. From the diagram itself, one can write differential equations for each compartment (i.e., for each state variable):

$$\frac{d\left(U_j(t)\right)}{dt} = \sum_{i \neq j} I_{i,j}(t) - \sum_{k \neq j} O_{j,k}(t) + in_j(t) - out_j(t) \tag{3}$$

where

$U_j(t)$ = value or level of the jth state variable

$\dfrac{d\left(U_j(t)\right)}{dt}$ = net rate of change in the level of the jth state variable

$I_{i,j}(t)$ = rate of flow into level j from the ith state variable at time t

$O_{j,k}(t)$ = rate of flow from the jth state variable into level k at time t

$in_j(t)$ = rate of flow into compartment j from outside the system

$out_j(t)$ = rate of flow out of compartment j, with the flow leaving the system

By applying this equation to each compartment in the system, a set of first order differential equations is obtained. The number of these first order differential equations, referred to as the dimensionality of the system model, is thus equal to the number of state variables. However, this is only the first step in developing the complete mathematical model. Each of the flows must be described using mathematical functions in terms of the system state variables, explanatory variables, and parameters. This is not a trivial task; it requires knowledge of what drives the process rates. Sometimes, this knowledge comes from basic principles, such as the flow of heat from an object of higher temperature into an object with lower temperature. However, in many cases, the flow equations are not easily derived from basic laws of physics and thermodynamics. This is true in most agricultural system models, partly due to the scale at which these systems are modeled (time and space) and the complexities of the processes involved. In these cases, model developers rely on simple functions to represent flows, based on empirical data and reasonable assumptions. This fact is one of the main reasons that multiple models for the same system are different; different model developers may use different empirical data, functional relationships, and assumptions about these process rates.

After all of the specific equations for rates are written, one can then include those terms to write the final set of equations that contain all variables needed to describe the system. Following this, a more complete Forrester diagram can be constructed showing all of the auxiliary variables that were used to finalize the mathematical equations. However, this is not always done, particularly for complex system models, because showing all of the variables in the diagram would make it difficult to read. The diagrams are very useful even if they do not include the complete list of auxiliary variables.

3.2 Example Development of a System Model

Here we present a simple system model that will demonstrate the use of the Forrester diagram to represent a system and to develop equations that describe the dynamic behavior of that system. For simplicity, we use a physical system example with example models of agricultural systems being given later in this chapter.

The system consists of two water storage tanks that are cylindrical in shape, as shown in the diagram in Figure 1.3. Water flows into the first tank from the external environment, from the first tank into the second tank through a horizontal pipe located at the bottom of the tank, and from the second tank into the environment from a horizontal pipe at the bottom of this tank. The cross-sectional areas (A_1 and A_2 in units of m^2) of the two tanks are different, but are known. Water in the first tank can be described by its height (m) in tank 1 or by its volume (m^3), and the same is true for the second tank. We are ultimately interested in how high the water rises in each tank, but we do not choose depth of water as our state variables because the flow variable is water volume per unit of time. We can compute the height of the water in each tank by knowing the water volume in each tank and the

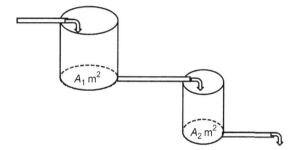

FIGURE 1.3   Schematic of the simple system in which water flows into and out of two tanks, which was used to demonstrate model development principles.

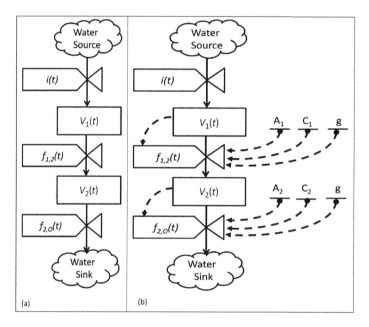

FIGURE 1.4   Forrester diagram for the two tank system model: (a) initial diagram used to develop general system dynamics equations; (b) final diagram showing the auxiliary variables that were used for the final system model.

cross-sectional area of each (e.g., $H_1(t) = V_1(t)/A_1$). Thus, our two state variables are $V_1(t)$ and $V_2(t)$; these are shown in a first drawing of a Forrester diagram for this simple system (Figure 1.4a).

Since we have two compartments (and two state variables) our system model will have two differential equations, which can first be written as:

$$\frac{d(V_1(t))}{dt} = i(t) - f_{1,2}(t) \ and \ \frac{d(V_2(t))}{dt} = f_{1,2}(t) - f_{2,o}(t) \qquad (4)$$

These two equations represent the overall net rates of change in water volume in the tanks, but they cannot yet be solved. Note that $i(t)$ is an explanatory variable (with units of $m^3 s^{-1}$) and must be known ahead of time for all time t. However, we do not yet have functional relationships for either $f_{1,2}(t)$ or $f_{2,0}(t)$, the flows from tank 1 and from tank 2. Knowledge is needed on how to model these flow rates. Here, we assume that the flow from each tank is controlled by an orifice, and use the equation for orifice flow rates (rate proportional to square root of pressure):

$$f_{1,2}(t) = C_1(2gH_1(t))^{1/2} \ and \ f_{2,0}(t) = C_2(2gH_2(t))^{1/2} \qquad (5)$$

Note that C_1 and C_2 are parameters that must have been estimated ahead of time for the orifice flow in this system and g is the gravitational constant. These flow rate equations have water heights instead of water volumes, so they are transformed since we know the relationships between heights and volumes in each tank. We can write the complete system model as:

$$\frac{d(V_1(t))}{dt} = i(t) - C_1\left[\left(\frac{2gV_1(t)}{A_1}\right)\right]^{1/2} \ and$$

$$\frac{d(V_2(t))}{dt} = C_1\left[\left(\frac{2gV_1(t)}{A_1}\right)\right]^{1/2} - C_2\left[\left(\frac{2gV_2(t)}{A_2}\right)\right]^{1/2} \qquad (6)$$

Figure 1.4b shows a revised Forrester diagram that contains more of the final relationships. Note that Equation (6) has the same form as Equation (1) for this two state variable system model. It has two first order, non-linear differential equations; each state variable must have an equation describing its dynamic behavior. (A linear differential equation has the form $dV/dt = a + bV$. Here the state variables appear as square roots rather than linear functions, so the equations are non-linear). System state variables appear on both sides of the equations and all explanatory variables and parameters appear only on the right-hand side ($i(t)$, C_1, C_2, A_1, A_2, and g).

Equation (6) has all of the information needed to simulate the dynamic changes in V_1 and V_2 vs. time, except initial conditions. In all dynamic models, initial values of all state variables are needed to solve the trajectories of state variables over time. In this example, we have two state variables ($V_1(t)$ and $V_2(t)$), and thus we must have initial values for each, $V_1(0)$ and $V_2(0)$, the amounts of water in the tanks at time $t = 0$. Note that we can also convert this set of equations into equivalent ones that contain heights of the water instead of water volumes in each tank. This will be left as an exercise.

When developing equations for system models, care must be taken to ensure dimensional consistency of the relationships, including consistency in the time dimension. Although time in continuous models can be set to any desired time unit (e.g., minutes, hours, or days), one must be sure that all of the rate variables on the left side of the equations (e.g., Equation (1) and

Equation (6)) use the same time scale and that all parameters and constants that appear on the right-hand side of the equations use this same basic time dimension. In this model, units on the left side of the equation are $m^3\,s^{-1}$, so model developers should confirm that the overall dimensions of each term on the right-hand side of the equation have the same units.

This example demonstrates the simplicity with which this approach can be used to obtain dynamic equations for system models. Complexities arise due to the complexities of the system being modeled, including the number of state variables involved, the number of parameters that must be estimated, and the determination of equations for modeling the different processes. Many choices have to be made by modelers in choosing what components to include in the system, what state variables are necessary to adequately represent the dynamic behavior of those components for particular purposes, what processes need to be included, and what functional relationships to use to represent the rates of those processes. Furthermore, some state variables are not continuous and cannot be represented by differential equations. These complexities are common in the development of agricultural systems models as will be demonstrated with examples presented later in this chapter.

4. OTHER FORMS OF SYSTEM MODELS

Before giving examples of agricultural system models, we want to present two additional formulations that are important for use with modern systems analysis tools. The first new formulation is based on the realization that the models are imperfect and include random elements. Equations (1) and (2) are deterministic in that uncertainty is not even considered. Such models always produce the same results when the same explanatory variables and parameters are used in their solution. The second formulation involves results of the model integration over some time period multiple times, such that the integrated results can be considered as a response function.

4.1 Random Elements in Dynamic Equations

We have written the dynamic model equations as perfect equalities in Equations (1) and (2). In practice, however, they are only approximations. The actual time evolution of a state variable in a system as complex as an agricultural system can depend on a very large number of factors. Most agricultural system models, such as crop models, generally include only a small number of factors; those considered to be the most important for the intended uses of the models. The form of the equation is also in general chosen for simplicity and may not be exact. Many modern applications of dynamic models have been developed to incorporate uncertainty in predictions, such as methods for parameter estimation, data assimilation, and uncertainty analysis that are described in later chapters in this book. Furthermore, many agricultural models have been used to provide information for policy makers who are increasingly demanding that model results include estimates of uncertainty.

Uncertainties in the models are caused by different sources, including the model equations, parameters, and explanatory variables. Thus, the equations of an agricultural system model should include random variables to account for these uncertainties. This can be done by adding a random variable to each system model equation as follows:

$$\frac{d(U_i(t))}{dt} = g_i(\boldsymbol{U}(t), \boldsymbol{X}(t), \boldsymbol{\theta}) + \eta_i(t), i = 1, \ldots, S \tag{7}$$

In discrete time form, the equations are written as:

$$U_i(t + \Delta t) = U_i(t) + g_i[\boldsymbol{U}(t), \boldsymbol{X}(t); \boldsymbol{\theta}]\Delta t + \eta_i(t), i = 1, \ldots, S$$

where the error $\eta_i(t)$ is a random variable that describes the uncertainty in the computation of the rate of change of state variable U_i at time t. Equation (7) in continuous or discrete time form is a stochastic dynamic equation. In this formulation, $\eta_i(t)$ summarizes all of the uncertainty for the given explanatory variables and parameters. As we shall see in later chapters, it is sometimes useful to consider uncertainty in model parameters, $\boldsymbol{\theta}$. Another major source of uncertainty in the dynamic equations comes from the explanatory variables, $\boldsymbol{X}(t)$. In particular, future climate is a major source of uncertainty when agricultural system models are used to analyze how crops will perform in simulated experiments for studying climate impacts, management alternatives, and new production technologies. For example, many applications of crop models have included uncertainty in climate by simulating crop growth and yield using multiple years of historical weather data as equally likely samples of future weather conditions (e.g., Thornton et al., 1995; Jones et al., 2000).

When these stochastic model equations (Equation (7)) are simulated, responses are distributions of state variables. This is true even when the same parameters and explanatory variables are used as model inputs.

4.2 A Dynamic System Model as a Response Model

We can integrate the differential equations or discrete time form of difference equations of the dynamic system (Equations (1) or (2), respectively) model. Often we talk of 'running' or 'simulating' the model when the equations are embedded in a computer program and integration is done numerically on the computer. For the difference equations, one simply starts with the initial values of all state variables at time $t = 0$, uses the dynamic equations (Equation (2)) to update each state variable to time $t = \Delta t$, uses the dynamic equations again to get the state variable values at $t = 2\Delta t$, etc., up to whatever ending time T that one has chosen.

The result of integration is to eliminate intermediate values of the state variables. The state variables at the chosen time T are then just functions of the explanatory variables for all times from $t = 0$ to $t = T - \Delta t$. Thus, after integration the state variables can be written in the discrete time form as:

$$U_i(T) = f_{i,T}[X(0), X(\Delta t), X(2\Delta t), X(3\Delta t), \ldots, X(T - \Delta t); \theta], i = 1, \ldots, S \quad (8)$$

In general, there are a limited number of model results that are of primary interest. We will refer to these as the model response variables. They may be state variables at particular times or functions of state variables. The response variables may include: variables that are directly related to the performance of the system such as yield, total nitrogen uptake or total nitrogen leached beyond the root zone; variables that can be used to compare with observed values, for example, leaf area index and biomass at measurement dates; variables that help understand the dynamics of the system, for example daily water stress.

We note a response variable Y. According to Equation (6) the equation for a response variable can be written in the form:

$$Y = f(X; \theta) \quad (9)$$

where X stands for the vector of explanatory variables for all times from $t = 0$ to whatever final time is needed and θ is the same parameter vector as in Equation (1). When we want to emphasize that the model is only an approximation, we will write $\hat{Y}$ in place of Y. Note that one can choose a number of response variables in this way, such that Y becomes a vector of response variables.

4.2.1 Random Elements in System Response Equations

Since the dynamic equations are only approximate, the response equations are also only approximate. Including error, the equation for a response variable can be written as:

$$Y = f(X; \theta) + \varepsilon \quad (10)$$

where ε is a random variable. For the moment we ignore the uncertainty in X. Since the response equations derive directly from the dynamic equations, ε is the result of the propagation of the errors in Equation (8). However, it is not obligatory to first define the errors in the dynamic equations and then derive the errors in the response equations. An alternative is to directly make assumptions about the distribution of ε. In this case, Equation (10) is treated as a standard regression equation for some purposes, such as estimating model parameters. If there are several response variables to be treated, then one is dealing with a (generally non-linear) multivariate regression model.

The error arises from the fact that the explanatory variables do not explain all the variability in the response variables, and from possible errors in the equations. In addition there may be uncertainties in the explanatory variables, in particular climate, when the model is used for prediction.

5. EXAMPLES OF DYNAMIC AGRICULTURAL SYSTEM MODELS

Examples of agricultural system models are presented here to demonstrate a range of model capabilities that include a simple crop growth model, a model of soil water content dynamics in a layered soil, a simple population dynamics model, and an insect predator−prey model with age structure. The simple crop model will be used throughout the book to demonstrate the different model applications presented in later chapters. Furthermore, these examples will explain how to use the compartment modeling approach presented above for systems that vary over space.

5.1 Simple Maize Crop Model

Here, we present a simple model of the growth of a maize crop. Our aim was to create the model such that it demonstrates some of the basic characteristics of many crop models, but is simple enough to be easily used along with methods presented in later chapters for applying dynamic models, such as for parameter estimation, uncertainty analysis, and sensitivity analysis. We also show the Forrester diagram of this model to further demonstrate system modeling principles from section 3 in this chapter.

The system consists of a maize crop growing in a homogenous area of a field, a single component, with three state variables. The three state variables are above ground biomass, the physiological age of the crop on any day of the season, and the crop's leaf area index (LAI), the area of leaves divided by the area of the homogenous ground area (1.0 m^2 in this case). The system environment consists of daily weather, in particular daily maximum and minimum temperatures and daily solar radiation values. All crop models include crop age, leaf area, and crop biomass state variables, but most models have many others.

A few simplifying assumptions were necessary to develop this model. First, it was assumed that water and nutrients are readily available and do not limit growth, and that there is neither damage from pests or diseases nor any competition from weeds growing in the field. Thus, it is assumed that this maize crop is cultivated under potential growth conditions for the particular temperature and solar radiation explanatory variables used to simulate crop growth. The implication of this assumption is that we do not need to include a soil component in the system, which greatly simplifies the task of developing the model. If we assume that water could be a limiting factor, then a soil component would be needed and the soil water availability to the crop would need to be modeled. This is the case with most published models.

We have used a time step (Δt) of one day. Thus, we will use the difference equation form to describe the model. This decision allows us to make use of readily available daily weather data. This choice also means that the time units in the rate equations used to model the changes in state variables must also have time units of days (d). The physiological age of the crop is assumed to be affected only by the time course of temperature, which allows us to use thermal time (degree-days above a base temperature) to compute the daily aging of the crop. Thermal time is used to model the development phases and duration of growth in most models developed to simulate crops cultivated in potential growth conditions, although some models also incorporate day length.

Another assumption is that we are only interested in above ground biomass of the crop, not grain yield. We included crop physiological age and *LAI* because these variables can have major effects on biomass growth. Furthermore, the biomass growth rate is assumed to be affected by light interception, which is dependent on *LAI* and daily solar radiation, and a radiation use efficiency (*RUE*) parameter. Many existing models use *RUE* to compute daily crop growth rate, although some models modify this *RUE* when temperature is outside an optimal range. Finally, we assume that leaf area expansion in the model depends on thermal time as well as the existing LAI of the crop; it does not depend on crop biomass in this simple model. This is not such a good assumption in that many factors control leaf area development, including temperature, solar radiation, and crop biomass. However, this approach provides a good first approximation to *LAI* development under potential growth conditions.

Figure 1.5 shows the Forrester diagram of this system, showing the three system state variables, the system boundary, the processes that cause these variables to change vs. time, parameters that were in the functions used to compute the rates of change of state variables, and the explanatory weather variables that must be known in order to simulate crop growth. Note that there are three rectangular boxes or levels to represent the three state variables. Note also that the rate of accumulation of thermal time age of the crop, $dTT(t)$, depends only on maximum and minimum daily temperatures and an explanatory variable, Tbase (Equation (11)). Thermal age of the crop is the accumulation (integration) of this rate.

$$dTT(t) = max\left[\frac{TMIN(t) + TMAX(t)}{2} - Tbase; 0\right] \qquad (11)$$

Equation (11) shows that dTT(t) is restricted to non-negative values. The rate of leaf area expansion depends on thermal age of the crop in that leaves expand only during a vegetative phase of crop growth, which is defined in this model by *TTL*, the thermal age of the crop when the crop ends its vegetative phase and begins reproductive growth. The rate of expansion is also reduced as *LAI* approaches *LAIMAX*, its maximum value; thus it depends on the current *LAI* as shown in Figure 1.5. *TTL*, *LAIMAX*, and *ALPHA* are variables that control the rate of LAI increase using the following empirical equation, a variation of the logistic equation.

$$dLAI(t) = ALPHA \cdot dTT(t) \cdot LAI(t) \cdot max[LAIMAX - LAI(t), 0] \, for \, TT(t) \le TTL$$

$$= 0 \, for \, TT(t) > TTL$$

$$(12)$$

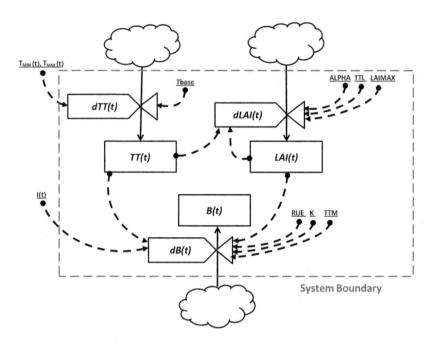

FIGURE 1.5 Forrester diagram of the simple maize crop growth model. The model has three state variables ($TT(t)$, $B(t)$, and $LAI(t)$) and three daily weather explanatory variables ($TMAX(t)$, $TMIN(t)$, and $I(t)$). See text for definitions of variables.

The rate of biomass accumulation per day is computed by the following equation, which computes the fraction of light intercepted depending on LAI and a light extinction parameter, K, the RUE parameter, and the solar radiation of the day, $I(t)$.

$$dB(t) = RUE \cdot [1 - e^{-K \cdot LAI(t)}) \cdot I(t) \, for \, TT(t) \leq TTM$$
$$= 0 \, for \, TT(t) > TTM \tag{13}$$

This equation also shows that daily biomass growth rate, $dB(t)$, continues only until a maximum thermal time is reached, TTM.

Table 1.1 gives the definition of variables and parameters used in this model. Note that there are three variables listed as explanatory variables, each from the environment (maximum and minimum daily temperature and solar radiation). It has seven parameters listed ($ALPHA$, $LAIMAX$, RUE, K, $Tbase$, TTL, and TTM), which are assumed to be unknown; we show in Chapter 6 how these can be estimated from observations of biomass and LAI of the crop. The vegetative phase starts at a $TT(t)$ of 0 and ends when $TT(t)$ reaches TTL, and biomass accumulation ends at physiological maturity, expressed as TTM degree days. TTL and TTM are assumed to be constant and must be quantified in order to simulate crop growth.

TABLE 1.1 Definition of Variables in the Simple Maize Crop Model

Name	Unit	Description
Explanatory Variables		
TMIN(t)	°C	Daily minimum temperature on day t
TMAX(t)	°C	Daily maximum temperature on day t
I(t)	MJ m^{-2} d^{-1}	Daily total solar radiation on day t
State Variables		
TT(t)	°C d	Thermal time age of the crop on day t
LAI(t)		Area of leaves per unit ground area, m^3 m^{-3}, or dimensionless
B(t)	g m^{-2}	Biomass of crop per square meter of ground area
Parameters		
K	–	Light extinction coefficient used to compute intercepted radiation
RUE	g MJ^{-1}	Radiation use efficiency
alpha	–	Relative rate of leaf area index increase for small values of leaf area index
LAImax	–	Maximum leaf area index
Tbase	°C	Base temperature below which the crop does not develop (age)
TTM	°C d	Thermal time threshold at which time the crop is mature
TTL	°C d	Thermal time threshold at which time the vegetative phase of growth ends

Using the discrete time form of the model from Equation (2) and a Δt of 1 day, we now present the model equations needed to simulate the maize crop.

$$TT(t+1) = TT(t) + dTT(t) \tag{14}$$

$$LAI(t+1) = LAI(t) + dLAI(t) \tag{15}$$

$$B(t+1) = B(t) + dB(t) \tag{16}$$

These three equations, along with values of the seven parameters, daily temperature, and solar radiation values, and initial values for each state variable are all that are needed to simulate the behavior of this crop model. Initial conditions are assumed to be: $TT(0) = 0.0$, $LAI(0) = 0.01$, and $B(0) = 0.0$. Note that the equations compute the values of state variables on a particular day based on the values of those state variables in the preceding day and the daily rates of change. Later, we will describe the development of the code in *R* for simulating this model and show example results.

5.2 Dynamic Soil Water Model and Drought Index

Soil water content changes daily in response to rainfall, irrigation, surface water runoff, evapotranspiration, and deep drainage. Atmospheric conditions greatly affect the rate at which crops use water. When water content of a soil is low, a crop growing in the soil is not able to take up enough water from the soil to meet the potential rate of water loss by leaves, which is determined by atmospheric conditions. When crop roots are unable to take up enough water from the soil to meet this atmospheric demand, plants experience what is referred to as water stress. Under these conditions, the crop will lose turgor and partially close its stomata. During periods of water stress, transpiration is lower than its potential rate, leaf area expansion decreases, and photosynthesis is reduced.

Many models have been developed to describe the dynamics of soil water, some of which are included in cropping system models. Under irrigated conditions, soil water models are useful for predicting when irrigation is needed and how much water to apply to avoid water stress that causes decreased growth and yield. In rain-fed crop production, soil water models are useful in estimating the level of stress that the crop experiences and estimating how much yield will be reduced due to that stress.

Woli et al. (2012) developed a simple soil water model for use in computing a drought index, referred to as ARID, the Agricultural Reference Index for Drought. In this section, we present this simple soil water model, then discuss how one could combine it with the simple maize crop model for simulating maize growth when water may be limited during all or parts of the growing seasons, thus predicting water-limited production. We will also explain how this simple model can be extended to account for variations in soil water contents vs. depth.

5.2.1 The ARID Soil Water Model

The system has a single component, the soil, which is assumed to be a homogenous volume with a depth of Z mm and 1 mm^2 in area (Note that we could also use units of m for the variable Z and m^2 for the homogeneous area, but we chose mm for this example because of the units usually used for precipitation, an important input to the model). This soil has a single state variable, soil water volume for a unit area of soil at time t ($W(t)$ mm^3 mm^{-2} soil area). Note that $W(t)$ thus has units of mm^3 mm^{-2}, or mm, and $W(t)$ is equal to the soil water volumetric concentration in the soil ($\omega(t)$ mm^3[water] mm^{-3}[soil volume]) multiplied by the soil depth, Z mm, for the unit area soil volume, thus $\omega(t) = W(t)/Z$. The soil is assumed to be well drained and that as water drains

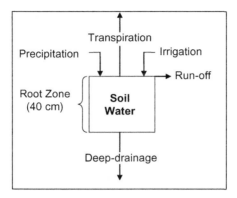

FIGURE 1.6 Diagram of the simple soil water model with one state variable (from Woli et al., 2012).

from the bottom of the soil volume, it leaves the system. The uniform cross-section soil volume is assumed not to be affected by the surrounding soil. It is assumed that a well-established, dense grass crop is growing in the soil, which completely covers the soil surface (Woli et al., 2012). Grass is used so that the assumptions used to compute reference evapotranspiration on day t, $ET_O(t)$ (Allen et al., 1998, 2005), can be used to compute potential evapotranspiration and the ARID reference drought index. The grass is assumed to be perennial, thus allowing ET_O to be computed throughout the year. The model does not compute growth of the grass.

The model is based on a daily time interval. Four processes are assumed to change soil water content on a daily basis: precipitation, $P(t)$, surface runoff, $RO(t)$, evapotranspiration, $ET(t)$, and deep drainage, $D(t)$ (Figure 1.6). Because the soil is completely covered by grass, it was further assumed that all of the evapotranspiration of the crop goes through the plants, thus that water evaporation from the soil surface is insignificant, so that transpiration, $T(t)$, is equal to $ET(t)$. Daily potential evapotranspiration is independent of the soil water, and thus can be used as an explanatory variable or it can be computed by using daily weather values. For example, $ET_O(t)$ can be computed using the FAO-56 Penman−Monteith model (Allen et al., 1998), which requires daily explanatory variables of net solar radiation ($RN(t)$), average daily temperature ($T_{av}(t)$), average wind speed at two m height ($u(t)$), and ambient vapor pressure ($e_a(t)$). Thus, we could use these daily weather variables to compute $ET_O(t)$, and this is typically what is done. However, here we use $ET_O(t)$ as an explanatory variable because it is easily computed and stored for model input, and this simplifies the presentation here. Thus, the soil water has two explanatory variables, $P(t)$ and $ET_O(t)$. We refer readers who are interested in details of this $ET_O(t)$ equation to Allen et al. (1998) and Woli et al. (2012).

Since we only have one state variable ($W(t)$), there will be only one equation that expresses the change in that variable using the discrete time formulation (Equation (2)). This simple equation is written using the discrete time form with $\Delta t = 1$ as:

$$W(t + 1) = W(t) + (P(t) - RO(t) - T(t) - D(t)) \qquad (17)$$

Since $P(t)$ values are known (it is an explanatory variable), we only have to compute the rates of surface runoff, transpiration, and deep drainage. The explanatory variables, state variables, and parameters for this model are given in Table 1.2.

To estimate surface runoff, ARID uses the runoff curve number method of the US Soil Conservation Service (SCS). This method is computationally efficient and is used by various soil water and crop models (Ritchie, 1998;

TABLE 1.2 Variables and Their Definitions for the Simple Soil Water Model

Name	Unit	Description
Explanatory Variables		
$P(t)$	mm d^{-1}	Daily precipitation on day t
$ET_O(t)$	mm d^{-1}	Daily potential evapotranspiration on day t
Z	mm	Depth of soil compartment (assumed to be 400 mm by Woli et al., 2012
ω_{WP}	–	Wilting point volumetric soil water concentration, lower limit for plant water uptake
ω_{FC}	–	Drained upper limit of soil water concentration, above which is subject to drainage
State Variables		
$W(t)$	mm	Water stored in the soil volume on day t per unit soil area
Computed Intermediate Variables		
$\omega(t)$		Volumetric concentration of soil water on day t, m^3 m^{-3}, or dimensionless
$ARID(t)$		Water stress variable (= $T(t)/ET_O(t)$) that affects crop growth processes
Parameters		
α	d^{-1}	Maximum fraction of plant available soil water that roots can take up in a day
β	d^{-1}	Fraction of soil water above field capacity that is drained per day

We use mm as the units for precipitation, soil depth, and potential evapotranspiration variables, and a Δt of 1 day.

Jones et al., 2003). It is based on the assumption that runoff occurs when the rate of rainfall is greater than that of infiltration. Runoff occurs after the initial demands of interception, infiltration, and surface storage have been satisfied. Conversely, runoff does not occur if precipitation is less than or equal to the initial abstraction (amount of rainfall that occurs before water begins to flow off of the soil surface). Using the Soil Conservation Service (1972) method, the daily runoff is estimated as (United States Department of Agriculture, 1986).

$$RO(t) = \frac{(P(t) - 0.2 \cdot S)^2}{P(t) + 0.8 \cdot S} \tag{18}$$

In the ARID model, S is computed by $S = 25{,}400/\eta - 254$, and η (the runoff curve number), which is assumed to be 65, a value for a well-managed grass. Thus, the only variable needed to compute $RO(t)$ is precipitation in this model.

Transpiration in the model is assumed to be limited either by $ET_O(t)$, the reference evapotranspiration rate, or by the maximum rate at which roots can absorb the soil water $RWU_M(t)$. This maximum rate of root water uptake depends on the amount of plant-available water in the soil on day t and a parameter that represents the maximum fraction (α) of that available water content that can be taken up by roots in a given day (Passioura, 1983; Dardanelli et al., 1997). Plant-available water is the amount of water above the wilting point soil water (ω_{WP}), the lower limit of water at which plants can uptake water. These two characteristics of the soil along with Z are listed in Table 1.2 as explanatory variables since they can be measured directly, and both α and β are assumed to be parameters that must be estimated, as shown in the table.

$$RWU_M(t) = \alpha \cdot (\omega(t) - \omega_{WP}) \cdot Z \tag{19}$$

This equation indicates that as $\omega(t)$ decreases and approaches $\omega_{FC}, RWU_M(t)$ will decrease and may drop to 0. Transpiration on day t is computed by:

$$T(t) = min(ET_O(t), RWU_M(t)) \tag{20}$$

Deep drainage depends on the amount of soil water in excess of the field capacity (ω_{FC}) and soil depth.

$$D(t) = \beta \cdot (\omega(t) - \omega_{FC}) \cdot Z \tag{21}$$

Equations (17)–(21) describe the soil water model, with $P(t)$, $ET_O(t)$, ω_{WP}, and ω_{FC} required as explanatory variables and with parameters Z, α, β. Woli et al. (2012) defined the drought index, ARID(t), to be:

$$ARID(t) = 1 - \frac{T(t)}{ET_O(t)} \tag{22}$$

This variable expresses the magnitude of stress a crop experiences when roots are unable to take up enough water to meet the atmospheric demand.

Thus, ARID(t) is a dynamic variable, but it is not a state variable; it can be computed from Equations (19) and (20). Woli et al. (2012) compared this index with other drought indices that are frequently used to represent agricultural drought and concluded that this simple index provides a more realistic indication of water stress that crops experience than the tested indices. The transpiration to potential transpiration ratio in Equation (22) is used by a number of crop models to reduce leaf expansion and biomass accumulation.

5.2.2 Combining Soil and Crop Models

Here, we discuss the structure of a crop model that combines the potential growth crop model from section 5.1 with the soil water model of section 5.2.1. Whereas assumptions in these two separate models eliminated the need to include dynamic interactions between soil water and crop growth, when these components are combined, one needs to define the important interactions and add relationships to quantify those interactions. However, there are important inconsistencies in the assumptions in these two particular models so that the models cannot just be combined as they are. This is generally true. When one wants to modify a particular model, care has to be taken to ensure that the assumptions are consistent or that modifications are made to one or both original models as needed. The key inconsistency between assumptions in these models is that the soil water model assumed that the soil was always covered by a grass so that evapotranspiration is dominated by transpiration. In the maize model, the soil is bare initially, then as the crop grows and LAI increases, most of the light is captured by the crop. Thus, during much of the season, evapotranspiration could include a significant amount of soil water evaporation.

Figure 1.7 shows the system diagram that combines the simple maize model with the soil water model. One can see from this diagram that there are four state variables—the three from the crop model and one from the soil. The diagram also shows four specific interactions between the crop and soil (see the double-dashed arrows). The figure shows that the soil water state variable, $W(t)$, is affected by daily rates of soil evaporation $E_S(t)$, a rate that was not needed in the simple soil model) in addition to transpiration, $T(t)$, deep drainage, $D(t)$, surface runoff, $RO(t)$, and precipitation, $P(t)$.

Interactions have to be identified and included in model equations. In this combined model, LAI affects the rates of actual soil evaporation and transpiration. In the figure, three new quantities (ω_{AD}, γ, and K_{EP}) are shown that were not in the soil water model. The ω_{AD} explanatory variable is the air dry volumetric soil water content, and γ is a parameter that reduces soil evaporation rate as the soil dries. The K_{EP} parameter is the solar radiation extinction coefficient that is used to partition potential evapotranspiration into the potential daily transpiration of the crop and the potential evaporation from the soil. This approach was first used by Ritchie (1972), and it is now included in a number of cropping system models (e.g., DSSAT, Jones et al., 2003). In this simple

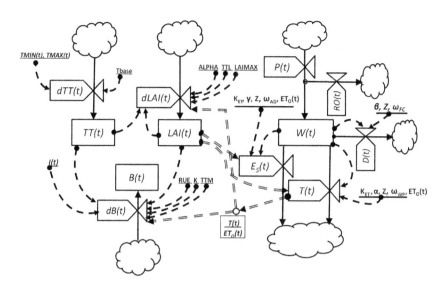

FIGURE 1.7   System diagram of a crop model that combines the simple maize crop model with the soil water dynamics model. The double-dashed lines show interactive relationships between the two components in which *LAI* expansion and biomass accumulation rates are affected by the water stress variable computed by the ratio of actual transpiration of the crop to the potential rate and where transpiration rate may also depend on *LAI* at time *t*.

model actual tranpsiration rate, $T(t)$, is computed as before using the minimum of the maximum root water uptake and the potential transpiration rate.

The status of the soil water also affects crop growth rates. In this diagram, we show the calculation of a water stress factor each day ($T(t)/ET_O(t)$), which is equal to $(1 - ARID(t))$ as defined by Woli et al. (2012). This stress index affects two processes in the crop, rate of leaf area expansion ($dLAI(t)$) and rate of biomass accumulation ($dB(t)$), as shown in Figure 1.7. This is the approach used in the *DSSAT* cropping system model to reduce daily leaf expansion and biomass growth rates.

Our purpose here was not to present the specific equations that one might use to model these expansions and interactions among the components. Most crop models include crop and soil components similar to what is shown in Figure 1.7, but they use different assumptions and functions to account for these dynamic interactions. Readers can get more details on how several of the major crop models model these interactions (e.g., Brisson et al., 2003; Jones et al., 2003; Keating et al., 2003; Kersebaum et al., 2007; Stockle et al., 2003).

5.2.3 Extending the Soil Water Model for Non-Homogenous Soils

We describe one additional extension to the simple homogenous soil water model presented in section 5.2.1 that represents a general issue that agricultural system modelers frequently face. Our models have assumed homogenous soil and crop components, meaning that state variables depend only on time and do not vary over space. However, one may need to explicitly incorporate spatial variability of soil state variables. An example is presented here in which soil water content (and soil properties) varies with soil depth. This variability can have a large effect on root water uptake as well as nutrient leaching and microbial processes in soils.

The homogenous model (section 5.2.1.) can be modified to account for variations in soil water content with depth. In fact, most cropping system models include soil models that include variability in soil properties and soil water with depth. However, we will see that this extension increases the dimensionality of the model (number of state variables) and leads to a requirement for incorporating additional processes, equations, and parameters. The soil component in the non-homogenous system is still assumed to have unit area and have a defined depth. The soil column is divided into a number of soil compartments or layers, each of which is assumed to be homogenous. Figure 1.8 shows how this will affect the system diagram of the simple soil water model (shown schematically on the right side of Figure 1.7 with only one compartment for the soil water). Figure 1.8 shows five compartments, each one representing the soil water content of a particular soil layer. This approach, which has soil water in each layer as state variables and has been referred to as a 'tipping bucket' approach, is used in a number of crop models.

A model of the system depicted in Figure 1.8 will have five state variables, and five discrete time equations are needed to compute changes in water for all of the layers. Each soil layer has its own vertical depth, and equations computing changes in soil water will include terms for downward movement of water from the layer above or to the layer below, upward flow from the layer below or to the layer above, and root water uptake, if roots are in the layer. At the soil surface and bottom boundaries, the rates of change need to take into account the assumptions about how the environment, or boundary conditions, affect flow across the boundaries. For example, one needs to define whether water can flow from the bottom layer or not and whether or not there is a water table that can result in water entering the soil profile from below. Furthermore, to simulate the system of five equations, initial conditions are needed for each state variable, the initial soil water content in each layer in this case.

Adding the capability to simulate vertical distribution of soil water has increased not only the number of state variables, but also the number of functions that are needed to describe the system dynamics and the number of input variables needed to simulate the model. For example, if a homogenous soil water model has three variables to describe the soil water limits, then the model with five soil layers will require 15 different soil characteristic variables, three for each layer.

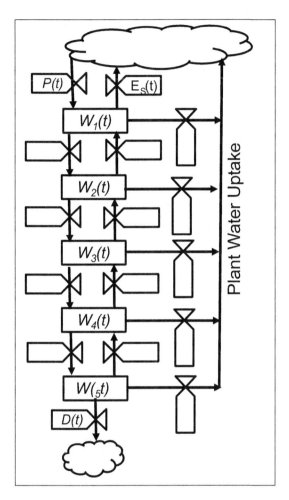

FIGURE 1.8   Initial Forrester diagram of a soil water model with multiple soil layers to represent vertical variations in soil water content.

This example demonstrates one way that agricultural models can be extended to be more realistic for some applications, as well as a giving a glimpse into the increases in model complexity when this is done. When spatial variability is important, the overall system can be divided into a number of homogenous compartments. These compartments typically interact with each other, as the vertical soil layers do in this example, where water flow from one compartment affects its adjacent layer. Furthermore, each of the compartments can have multiple state variables. In the vertical soil example, each soil layer could have soil inorganic nitrogen content and root length density that change with time in addition to soil water, and these variables can also interact. The structure for modeling various spatially-varying elements provides the basis for determining the dimensionality of the model in that state variables for each element could exist in each compartment.

There is another important issue in developing spatial models in this way that model developers should be aware of. The number of layers used to represent the vertical variations in soil water and other variables can make a big difference in numerical results. Just as the discrete-time form of system models is an approximation of processes that may change continuously over time, breaking the soil into multiple imaginary homogenous layers is an approximation in that variables can change continuously over very short vertical distances. This is particularly true near the soil surface where gradients in soil water content can be very high, and soil water can change rapidly. The overall issue is that the discrete soil layers provide only a coarse approximation to the variations in soil water vs. soil depth, depending on the thickness of layers. Jones and Luyten (1998) explained that dividing the soil into very thin layers and using a very small time step can be used to obtain an approximate numerical solution to the well-known soil water diffusion second order partial differential equation. It is well known that accurate solutions to those types of equations require careful attention to the choice of discrete time and space steps used in the numerical solutions. This will be discussed in Chapter 4.

5.3 Population Dynamics Models

Many models have been developed for describing population dynamics of different types of biological organisms, including cells, plants, animals, insects, and humans. The populations are typically assumed to be homogenous with one or more state variables representing the numbers of individuals in the population in a given area or volume, thus the density of individuals. In this type of model, the processes that affect population dynamics are birth and mortality rates as well as movement into and/or out of the homogenous area. These models are typically written as differential or difference equations, and are thus consistent with the approach presented in this chapter and summarized by Equations (1) and (2). Many of the population dynamics models build on the early pioneering work of Lotka (1925).

Typically, differences in age and sex are assumed to be unimportant in the studies. However, these differences among individuals in a population can be taken into account by including different variables for each of several groupings of individuals, each of which is assumed to be homogenous in the area or volume being studied. There are practical reasons for breaking the populations into age classes, for example, noting that reproductive and mortality rates may be highly dependent on ages of individuals. Conceptually, breaking a homogenous population into age classes is similar to dividing a soil into different layers, each of which is assumed to be homogenous within the layer or age class. If we divide the population into age classes between 0 and 1 years, 1 and 2 years, etc., and define 20 age classes, then the model will have 20 state variables, one for each age class, and it will have 20 differential or difference equations. Whereas the simple models with only one homogenous population are useful for some purposes, models that include more details about individuals have much more interesting behavior, at the cost of requiring more data and functional relationships that quantify the class-specific processes and interactions.

Many population dynamics models include competition among species, beginning with the classical Lotka-Volterra model (Hutchinson, 1978) of prey and predator densities and their interactions. Although this is a simple model, it has been used as a basis for developing more complex formulations by many authors. Furthermore, it is easy to see that population dynamics models can include competition among several species, each with its own age, sex, and other classes. Finally, just as with the soil water example presented earlier, populations may not be homogenous over the area or within the volume one is studying. Thus, an area that is being studied can be subdivided into homogenous areas, each with its own population that evolves over time and interacts with populations over space. The number of state variables in population dynamics models can be very large if the populations are divided into different classes, include species interactions, and vary over space. We encourage caution to those who aim to include all of these classes in a model, not because of the problems of solving them using computer methods like those in this book, but due to the amount of quantitative data needed and understanding of relationships needed to develop and adequately parameterize the model.

In this section, we present a basic model for a single homogenous population, then demonstrate how this model can be expanded to include age classes, and finally show the expansion of this latter model into a predator–prey model.

5.3.1 Homogenous Population with Limited Food Supply

We first assume that we have a single homogenous population with a reproductive rate proportional to the density of individuals in the population. Furthermore, we assume that the organism lives in a habitat that has limited food supply, such that as the population reaches an environment-dependent 'carrying capacity', the rate of reproduction approaches zero. In this simple population, there is one state variable, $A(t)$, which increases with reproduction and decreases due to deaths. A simple equation can be used to represent the population changes with time:

$$\frac{dA(t)}{dt} = r \cdot A(t)\left[1 - \frac{A(t)}{K}\right] \qquad (23)$$

where

$A(t) =$ population density, number ha^{-1}
$t =$ time in days, d
$r =$ rate of reproduction per individual in the population, d^{-1}
$K =$ carrying capacity of the environment, number ha^{-1}

Note that the equation has a positive rate of increase in $A(t)$ as well as a negative term; this negative term increases as $A(t)$ approaches K, the maximum population. The solution to this equation shows an approximate exponential rate of increase in $A(t)$ at low population densities followed by $A(t)$ asymptotically approaching K over time. Although this is a simple equation and can be solved analytically when K and r are constant (the logistic equation), we leave the solution of this equation as a student exercise.

If K or r vary over time or in response to some other environmental variable like temperature, the behavior of this equation becomes very interesting. Temperature, for example, has a major influence on the reproductive and development rates of insects and other organisms. In this simple model, the parameter r could depend on temperature (T), an environmental variable that may also vary with time, $r = r(T(t))$. If K is constant, then Equation (23) becomes

$$\frac{dA(t)}{dt} = r(T(t)) \cdot A(t) \left[1 - \frac{A(t)}{K} \right] \qquad (24)$$

In order to solve this equation, one needs to develop an explicit equation to describe $r(T(t))$ and to have data on how temperature varies with time (i.e., $T(t)$ is needed for the duration of time t for which a solution is needed). Whereas Equation (23) can be solved to obtain an expression of $A(t)$ as a function of time, Equation (24) must be solved numerically, for example by using the difference form of the model.

$$A(t + \Delta t) = A(t) + r(T(t)) \cdot A(t) \cdot \left[1 - \frac{A(t)}{K} \right] \cdot \Delta t \qquad (25)$$

The time step, Δt, is an important consideration in this type of model. Many population dynamics models use a time step of one day and thus use a daily average temperature. However, since the temperature effects are highly nonlinear, the use of daily average temperatures may not be adequate (Purcell, 2003). If a Δt of one hour is used, the effects of temperature variations in the environment are more reliable, but for this time step, hourly values of temperature are needed to simulate $A(t)$ using Equation (25). If only daily maximum and minimum temperatures are collected, there are methods of interpolating the daily time variations in temperature, thus to generate hourly values necessary to simulate the population dynamics.

5.3.2 Population Dynamics Model with Age Classes

In homogenous population models, such as the one presented above, a number of processes that affect populations are not directly accounted for. The reproduction, development or aging, migration, and mortality are not explicit. They can be included with a fairly straightforward extension of the homogenous population assumption—the incorporation of age classes of the population. One way to model populations is to divide the population into discrete age classes, with each age class assumed to be homogenous. For example, one can divide a population into age classes of individuals between 0 and 5 days of age, between 5 and 10 days, etc., with the number of individuals in each age class in a given area of land being state variables. If we divide the population into 10 age classes, then the model will have 10 population density state variables, which can be represented with a Forrester diagram. This formulation allows the modeler to vary inflow and outflow processes for each state variable. For example, animals may have to reach a certain age before they reproduce, mortality and migration rates may vary with age, and environmental effects on population processes may also vary. The number of age classes can be important in the overall population dynamics.

Some populations have specific life stages that are frequently included in population models. For example, insects have egg, larvae, pupae, and adult stages. These stages generally represent different ages in the life of an insect, each having specific processes that result in changes in populations of individuals in each stage. It is also possible to divide any of the stages into age classes if the processes vary with age within a life stage. We demonstrate this type of model using an insect model with egg, four larval, pupae, and adult stages (a total of seven age classes), assumed to represent the population dynamics of ladybeetles per m^2 area in a field. Figure 1.9 shows an initial Forrester diagram with seven state variables, rate of egg laying, influx/efflux of adult ladybeetles, mortality of each class of insects, and development rates from stage and age classes. This example is based on the ladybeetle and aphid model published rates of development by Bianchi and van der Werf (2004), although variable names are different.

The figure depicts the 'flow' of individuals from one stage or age class into the next. At any instant, for example, the number of eggs in the area of land can increase due to new egg laying by adult ladybeetles, with r_b defined as the rate of 'birth' or eggs laid per adult per unit area (units of d^{-1}), by a relative mortality rate (m_E, with dimensions of d^{-1}), and by a development rate (the fraction leaving a stage or age class per day, r_E with units of d^{-1}). The fraction leaving any particular life stage is typically approximated as the inverse of the time required by that stage to develop. For example, if 5 days are required for a newly-laid egg to develop into a new larvae, then r_E would be equal to 1/5, meaning that 20% of the eggs develop into new larvae per day. For the larvae age classes, if the duration of each age class is 10 days,

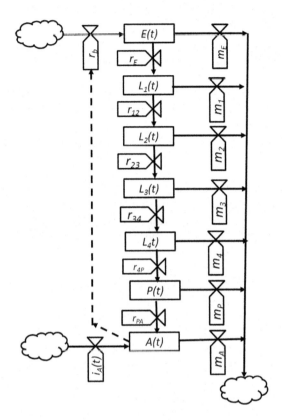

FIGURE 1.9 Initial Forrester diagram of a system model for a population of ladybeetles per m² area, showing compartments for life stages as well as four age classes of larvae. There are seven state variables (E, L_1, L_2, L_3, L_4, P, and A, for eggs, larvae in each of four age classes, pupae, and adults) each with processes that affect changes in population density of those compartments.

then 10% of the larvae in an age class would leave that class and be put into the next age class per day. In addition, this model includes the possibility of adult insects flying into the area on any day t, $i_A(t)$, with units of (number m^{-2} d^{-1}).

The mathematical model for this system has seven equations, which can be written as differential equations or difference equations. The differential equations are:

$$\frac{dE(t)}{dt} = r_b \cdot A(t) - r_E \cdot E(t) - m_E \cdot E(t)$$

$$\frac{dL_1(t)}{dt} = r_E \cdot E(t) - r_{12} \cdot L_1(t) - m_1 \cdot L_1(t)$$

$$\frac{dL_2}{dt} = r_{12} \cdot L_1(t) - r_{23} \cdot L_2(t) - m_2 \cdot L_2(t)$$

$$\frac{dL_3(t)}{dt} = r_{23} \cdot L_2(t) - r_{34} \cdot L_3(t) - m_3 \cdot L_3(t) \qquad (26)$$

$$\frac{dL_4(t)}{dt} = r_{34} \cdot L_1(t) - r_{4P} \cdot L_4(t) - m_4 \cdot L_4(t)$$

$$\frac{dP(t)}{dt} = r_{4P} \cdot L_4(t) - r_{PA} \cdot P(t) - m_P \cdot P(t)$$

$$\frac{dA(t)}{dt} = r_{PA} \cdot P(t) - m_A \cdot A(t) + i_A(t)$$

The solution of these equations provides values of all seven state variables vs. time, starting at some arbitrary time when the initial values of each of these state variables are provided. To do this, one needs to write these as difference equations and set a specific value for Δt (for example, 1 day). The rate coefficients (r's) and mortality coefficients (m's) may be constant and may be assumed not to vary with time or state of the system. However, the influx of adult ladybeetles, $i_A(t)$, is a time-dependent variable that must be known for all time t in order to simulate the population dynamics in the specific unit field area that is modeled.

In contrast to the simple homogenous population model with one state variable (Equation (23)), there is not an explicit carrying capacity. However, the mortality rates of any life stage or age class can easily be modeled such that mortality increases as the population in that class approaches a maximum density. For example, the mortality of the ladybeetle adults could be dependent on a food-carrying capacity, for example:

$$m_A = m_A' \cdot \frac{A(t)}{K} \qquad (27)$$

Also, the rate and mortality coefficients in Equation (26) do not include the effects of temperature or other environmental factors on reproductive, development, and mortality processes. In particular, temperature has major effects on insect population dynamics, and most models include those effects. This can be done by representing each of the rate and mortality coefficients as functions of temperature, similar to what was done in the simple one state-variable model shown in Equation (24). In addition, the carrying capacity can be affected by temperature and other factors in the environment. System models may also include the population dynamics of predators in addition to the insect population. This example ladybeetle population dynamics model will be extended by assuming that the ladybeetles are predators feeding on a prey population of aphids.

5.3.3 Predator–Prey Population Dynamics Model

Consider that the ladybeetles in an area feed on aphids, which are also increasing and decreasing due to the aphid population's own reproductive and mortality processes in addition to mortality due to the adult population of ladybeetles feeding on the aphids. For this system model, we will use Equations (26) and (27) to represent the population dynamics of ladybeetles and a modification of Equation (23) to model the aphid population, treating aphids as a homogenous population. We define $H(t)$ as the population density of aphids in the same area as ladybeetles, K_H as the carrying capacity for aphids, r_H as the aphid relative rate of reproduction, and m_H is a predation rate, relative to populations of both ladybeetles and aphids.

$$\frac{dH(t)}{dt} = r_H \cdot H(t) \cdot \left[1 - \frac{H(t)}{K_H}\right] - m_H \cdot A(t) \cdot H(t) \tag{28}$$

$$\frac{dE(t)}{dt} = r_b \cdot A(t) \cdot H(t) - r_E \cdot E(t) - m_E \cdot E(t) \tag{29}$$

These equations, when combined with Equation (26), are an example of a predator (ladybeetle)–prey (aphids) model. These population dynamics models demonstrate how models can be very simple and also be expanded to include more details about the populations and the processes that change the dynamics of one population. We demonstrated how age structure and temperature effects can be added as well as how one can include interactions among species. We point out that the structure of this type of model is not very complex. However, the model may contain a large number of parameters that must be quantified in order to simulate the populations. Population dynamics parameters for the development rates of organisms can be estimated using laboratory studies in many cases. However, the relative reproductive and mortality rates are affected by many factors and have to be estimated from observed population datasets. In addition, migration of populations into a study area is needed unless the area is isolated from other populations.

5.3.4 Modeling Spatial Variations in Population Dynamics

Populations may vary considerably over space, and one may be interested in studying spatial and temporal changes in populations. Generally, spatially-variable populations are modeled by dividing a selected study area into a number of areas, each with its own populations. For example, an area may be divided into square grid cells, each having four adjacent grid cells. Each of the grid cells is assumed to be homogenous. The size of these grid cells depends on the characteristics of the populations being studied and on the study's objectives. Each grid cell will have its own population dynamics equations, similar to those above, with or without age structure. However, the mathematical model in each grid cell will contain additional terms for the movement into and out of each grid cell. It is easy to see that the number of state variables (the dimensionality) of the model could considerably increase, depending on the number of grid cells. Whereas it is not difficult to simulate the additional complexity associated with spatially variable populations, it is a challenge to parameterize these models. See Bianchi and van der Werf (2004) for an example where the ladybeetle–aphid model is extended to study variations in populations over space.

There is considerable literature on modeling population dynamics of many types of organisms and in many fields of study. Some crop models also contain components that are modeled using population dynamics models similar to those presented in this section. For example, crop models may have state variables of leaves and fruit that are 'born', develop, and senesce. Temperature affects the rates at which they are created, develop toward maturity, and die. Therefore, even though the examples presented in this chapter are simple, understanding how these types of models are developed is highly useful across a wide range of different types of systems. Typically, system models will have interacting components with equations that will contain different functions describing the transport, transformation, and development of important system variables. Frequently, such system models include features that are modeled using concepts presented here.

EXERCISES

Easy

1. Extending the water tank model.
   a. Write the model equations for the water tank system in terms of heights of the water in each of the two tanks, completely eliminating the tank water volumes from the equations.
   b. Add a third tank in the system. This tank is connected to tank #2 with its base at exactly the same elevation of that of tank #2. A single pipe connects tank #3 to tank #2, and this pipe is the only source of water flowing into or out of the new tank #3. The flow coefficient for this tank is C_3 and the cross-section area is A_3.
   i. Draw the Forrester diagram for the system with three tanks.

ii. Write the equations for this revised system. Note that flow between tanks #2 and #3 is proportional to the difference in water heights between the two tanks.

2. Dynamic system models are frequently used as response models.

   a. Equations (24) and (25) describe the population dynamics of a simple homogenous population. Write this model as a response function for time $t = 2$ days with $\Delta t = 1$ day, noting that at time $t = 0$, $A(0)$ is known.

   b. Equation (6) consists of differential equations. Write these equations using a discrete time step approximation. Then assuming that $V_1(0) = V_2(0) = 0.0$ and that $i(t) = I$ is a constant for all time t, write the response equations for V_1 and V_2 at time $t = 2\Delta t$.

   c. Equations (17)–(22) describe soil water changes over time. Write the response function for W at time $t = 3$ days, assuming that W at time $t = 0$ is equal to W_0.

3. Write the overall model equations for the system shown in the Forrester diagram in Figure 1.8. What additional information do you need to complete the equations? Discuss the implications of adding soil layers vs. having only one soil volume.

4. You are asked to write the analytical solution to the equation for a homogenous population, given in Equation (23) by:

$$\frac{dA(t)}{dt} = r \cdot A(t)\left[1 - \frac{A(t)}{K}\right]$$

Please list all of the assumptions needed to obtain this form of the Logistic equation.

Moderate

5. A schematic of a biological reactor is shown below. In this system, there is a flow of a mixture of nutrients and water into the reactor vat (F liters per hour, $1\ h^{-1}$), with the concentration of the nutrients of $[S_i]$ (with units of mg of nutrient per liter of solution). Initially, the reactor vat is filled with water along with enzymes that can convert nutrients in the vat, S(t) mg of nutrients, into a desired product denoted by B(t). Units of B(t) are mg of product. Note that the conversion of one mg of nutrients results in a larger mass of product due to the amount of nutrient in the composition of the product. V is the volume of the reactor vat, which is assumed to be full of a mixture of water, nutrients, and product. Flow from the vat also occurs at a rate of F liters per hour. Note that the concentration of S is $[S]$.

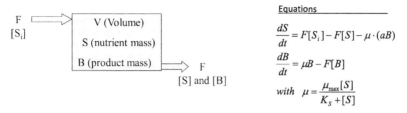

Equations

$$\frac{dS}{dt} = F[S_i] - F[S] - \mu \cdot (aB)$$

$$\frac{dB}{dt} = \mu B - F[B]$$

$$\text{with} \quad \mu = \frac{\mu_{max}[S]}{K_S + [S]}$$

F = volume flow rate of fluid into (and out from) reactor vessel, liters s^{-1}
$[S_i]$ = concentration of nutrient flowing into vessel
$[S]$ = concentration of nutrient in vessel
B = product of reaction, mg
$[B]$ = concentration of product in vessel, mg l^{-1}
V = volume of vessel, liters
μ = specific growth rate at which B is produced from S
a = fraction of product that is nutrient (S)
K_S = concentration of [S] when reaction rate is 0.5 of μ_{max} (maximum reaction rate)
Also, note that S = V[S] and B= V[B] and that the contents are well mixed

a. You are asked to draw a Forrester diagram of this system, showing all flows, explanatory variables and parameters.

b. Now, you are asked to modify the model to include the effects of temperature on the conversion process, μ such that μ now depends both on the concentration of S as well as temperature. Also assume that K_s does not depend on temperature, but that μ_{max} does. You are asked to look in the literature for an appropriate temperature function, then use it to modify the equations above such that the performance of the bioreactor will depend on temperature in addition to what was presented above. (Later, you will be asked to simulate this system, in Chapter 4).

6. You are asked to create a Forrester diagram for a population dynamics model that describes fox populations, some of which have rabies. This model was published in the journal *Nature*, volume 289, in 1981, on pages 765–771.

Difficult

7. Many biological models contain components that are described by the diffusion of gases across membranes or through a continuous medium. In this exercise, you are asked to develop a Forrester diagram and equations for flow of gases across membranes and through a continuous medium.

a. For the first case, assume that you have five compartments connected end to end, and that the center compartment initially has a concentration of gas that may flow across membranes at both ends of the compartment. Further, you can assume that the concentration of the gas in the surrounding air environment is 0.0, and that the cylindrical exterior of the compartments is solid, preventing gas from diffusing from each of the five compartments. In other words, the flow only occurs horizontally across membranes at each end of each compartment, including membranes at both ends of the compartments that are placed end to end horizontally. Draw the Forrester diagram for this system and write the dynamic differential equations for diffusion from each compartment and into the atmosphere. State all assumptions that you make to model this system.

b. Continuing with the diffusion across membrane problem from Question 7.a, now assume that the compartment on one end has a solid cover that prevents gas diffusion. For this case, there is only one exit pathway for the gases in the 5 compartments.

c. Now, assume that the compartments have leaky external boundaries, in addition to the membranes at each end. Draw the Forrester diagram for this system and write the equations.

d. As a final diffusion problem, you will consider the diffusion of a gas in a long hollow pipe with a constant diameter. Initially, the pipe has no gas (e.g., [CO_2]), but that at time 0.0, the ends of the pipe are opened (at $x = 0$ at one end and $x = X$ at the other end) and gas begins to diffuse along the horizontal pathway of the pipe in response to the concentration of CO_2 ($C(x,t)$) in the pipe. The diffusion coefficient (D) is assumed to be constant. [CO_2] in the atmosphere is assumed constant.

$$\frac{\partial C(x,t)}{\partial t} = D \frac{\partial^2 C(x,t)}{\partial^2 x}$$

You are asked to develop a difference equation to approximate this continuous diffusion problem. Assume that you will divide the diffusion pathway into 20 compartments. Number the compartments starting using numbers (1,2,3, etc.) but note that distance along the flow path is (compartment number $*$ Δx). Simulate the diffusion along a linear gradient (one dimension) of 20 compartments. Draw a Forrester diagram of this system and write the approximating difference equations.

REFERENCES

Addiscott, T.M., Wagenet, R.J., 1985. Concepts of solute leaching in soils: A Review of Modeling Approaches. J. Soil Sci. 36, 411−424.

Allen, R.G., Pereira, L.S., Raes, D., Smith, M., 1998. Crop evapotranspiration−guidelines for computing crop water requirements. Irrigation and Drainage Paper No. 56. FAO, Rome.

Allen, R.G., Walter, I.A., Elliott, R., Howell, T., 2005. The ASCE standardized reference evapotranspiration equation. ASCE, Reston, VA.

Asseng, S., Ewert, F., Rosenzweig, C., Jones, J.W., Hatfield, J.L., Ruane, A.C., et al., 2013. Uncertainty in simulating wheat yields under climate change. Nature Clim. Change, 3, 827−832. doi:10.1038/nclimate1916.

Bianchi, F.J.J.A., van der Werf, W., 2004. Model evaluation of the function of prey in non-crop habitats for biological control by ladybeetles in agricultural landscapes. Ecological Modeling 171, 177–193.

Brisson, N., Gary, C., Justes, E., Roche, R., Mary, B., Ripoche, D., et al., 2003. An overview of the crop model STICS. European J. Agron 18, 309–332.

Dardanelli, J.L., Bachmeier, O.A., Sereno, R., Gil, R., 1997. Rooting depth and soil water extraction patterns of different crops in a silty loam Haplustoll. Field Crops Research 54 (1), 29–38.

Dent, J.B., Blackie, M.J., 1979. System Simulation in Agriculture. Applied Science Publishers, London.

Forrester, J.W., 1961. Industrial Dynamics. Wiley, New York.

Forrester, J.W., 1971. Principles of Systems. Wright-Allen, Cambridge, MA.

Hutchinson, G.E., 1978. An Introduction to Population Ecology. Yale University Press, New Haven, Conn.

Jones, J.W., Luyten, J.C., 1998. Simulation of biological Processes. In: Peart, R.M., Curry, R.B. (Eds.), In Agricultural Systems Modeling and Simulation. Marcel Dekker, Inc., New York, pp. 19–62.

Jones, J.W., Hansen, J.W., Royce, F.S., Messina, C.D., 2000. Potential benefits of climate forecasting to agriculture. Agr. Ecosystems & Env 82, 169–184.

Jones, J.W.G., Hoogenboom, C.H., Porter, K.J., Boote, W.D., Batchelor, L.A., Hunt, P.W., et al., 2003. The DSSAT cropping system model. European Journal of Agronomy 18 (3–4), 235–265.

Keating, B.A., Carberry, P.S., Hammer, G.L., Probert, M.E., Robertson, M.J., Holzworth, D., et al., 2003. An overview of APSIM, a model designed for farming systems simulation. Eur. J. Agron. 18, 267–288.

Kersebaum, K.C., Hecker, J.-M., Mirschel, W., Wegehenkel, M., 2007. Modelling water and nutrient dynamics in soil–crop systems: a comparison of simulation models applied on common data sets. In: Kersebaum, K.C., et al., (Eds.), Modelling Water and Nutrient Dynamics in Soil–Crop Systems. Springer, pp. 1–17.

Odum, H.T., 1973. An energy circuit language for ecological and social systems. In: Systems Analysis and Simulation in Ecology. Vol II, B.C. (Ed.), Patten. Academic Press, New York, pp. 140–211.

Passioura, J.B., 1983. Roots and drought resistance. Agric. Water Management 7 (1–3), 265–280.

Penning de Vries, F.W.T., van Laar, H.H. (Eds.), 1982. Simulation of Plant Growth and Crop Production. PUDOC, Wageningen, The Netherlands.

Purcell, L.C., 2003. Comparison of thermal units derived from daily and hourly temperatures. Crop. Sci. 43 (5), 1874–1879.

Ritchie, J.T., 1998. Soil water balance and plant water stress. In: Y.T. Gordon et al. (ed.) Systems approaches for sustainable agricultural development—Understanding options for agricultural production. vol. 7. Kluwer Academic Publishers, Dordrecht, The Netherlands, pp. 41–54.

Rosenzweig, C., Jones, J.W., Hatfield, J.L., Ruane, A.C., Boote, K.J., Thorburn, P., et al., 2013. The Agricultural Model Intercomparison and Improvement Project (AgMIP): Protocols and pilot studies. Agr. Forest. Meteorol. 170, 166–172.

Rotter, R.O., Nendel, C., Olesen, J.E., Patil, R.H., Ruget, F., Takac, J., et al., 2012. Simulation of spring barley yield in different climatic zones of northern and central Europe. Field Crops Research 133, 23–36.

Stockle, C.O., Donatelli, M., Nelson, R., 2003. CropSyst, a cropping systems simulation model. European Journal of Agronomy 18, 289–307.

Thornton, P.K., Hoogenboom G., Wilkens P.W., Bowen W.T., 1995. A computer program to analyze multi-season crop model outputs. Agron. J. 87:131–136.

United States Department of Agriculture, 1986. Urban hydrology for small watersheds. Technical Release 55 (TR-55) (Second Edition ed.). Natural Resources Conservation Service, Conservation Engineering Division. ftp://ftp.wcc.nrcs.usda.gov/downloads/hydrology_hydraulics/tr55/tr55.pdf.

Woli, P., Jones, J.W., Ingram, K.T., Fraisse, C.W., 2012. Agricultural Reference Index for Drought (ARID). Agron. J. 104, 287–300.

Chapter 2

Statistical Notions Useful for Modeling

1. INTRODUCTION

In many ways, a systems model is just a complex nonlinear regression model. It relates response variables to explanatory variables. This places systems models firmly in the realm of statistics, where regression models have been extensively studied. Essentially all the chapters of this book make use of statistical notions, and several of the chapters rely heavily on statistical reasoning. The problems of parameter estimation and prediction error, which are central to regression in general and systems models in particular, are essentially statistical problems.

As far as possible, each chapter in this book is self-contained if any advanced statistical methods are used. However, it is assumed that basic statistical notions are known. The purpose of this chapter is to briefly review those notions.

This chapter is not meant as an all-purpose introduction to statistics. It focuses on the basic notions that are important for this book. Some topics that are very important in general statistics are ignored here. An example is statistical hypothesis testing, which is not treated at all in this chapter.

1.1 In This Chapter

We first define the notion of a random variable, and then the notion of a probability distribution function. Then we consider multiple random variables, and the ways of describing distributions in this case. We are particularly interested in conditional distributions, since a regression model is in fact an approximation to the conditional expectation of the response variable, given the explanatory variables. Next we consider sampling and samples. Since our information about the random variables of interest comes from sampling, it is essential to show how sampling quantities are related to population quantities. We discuss simple random sampling and also sampling

Working with Dynamic Crop Models. DOI: http://dx.doi.org/10.1016/B978-0-12-397008-4.00002-2

schemes more relevant to agronomy. Next we present regression models. The last section is a brief introduction to Bayesian statistics.

2. RANDOM VARIABLE

A random variable, say X, is a function that maps the set of possible outcomes of an experiment involving some degree of randomness to numerical values. A classic example is a throw of a die, with X being the number of dots on the upturned side. Another example would be $X =$ yield in a randomly chosen French corn field next year. This X is a random variable both because the choice of field is random and because yield is not known in advance.

It is important to identify the range, or 'population', of experiments and observations which give rise to a random variable. The die throwing experiment concerns some particular die or similar dice and throws that make the die turn over several times before it lands. For the random variable $X =$ yield in a randomly chosen French maize field next year, the population is all fields planted with maize in France next year. If we restricted the population to fields with conventional practices (eliminating, for example, organic farmers), that would give rise to a different random variable.

The result of an observation of a random variable is called a realization of the random variable. We will use here small case letters to indicate a realization (x) and capital letters to indicate random variables (X).

Random variables can be discrete or continuous or a mixture of the two. A discrete random variable takes only a finite number of possible values. The result of tossing a die is one of the set $\{1,2,3,4,5,6\}$; this is a discrete random variable. A continuous random variable can take values on a segment of the real line. $X =$ yield in a randomly chosen French maize field next year is a continuous random variable because possible yield values are not restricted to a set of discrete values.

In general we are interested in continuous random variables in this book, so in many cases below we only give results for such variables. The results for discrete random variables can usually be obtained by replacing integrals with sums.

3. THE PROBABILITY DISTRIBUTION OF A RANDOM VARIABLE

3.1 Cumulative Distribution and Density Functions

A random variable is described by the probability of obtaining different possible values. The basic function which describes these probabilities is the cumulative distribution function (CDF), denoted by $F_X(x)$, defined as the probability that X has a value less than or equal to x:

$$F_X(x) = P(X \leq x)$$

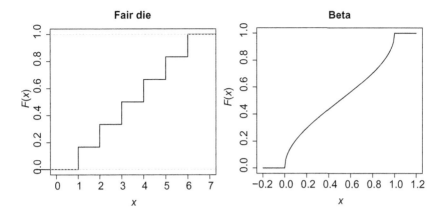

FIGURE 2.1   Cumulative distribution function for a discrete (left) and a continuous (right) random variable.

This function is well defined for both discrete and continuous random variables. Figure 2.1 shows the CDF for a discrete random variable (a fair die with probability 1/6 on each of the integers 1,...,6) and for a continuous random variable (a beta distribution with parameters $\alpha = \beta = 0.5$).

We can restate the definition of discrete and continuous random variables in terms of the CDF. A random variable X whose CDF is constant except for a finite number of jumps is a discrete random variable. A random variable X whose CDF is continuous as a function of x is a continuous random variable.

Often it is more convenient to work with the probability or probability density at each possible value of x. For a discrete random variable the probability density function (PDF), denoted by $f_X(x)$ or $P(X = x)$, gives the probability of each possible value. For a fair die, $f_X(x) = 1/6$ for x = 1,...,6 and 0 elsewhere.

For a continuous random variable the probability density function $f_X(x)$ is proportional to the probability that X takes values in a vanishingly small neighborhood around x. That is, the probability that X is in the range $(x, x + dx)$ is $f_X(x)dx$ as dx approaches 0. $f_X(x)$ is the derivative of the CDF.

To indicate that a random variable has the PDF $f_X(x)$ we write

$$X \sim f_X(x)$$

which is read as 'X is distributed as $f_X(x)$'.

We will often refer to the 'distribution' or 'probability distribution' of a random variable. This could be defined by its CDF or, equivalently, by its PDF.

3.2 Expectation, Variance, and Quantiles of a Random Variable

Two important values that characterize a distribution are the expectation and the variance. For a discrete random variable, the expectation is the sum of the possible values, each weighted by the probability of that value.

$$E(X) = \sum_{i=1}^{n} x_i f_X(x_i)$$

where the x_i are the values with non-zero probability. The analogous expression for a continuous random variable replaces the sum by an integral over the PDF

$$E(X) = \int_{-\infty}^{\infty} x f_X(x) dx$$

The expectation of a function of a continuous random variable, say $g(X)$, is defined by

$$E(g(X)) = \int_{-\infty}^{\infty} g(x) f_X(x) dx$$

The variance of a distribution is related to the variability around the expectation. It is defined by

$$var(X) = E(X - E(X))^2$$

The square root of the variance is called the standard deviation.

The qth quantile of a continuous random variable X with CDF $F_X(x)$ is defined by $F_X(x) = q$. For example, the 1/4th quantile (also called the lower quartile), is the value of x such that the probability that $X \leq x$ is one fourth. The 1/2th quantile is called the median. For a discrete random variable, the qth quantile is any x value that satisfies $F_X(y) \leq q \leq F_X(x)$ for any $y < x$.

3.3 Best Predictor of Y Using a Constant

Suppose that we want to predict a future value of X using some fixed value g. Suppose that our criterion is mean squared error, so we want to minimize

$$E(X - g)^2$$

The best predictor (smallest mean squared error) is $g = E(X)$. The mean squared error of that predictor is var(X). The proof is as follows:

$$\begin{aligned} E(X-g)^2 &= E(X-E(X)+E(X)-g)^2 \\ &= E(X-E(X))^2 + (E(X)-g)^2 \qquad (1) \\ &= \text{var}(X) + (E(X)-g)^2 \end{aligned}$$

In the first line, we simply add and subtract $E(X)$. In going from the first to the second line, the cross term in developing the square is $2 * E((X - E(X))(E(X) - g))$. This is 0 because $(E(X) - g)$ is just a constant and $E(X - E(X)) = 0$. The sum in the last line is minimized when $g = E(X)$, and then $E(X - g)^2 = \text{var}(X)$.

If the criterion of prediction quality is mean absolute error,

$$MAE = E\left(\left|X - g\right|\right)$$

then the best predictor is g = median of X.

3.4 Particular Distributions

We describe here some particular distributions that are used in this book. They are summarized in Table 2.1 and the corresponding CDFs and PDFs are shown in Figures 2.2 to 2.5.

A very simple probability distribution is the uniform distribution (Figure 2.2). It is defined by two parameters, a (the lower bound) and b (the upper bound). It is noted $U(a,b)$. If $X \sim U(a,b)$, then X has equal probability of taking any value in the range (a,b) and zero probability of being outside this range.

A very commonly encountered distribution is the normal distribution (Figure 2.3). This distribution is determined by two parameters, μ (the expectation) and σ (the standard deviation). It is noted $N(\mu,\sigma)$. The distribution is symmetric around μ. σ measures the spread in the distribution.

The gamma distribution function is useful because it only has non-zero probability for $x \geq 0$ (Figure 2.4). It has two parameters, a shape parameter α and a scale parameter β. It is noted $G(\alpha, \beta)$.

In some cases we can sample or generate values from a distribution, but we don't know the exact form of the distribution. In that case we can define an empirical distribution function, which puts equal weight on each of the sampled or generated values. For example, if we have 50 values of a random variable, our empirical distribution function would have a probability of 1/50 for each of those values. The type of CDF and histogram one might obtain are shown in Figure 2.5.

TABLE 2.1 Characteristics of Some Common Probability Distributions

Name of Distribution	PDF	Parameters	Expectation	Variance
Uniform	$\frac{1}{b-a}$ $a \le x \le b$	a = lower bound b = upper bound	$(b+a)/2$	$(b-a)^2/12$
Normal	$\frac{1}{\sqrt{2\pi\sigma^2}}\, e^{\frac{-(x-\mu)^2}{2\sigma^2}}$ $-\infty < x < \infty$	μ (expectation) σ (standard deviation) $\sigma > 0$	μ	σ^2
Gamma[1]	$\frac{1}{\Gamma(\alpha)\beta^\alpha} x^{\alpha-1} e^{-x/\beta}$ $0 \le x < \infty$	α (shape) β (scale) $\alpha, \beta > 0$	$\alpha\beta$	$\alpha\beta^2$
Empirical distribution function	x_i, $I = 1,\dots,n$ are sampled or generated values $f_X(x) = 1/n$ $x = x_i$ $f_X(x) = 0$ otherwise		$(1/n)\sum\limits_{i=1}^{n} x_i$	$(1/n)\sum\limits_{i=1}^{n} [x_i - E(X)]^2$

[1] In the PDF, $\Gamma(\alpha)$ is the gamma function evaluated at α.

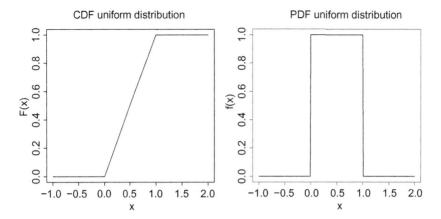

FIGURE 2.2 Cumulative distribution function (left) and probability density function (right) for a uniform distribution with lower bound = 0, upper bound = 1.

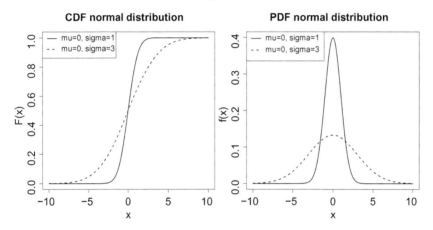

FIGURE 2.3 Cumulative distribution function (left) and probability density function (right) for two normal distributions.

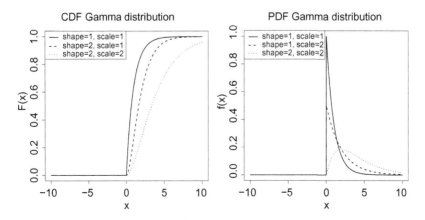

FIGURE 2.4 Cumulative distribution function (left) and probability density function (right) for three gamma distributions.

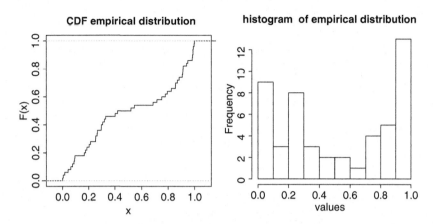

FIGURE 2.5   Cumulative distribution function (left) and histogram (right) for an empirical distribution with 50 values.

4. SEVERAL RANDOM VARIABLES

Modeling is in essence a study of the relationship between random variables. Thus we need to consider not only single random variables but also vectors of random variables.

We now let X represent a vector of random variables with k components $X_1, X_2,\ldots,X_k$. In cases where we are specifically interested in one response variable (for example yield), we will let Y represent the response variable and X the vector of other variables. Here are a few examples:

- We could look at plant biomass (Y) and total intercepted radiation (X_1) in maize fields. We would be interested in the relationship between them.
- An example of interest for systems models would be the relation of yield (Y) to daily weather, soil, initial conditions, and management (X). In this case X can have several hundred components, since it includes the weather variables each day.

4.1 Joint Distribution

The full description of multiple random variables is given by their joint CDF. For a random vector X with components $X_1,\ldots,X_k$ this is

$$F_X(x) = P(X_1 \leq x_1, X_2 \leq x_2, \ldots, X_k \leq x_k)$$

This is the probability that simultaneously $X_1 \leq x_1, X_2 \leq x_2$, etc.

4.2 Marginal Distribution

Suppose that we are interested in properties of just some of the random variables that we have defined irrespective of the other random variables in the population. The marginal distribution of any subset of X is the distribution of that subset in the population, where we don't concern ourselves with the values of the other random variables.

The marginal distribution is related to the joint distribution. It is obtained by integrating (or summing for discrete random variables) the joint distribution over the random variables that are not of interest.

As an example, suppose that we have defined two random variables, X and Y. The marginal density of Y is related to the joint distribution by

$$f_Y(y) = \int_{-\infty}^{\infty} f_{X,Y}(x, y)dx \tag{2}$$

That is, the marginal density of Y is obtained by integrating the joint density over all the possible values of X.

4.3 Conditional Distribution and Independence

The conditional distribution of a random variable is the distribution of that variable, when the values of other random variables are fixed. The conditional density of Y, given that $X_1 = x_1$, ..., $X_k = x_k$, is noted

$$f_{Y|X_1,\ldots,X_k}(y|x_1,\ldots,x_k)$$

There will be a different conditional PDF for every set of values $(x_1,\ldots,x_k)$. To indicate conditional random variables we will use the notation $Y|X$, which should be read as 'Y given X'. This is a family of random variables, since each value of X results in a different distribution of Y.

The conditional distribution of random variables is of major interest in modeling. The whole point of modeling is to predict some variable Y, the 'response' variable, for given values of other variables X, the 'explanatory' variables. The conditional distribution $f_{Y|X_1,\ldots,X_k}(y|x_1,\ldots,x_k)$ specifies exactly how X and Y are related.

Example 1

Lindquest et al. (2005) studied the relation of maize biomass (Y) to total intercepted radiation (X). Consider $Y|(X = 400 \text{ MJ/m}^2)$, which is maize biomass given that total intercepted radiation is $X = 400 \text{ MJ/m}^2$. This random variable has a distribution which is approximately centered around 1500 g/m². The random variable $Y| (X = 600 \text{ MJ/m}^2)$, which is yield given that total intercepted radiation is 600 MJ/m², is a different random variable, which is approximately centered around 2300 g/m².

The conditional distribution is related to the joint and marginal distributions as follows

$$f_{Y|X}(y|x) = \frac{f_{X,Y}(x, y)}{f_X(x)} \tag{3}$$

where X and/or Y can be vectors of random variables.

Equation (3) is easy to understand in the case of discrete variables with equal probability for each value. The denominator on the right-hand side is the number of cases where $X = x$, divided by n, the total number of possibilities. The numerator is the number of cases where both $X = x$ and $Y = y$, divided by n. The ratio is the fraction of those cases where $X = x$ that also have $Y = y$ or, in other words, the probability that $Y = y$ given that $X = x$.

The random variable Y is said to be independent of the random variable X if the distribution of $Y|X$ is the same as the distribution of Y for all values of X, so that

$$f_{Y|X}(y|x) = f_Y(y).$$

That is, knowing the value of X makes no difference to our knowledge of Y.

It is easy to show that independence implies that the joint distribution is equal to the product of the individual PDFs. Replacing $f_{Y|X}(y|x)$ with $f_Y(y)$ in Equation (3) gives

$$f_Y(y) = \frac{f_{X,Y}(x, y)}{f_X(x)} \Rightarrow f_{X,Y}(x, y) = f_Y(y)f_X(x)$$

4.4 Covariance and Correlation

The covariance measures to what extent two random variables tend to vary together. The definition is

$$\text{cov}(X, Y) = \int_{-\infty}^{\infty} \int_{-\infty}^{\infty} [x - E(X)][y - E(Y)]f_{X,Y}(x, y)dxdy \qquad (4)$$

Note that $\text{cov}(X,Y) = \text{cov}(Y,X)$, and that $\text{cov}(X,X) = \text{var}(X)$.

When we have multiple random variables, there is a covariance between every pair. These covariances can be arranged in a matrix called the variance–covariance matrix and noted Σ. For k random variables $X_1, \ldots, X_k$ we have

$$\Sigma = \begin{pmatrix} \sigma_1^2 & \sigma_{1,2} & \cdots & \sigma_{1,k} \\ \sigma_{2,1} & \sigma_2^2 & \cdots & \sigma_{2,k} \\ \vdots & \vdots & \ddots & \vdots \\ \sigma_{k,1} & \sigma_{k,2} & \cdots & \sigma_k^2 \end{pmatrix} \cdots$$

where $\sigma_{i,j}$ is the covariance between X_i and X_j and σ_i^2 is the variance of X_i.

The correlation between two random variables X and Y, noted r, is defined as

$$r = \frac{\sigma_{X,Y}}{\sqrt{\sigma_X^2 \sigma_Y^2}}$$

It can be shown that $-1 \leq r \leq 1$. A negative value indicates that the random variables vary in opposite directions. If the random variables are independent, then $\sigma_{ij} = 0$. The converse is not necessarily true ($\sigma_{ij} = 0$ does not necessarily imply independence).

4.5 Expectation and Variance for Multiple Random Variables

If X and Y are two random variables and a and b are constants, then

$$E(aX + bY) = aE(X) + bE(Y)$$

The variance of a sum of two random variables is:

$$\text{var}(aX + bY) = a^2\text{var}(X) + b^2\text{var}(Y) + 2ab\text{cov}(X, Y) \qquad (5)$$

If X and Y are independent then

$$\text{var}(aX + bY) = a^2\text{var}(X) + b^2\text{var}(Y)$$

It is sometimes useful to have expressions for expectations and variances in terms of conditional variables. The expectation of Y can be written

$$E(Y) = E_X\left[E_{Y|X}(Y|X)\right] \qquad (6)$$

We first fix X and take the expectation over $Y|X$. The result is a function of X. Then we take the expectation over X. For clarity we have written explicitly the random variables over which we take the expectations. We can show this using equations (2) and (3). We have that

$$f_Y(y) = \int f_{Y|X}(y|x)f_X(x)dx$$

Then

$$E(Y) = \int y f_Y(y)dy = \int\int y f_{Y|X}(y|x)f_X(x)dxdy$$
$$= E_X\left(\int y f_{Y|X}(y|x)dy\right) = E_X E_{Y|X}(y|x) \qquad (7)$$

The variance of Y can be expressed as a sum:

$$\text{var}(Y) = \text{var}[E(Y|X)] + E[\text{var}(Y|X)]$$

In words, the total variance is the sum of two terms: the variance of the conditional expectation and the expectation of the conditional variance.

4.6 Best Predictor of Y Using a Function of X

Suppose that we want to predict Y using a function that depends on X, say $\pi(X)$. This is obviously of interest, because in modeling we want to predict a response as a function of explanatory variables. Suppose that our criterion is mean squared error, so that we want to minimize

$$MSE = E_{X,Y}\{[Y - \pi(X)]^2\}$$

We can use the relation in Equation (6) to express the expectation over the joint distribution as follows:

$$MSE = E_X\{E_{Y|X}[(Y|X - \pi(X))]^2\} \qquad (8)$$

For each fixed value x of X, the quantity in curly brackets is just MSE for predicting a random variable, $Y|(X = x)$, using a constant, $\pi(x)$. As we have seen, this is minimized by $\pi(x) = E(Y|(X = x))$. So the prediction function that minimizes MSE is

$$\pi(X) = E(Y|X)$$

For this predictor,

$$MSE = E_X \mathrm{var}(Y|X),$$

which is the variability of Y for each value of X, averaged over X.

The same argument holds if X is not a random vector but rather a vector of fixed explanatory values. For example, suppose that in a planned experiment, the N fertilizer dose (X), takes the values 0, 100, and 200 kg N/ha for every site-year. The best predictor of Y for each value of X would be $E(Y|X)$. Its squared error would be the variance of $Y|X$ averaged over the three X values. X could also be a mixture of random and fixed variables (for example, weather and fixed fertilizer doses).

The above shows that once the explanatory variables of a model are chosen, the best possible model is $E(Y|X)$. However, we still have to treat the problem of approximating this function, and in particular of estimating the parameters that appear in the model.

4.7 The Multivariate Normal Distribution

Consider a random vector with k components, $X = (X_1, \ldots, X_k)^T$. X has a multivariate normal distribution if its PDF is

$$f_X(x) = \frac{1}{(2\pi)^{k/2}|\Sigma|^{1/2}} e^{-(x-\mu)^T \Sigma^{-1}(x-\mu)/2}$$

where

$$\mu = (E(X_1), \ldots, E(X_k))^T$$

is the vector of expectations, Σ is the variance–covariance matrix and $|\Sigma|$ signifies the determinant of Σ.

Suppose that we partition X into two vectors: Z_1 of length n_1 and Z_2 of length n_2. The variance–covariance matrix is then partitioned as

$$\Sigma = \begin{pmatrix} E\left[(Z_1 - E(Z_1))(Z_1 - E(Z_1))^T\right] & E\left[(Z_1 - E(Z_1))(Z_2 - E(Z_2))^T\right] \\ E\left[(Z_2 - E(Z_2))(Z_1 - E(Z_1))^T\right] & E\left[(Z_2 - E(Z_2))(Z_2 - E(Z_2))^T\right] \end{pmatrix}$$

$$= \begin{pmatrix} \Sigma_{11} & \Sigma_{12} \\ \Sigma_{21} & \Sigma_{22} \end{pmatrix}$$

One can show that the conditional distribution of Z_2, given that $Z_1 = z_1$, is a normal distribution with expectation and variance as follows:

$$E(Z_1 | Z_2 = z_2) = E(Z_1) + \Sigma_{12}\Sigma_{22}^{-1}[z_2 - E(Z_2)]$$
$$\text{var}(Z_1 | Z_2 = z_2) = \Sigma_{11} - \Sigma_{12}\Sigma_{22}^{-1}\Sigma_{21}$$

$$(9)$$

The conditional expectation is a linear function of z_2. In the expression for the conditional variance, suppose that Z_1 is a single random variable. Then $\text{var}(Z_1 | Z_2 = z_2)$ is a number, which is less than or equal to $\text{var}(Z_1)$, with equality only if the correlation between Z_1 and every element of Z_2 is 0. This is the statistical statement of the fact that there is less variability in Z_1 if we restrict ourselves to cases where some related random variables Z_2 are fixed.

5. SAMPLES, ESTIMATORS, AND ESTIMATES

5.1 Simple Random Samples

A sample is a set of observations drawn from a population by some defined procedure. It is through the values of the random variables in a sample that we learn about these random variables in the population. So we need to consider the distributions of sample quantities and how these are related to distributions of the random variables.

The distribution of sample quantities depends on how the sample is obtained. The simplest case is that of a simple random sample. Each individual in the sample is drawn at random from the population, independently of all the other individuals in the sample. Let Y_i be the value of the variable Y measured for the ith individual in the sample. Since Y_i is drawn at random from the population, $f_{Y_i}(y_i) = f_Y(y)$. The distribution of Y_i is the same as the distribution of Y in the population. This is true for every individual in the sample. The Y_i in the sample are said to be independent and identically distributed (iid). The distribution would be written

$$Y_i \overset{iid}{\sim} f_Y(y)$$

The joint distribution of $Y_1, \ldots, Y_n$, where n is the sample size, taking into account independence, is

$$f_{Y_1, \ldots, Y_n}(y_1, \ldots, y_n) = f_Y(y_1) f_Y(y_2) \ldots f_Y(y_n)$$

The sample mean is

$$\frac{1}{n} \sum_{i=1}^{n} Y_i$$

Consider the expectation of the mean of the Y_i for a simple random sample. It is easily seen that

$$E\left(\frac{1}{n} \sum_{i=1}^{n} Y_i\right) = E(Y) \tag{10}$$

The expectation of the sample mean is equal to the expectation of Y in the population. We say that the sample mean provides an unbiased estimate of the population mean $E(Y)$.

The sample variance is defined by

$$s^2 = \frac{1}{n-1} \sum_{i=1}^{n} (Y_i - \bar{Y})^2$$

It can be shown that

$$E(s^2) = \sigma^2 \tag{11}$$

where $\sigma^2 = \text{var}(Y)$, and so the sample variance is an unbiased estimator of the population variance $\text{var}(Y)$.

It is also of interest to have an expression for the variance of the sample mean. For a simple random sample this is

$$\text{var}\left(\frac{1}{n} \sum_{i=1}^{n} Y_i\right) = \frac{1}{n} \sigma^2 \tag{12}$$

The variance of the mean decreases as 1 over the sample size.

5.2 Sampling in Agronomy

The sampling underlying systems models in agronomy is often more complex than simple random sampling. Consider crop models. Often the population consists of some large number of sites and years, where the year determines weather and history of the site. Management practices or strategies might be random variables determined by site and year (in the case of farmer fields) or might be fixed with the same strategies for each site-year combination (in planned experiments).

Sampling then involves choosing sites and years from the population. A simple but not totally unrealistic sampling strategy would be to choose sites at random from the sites in the population and years at random from possible years (or rather, yearly conditions at random from possible conditions). Then the sample might have various combinations of sites and years, with the same sites and the same years repeated several times in the sample. For example, if there are s sites and y years, then the sample might have all s*y combinations of the chosen sites and years. If these are planned experiments, then we might have the same set of management strategies repeated for each site-year combination.

The difference with random sampling is that the site-year combinations are not all chosen independently. Once a site is chosen it is reused several times, and the same for years.

In this case, the probability distribution for any variable of the sample is still equal to the probability distribution of that variable in the population, since each site and year is chosen at random from the population. Thus Equation (10) still holds, so that sample means are still unbiased estimators of population expectations. However, we do not have independence between elements of the sample. Equations (11) and (12) are no longer valid. We do not now have simple expressions for the variability of the sample quantities.

5.3 Estimators and Estimates

A point estimator is any function of sample measurements that provides a value for the quantity to be estimated. For example, the sample mean is a point estimator of the expectation of Y in the population.

An estimator is itself a random variable, because the values obtained in a sample are random, with some PDF. We are also interested in the numerical value obtained when an estimator is evaluated for a particular sample. We will refer to this as an 'estimate'. If the estimator of $E(Y)$ is the sample mean $(1/n)\sum_{i=1}^{n} Y_i$, and for a particular sample the mean is 4.0, then 4.0 is the estimate of $E(Y)$.

It is important to quantify the quality of an estimator. Suppose that we want to estimate a population quantity, say μ, using an estimator $\hat{Y}$. We use mean squared error as our criterion. Then

$$MSE = E\left[\left(\mu - \hat{Y} \right)^2 \right]$$

The expectation is over the distribution of $\hat{Y}$. *MSE* can de decomposed into two terms as follows

$$MSE = E\left[\mu - E(\hat{Y}) + E(\hat{Y}) - (\hat{Y})^2\right]$$
$$= \left[(\mu - E(\hat{Y}))^2\right] + E\left[\left(E(\hat{Y}) - \hat{Y}\right)^2\right] \tag{13}$$
$$= \text{bias}^2 + \text{variance of estimator}$$

We are particularly interested in the estimation of $E(Y|X)$ using some $\hat{Y}$ which will, in general, be a function of X. Then

$$\text{bias}^2 = E\left[\left(E(Y|X) - E(\hat{Y}|X)\right)^2\right]$$
$$\text{variance} = E\left[\left(E(\hat{Y}|X) - \hat{Y}|X\right)^2\right]$$

The first term in MSE is 0 if the estimator is unbiased, i.e., if

$$E(\hat{Y}|X) = E(Y|X).$$

The second term is the variance of the estimator. It measures how variable the estimator is for different possible samples. A good estimator has small bias and small variance. Being unbiased is not, by itself, enough to ensure that an estimator has a small error.

5.4 Effective Sample Size

We have seen that samples in agronomy are often not independently distributed. What are the consequences of this for the mean squared error of estimators? Put another way, how much information is there in a non-iid sample, compared to an iid sample?

We frame this question in terms of effective sample size (ESS). The ESS of a non-iid sample is the size of an iid sample that would give the same variance of the estimator under consideration.

As the following example shows, positive correlation reduces effective sample size.

Example 2

Suppose we are interested in using the sample mean to estimate a population expectation. Suppose further that in the sample, of size n,

$$E(Y_i) = E(Y) \quad \text{all } i$$
$$\text{var}(Y_i) = \text{var}(Y) = \sigma^2 \quad \text{all } i$$
$$\text{cov}(Y_i, Y_j) = r\sigma^2 \quad 0 \leq r \leq 1 \quad \text{all } i \neq j$$

TABLE 2.2 Effective Sample Size (ESS) for Samples of Size 5 or 10 with Various Levels of Correlation Between All Pairs of Random Variables

r	ESS for n = 5	ESS for n = 10
0	5	10
0.2	2.8	3.6
0.6	1.5	1.6
1.0	1.0	1.0

The difference between this sample and a simple random sample is that here the sample variables are correlated if $r > 0$. We have assumed a very simple correlation structure (all pairs of variables have the same covariance).

Since $E(Y_i) = E(Y)$, the sample mean is an unbiased estimator of $E(Y)$. The variance of our estimator, using Equation (5), is

$$\text{var}\left(\frac{1}{n}\sum_{i=1}^{n} Y_i\right) = \frac{1}{n^2}(n\sigma^2 + n(n-1)r\sigma^2) = \frac{\sigma^2}{[n/(1 + (n-1)r)]}$$

This is the same variance as a simple random sample with $n = [n/(1 + (n-1)r)]$. Therefore

$$ess = \frac{n}{1 + (n-1)r}$$

Results for several combinations of n and r are shown in Table 2.2. If the individuals in the sample have $r = 0$, this is simple random sampling and effective sample size is equal to n. If $r = 1$, all sample values are perfectly correlated and effective sample size is 1. This is the same as having just a single measurement. Correlations between these extremes lead to effective sample sizes between 1 and n.

In system modeling the sample variables may have a complex correlation structure, with unknown correlation parameters. In that case we cannot exactly calculate effective sample size. However, we do know that positive correlations will make the effective sample size smaller than the number of individual measurements.

6. REGRESSION MODELS

The general form of a regression model is

$$Y = h(X; \theta) + \varepsilon \tag{14}$$

where $h(X; \theta)$ is a function of X that in general contains unknown parameters θ, and ε is the residual error which is described by a PDF. The PDF of ε may also contain unknown parameters. Often, for example, we assume that $\varepsilon \sim N(0, \sigma^2)$ for all X, where σ^2 is an unknown parameter. This assumption says that $E(\varepsilon) = 0$, which implies that $E(Y|X) = h(X;\theta)$. This assumption further says that the variance of $Y|X$ is the same for all X. Of course other assumptions are possible.

We have already seen that the best predictor of Y based on X is $E(Y|X)$. According to the above assumptions, this is equal to the regression function $h(X;\theta)$. However, we still have to estimate the parameters in this function. We also want to estimate σ^2. These are the main objectives of regression analysis. We will discuss this in some detail in Chapter 6.

7. BAYESIAN STATISTICS

7.1 The Difference Between Bayesian and Frequentist Statistics

All of the previous material has been based on 'frequentist' statistics. That is the dominant school of thought in statistics, and most statistics courses are largely or wholly based on that approach. However, there is another major, and increasingly important, school of thought in statistics known as 'Bayesian' statistics. Here we briefly introduce some of the principles of Bayesian statistics. Chapter 7 gives details about how to estimate parameters using Bayesian statistics.

There are major differences between parameter estimation using frequentist statistics (see Chapter 6) and using Bayesian statistics (see Chapter 7). In frequentist statistics, a parameter is treated as a fixed quantity and one concentrates on obtaining an estimator of the parameter, based on the data in a sample. Maximum likelihood estimators are an example. We have explained that estimators are random variables because different samples would give different results. Frequentist statistics concentrates on the distribution of the estimator, in particular its bias and variance. The important point here is that random variability enters through the fact that the estimated value would vary if the sampling were repeated. The name 'frequentist' comes from the emphasis on the frequency of different possible outcomes of sampling or of an experiment.

Bayesian statistics on the other hand treats the parameter as a random variable whose distribution represents our knowledge about the parameter. There is no distinction here between the true value of the parameter and the estimator. There is just the distribution of the parameter. The second major difference with frequentist statistics is that the reasoning does not consider multiple possible samples. All the reasoning is based on the actual values that were obtained in the sample or the experiment. Uncertainty enters because, based on our sample or experiment, we do not have enough information to know perfectly the values of the parameters.

7.2 Basic Ideas of Bayesian Statistics

The term 'Bayesian' statistics derives its name from the fact that the basic theorem was published by Thomas Bayes (in 1764). The theorem says that, for two random variables A and B,

$$P(A|B) = P(B|A)P(A)/P(B) \qquad (15)$$

This is a basic theorem of probability, and by itself is neither specifically frequentist nor specifically Bayesian. It follows quite directly by noting that

$$f_{A,B}(a, b) = f_{B,A}(b, a).$$

According to Equation (3) this implies that

$$f_{A|B}(a|b)f_B(b) = f_{B|A}(b|a)f_A(a).$$

Rearranging and using the notation for discrete variables gives Equation (15). Equation (15) shows how to invert conditional probabilities, which can often be quite useful.

Example 3

This example is from Efron (2005). A woman is pregnant with twins, and a sonogram reveals that both are boys. The woman would like to know the probability that she is carrying identical twins. Here A is type of twin ($A = F$ for fraternal twins or $A = I$ for identical twins) and B is the sex of the two fetuses ($B = MM$, MF, FM or FF where $M =$ male and $F =$ female). The probability of having two boys if the twins are identical is well known; it is very close to 0.5. This is $P(B = MM|A = I)$. We want the opposite conditional probability, $P(A = I|B = MM)$. This is not immediately obvious, but we can use Bayes theorem to calculate it:

$$P(A = I|B = MM) = P(B = MM|A = I)P(A = I)/P(B = MM)$$

For the calculation we need $P(A = I)$ and $P(B = MM)$.

$P(A = I)$ is the marginal probability (i.e. the probability in the overall population) that twins are identical. This is currently about $P(A = I) = 0.12$. Fertility drugs have greatly increased the chances of having twins but don't affect the probability of identical twins, so the probability that twins are identical has been decreasing. (Efron, (2005) uses a probability of $P(A = I) = 1/3$.)

$P(B = MM)$ is the total probability (among both identical and fraternal twins) of having twins that are both males. It is not immediately known, but can be calculated. Equations (2) and (3) together imply that

$$f_Y(y) = \int f_{Y|X}(y|x)f_X(x)dx.$$

Here we use the discrete version, replacing the integral by a sum (and changing notation so that $Y = B$ and $X = A$):

$$P(B = MM) = P(B = MM|A = I)P(A = I) + P(B = MM|A = F)P(A = F)$$

$$= 0.5 * 0.12 + 0.25 * 0.88 = 0.28$$

We can now calculate

$$P(A = I|B = MM) = \frac{P(B = MM|A = I)P(A = I)}{P(B = MM)} = 0.5 * 0.12/0.28 = 0.21$$

Three aspects of the calculation in the example should be noted. First, the calculated probability that the twins are identical does not just depend on the measurement (that there are two boys). Rather, we have injected additional information, that the overall fraction of identical twins in the population is 0.12. This is termed 'prior' information. It is the information that we had before doing the measurement. Prior information plays an essential role in Bayesian statistics.

Secondly, note the interplay between prior information and measurement information. In the overall population there is a 12% chance that twins are identical, and that would be the probability that would be announced to the woman before the sonogram. As a result of the sonogram, the probability is updated to 21%. This is a result of combining the prior information and the measurement information.

Thirdly, the calculation is quite simple, except for the denominator $P(B = MM)$. For continuous distributions we need to do an integral in place of a sum to obtain the denominator, and the dimension of the integral is the dimension of B. If this is high (which is often the case), the integration presents difficult or insurmountable problems. There are, however, algorithms that allow one to do a Bayesian analysis without actually calculating that integral.

7.3 Bayesian Parameter Estimation in Modeling

We now show how to apply the Bayesian approach to model parameter estimation. We assume we have a model as in Equation (14), with unknown parameter vector θ. We assume also that we have sampled from the population, thus obtaining a sample of n values of Y and corresponding X.

For parameter estimation our two random variables are Y = results of measurements on sample and θ = parameter values. Bayes' theorem is then

$$f_{\Theta|Y}(\theta|y) = \frac{f_{Y|\Theta}(y|\theta)f_{\Theta}(\theta)}{f_Y(y)} = \frac{L(\theta|y)\pi(\theta)}{f_Y(y)}$$

The second expression on the right is the same as the first, but with the standard Bayesian notation.

The left hand side is the probability distribution of the parameters, given our measured values. This is called the posterior distribution. That is what we want; it contains all the information about the parameters, both from the experiment and prior to the experiment.

On the right hand side, the distribution $f_{Y|\Theta}(y|\theta)$ is the probability of observing y if the parameter values are equal to θ. This is known as the likelihood, noted $L(\theta|y)$, and is considered as a function of the parameters, given the data.

The distribution $f_{\Theta}(\theta)$, generally noted $\pi(\theta)$ in Bayesian analysis, is termed the 'prior' distribution of θ. It is the PDF of the parameters as far as we know, before having the results of the measurements Y. If one has essentially no prior information, one can use a 'non-informative' prior. For example, for one parameter one could use as prior a uniform distribution $U(-1000000,1000000)$ which simply says that the value has equal probability of being anywhere between -1000000 and $+1000000$.

The denominator on the right-hand side, $f_Y(y)$, is a constant which does not depend on θ. As we said, this is often very difficult to calculate, and in most cases one uses algorithms that do not require this calculation.

It should be noted in the limit of a very large amount of data, or a very uninformative prior, the prior information has a negligible effect on the posterior distribution. In this case, the posterior distribution is essentially determined by the likelihood, so the maximum of the posterior is the same as the maximum likelihood estimate. In fact, Bayesian and frequentist results are often similar or identical. However, when the prior has a large effect this will in general not be the case.

7.4 Frequentist or Bayesian?

A major argument that Bayesians make against frequentists is that frequentist reasoning is based on predicting what would happen if the experiment or the observations were repeated many times. In fact, often one has no intention of repeating an experiment. It seems then unrealistic to base conclusions on that unlikely or impossible repetition. Bayesian calculations, on the other hand, are all done conditionally on the results of the experiment that was actually carried out.

The major argument of frequentists against a Bayesian approach is the need for prior information. The Bayesian approach has an absolute need for a prior distribution. Frequentists argue that this introduces subjectivity into the analysis. Different people could have different information about the prior distribution, or different ways of interpreting the available information. Thus, even though faced with the same measured values, different Bayesians could come to different conclusions. (The argument does not include cases where one has solid quantifiable prior information. For example, if we had a previous series of measurements of Y, then both frequentists and Bayesians would take it into account. The problems arise in translating rather vague information into a prior.)

Bayesians, in rebuttal, note that there is always subjectivity in statistical analyses. Frequentists also have to make assumptions, for example about distributions and response functions, and different statisticians might make different assumptions. Even frequentists, faced with the same data, could come to different conclusions. Also, often there is real prior information available even if it is somewhat vague. It seems inefficient to ignore it.

In systems modeling there is in general information about many of the parameters from studies of the underlying processes. This would be the prior information, to be combined with the information contained in measurements on the overall system. It is tempting to use a Bayesian approach to take advantage of the prior information.

EXERCISES

1. Let $F_X(x)$ be the cumulative distribution function of some random variable X.
   a. For this variable, $F_X(3) = 0$. What can we say about the range of values where this variable has non-zero probability? What does this imply for the probability density function?
   b. For this variable, $F_X(8) = 1$. What can we say about the range of values where this variable has non-zero probability? What does this imply for the probability density function?
   c. For this variable, $F_X(x_1) = F_X(x_2)$ for $x_2 > x_1$. What can one say about the probability that X is in the range $x_1 < x \leq x_2$? What does this imply for the probability density function?

2. Let X and Y be two random variables with probability distribution functions $f_X(x)$ and $f_Y(y)$, respectively. We have $f_X(x) = f_Y(x)$ for $3 \leq x \leq 5$. What can we deduce about the cumulative distribution functions $F_X(x)$ and $F_Y(y)$?

3. X is a random variable with expectation 4 and variance 2.
   a. What is the best (minimum mean squared error) constant for predicting a future value of X?
   b. What is the average squared difference between this predictor and the true value?

4. X is distributed as a gamma distribution with parameters shape = 2, scale = 1. Y is distributed as a normal distribution with expectation 2 and

variance 3. Z is distributed as a uniform distribution with lower bound 0 and upper bound 4.

    a. What is the best (minimum mean squared error) constant for predicting a future value of X? Of Y? Of Z?

    b. For which variable is the prediction error (i) largest? (ii) smallest?

    c. One decides to use 3 as the predictor of future values of Z. What is the mean squared error of this prediction?

5. In the region of interest, the joint distribution of rainfall (X, in mm during the growing season) and maize yield (Y in t/ha) is treated as a discrete distribution, with a probability of 0.1 for each of the pairs of values of Table 2.3.

    a. What is the expectation of the marginal distribution of X?

    b. What is the expectation of the marginal distribution of Y?

    c. What is the expectation of the conditional distribution $Y|X = 400$?

    d. What is the expectation of the conditional distribution $Y|X = 500$?

    e. Calculate the covariance and correlation of X and Y. Are X and Y independent random variables?

    f. Use Equation (2) to evaluate $f_{Y|X}(7|600)$ and $f_{X,Y}(600|7)$. Show that Equation (15) holds for $f_{X,Y}(600|7)$.

TABLE 2.3 Joint distribution of rainfall (X) and maize yield (Y)

X	300	300	400	400	500	500	500	600	600	600
Y	4.0	2.5	4.5	6.5	2.7	4.2	7.0	6.0	5.0	7.0

6. In problem (5), we want to predict Y using a function of X.

    a. For each value of X, what is the best predictor?

    b. For each value of X, what is the squared error of prediction for the best estimator? Indicate the units.

7. Suppose that rainfall and maize have a bivariate normal distribution. The expectation and variance for rainfall (X) are 500 and 11,000, respectively, in units of mm or mm^2 during the growing season. The expectation and variance for yield (Y) are 4.8 and 1.7, respectively, in units of t/ha or (t/ha)^2. The covariance is $\mathrm{cov}(X,Y) = 0.98$. In Equation (9) then $\Sigma_{12} = \Sigma_{21} = 0.98$, $\Sigma_{11} = 11,000$ and $\Sigma_{22} = 1.1$.

    a. What is the mean of the conditional distribution $Y|X = 400$?

    b. What is the variance of the conditional distribution $Y|X = 400$?

    c. What is the best predictor of Y that is a function of X?

    d. What is the squared prediction error of this predictor?

8. A simple random sample size of 6 from the distribution of some variable Y is shown in Table 2.4.

a. What is the sample mean of Y?
b. What is the sample variance of Y?
c. What is the estimated standard deviation of the sample mean of Y?
d. We consider continuing the sampling, to double the sample size. Approximately how large do you expect the standard deviation of the sample mean to be for the doubled sample size?
e. Suppose that this is not the result of simple random sampling, but rather that there is a correlation $r = 0.2$ between each pair Y_i, Y_j for $i \neq j$. What then is the estimated standard deviation of the sample mean of Y?
f. What is the effective sample size?

TABLE 2.4 Sample values for some variable Y

i	1	2	3	4	5	6
Y_i	6.5	7.0	6.5	4.5	6.0	2.7

9. Suppose that Y, X_1, X_2, and X_3 are jointly distributed random variables. The conditional distribution of Y is

$$Y|X_1, X_2, X_3 \sim N(\theta_0 - e^{\theta_1 X_1} + \theta_2 X_2 + \frac{\theta_3}{X_3}, 2)$$

$$\theta_0 = 1, \quad \theta_1 = -0.03, \quad \theta_2 = 0.01, \quad \theta_3 = 0.05$$

a. What is the best model for predicting Y from X_1, X_2, and X_3?
b. What is the mean squared error of that prediction?
c. Suppose that the parameters $\theta_0, \theta_1, \theta_2, \theta_3$ need to be estimated from a sample.
   i. What is the best functional form for the model?
   ii. Can you be sure?
   iii. Might it depend on the sample? Explain your answer.

10. Among students in a certain class, it is found that there is a correlation between time devoted to homework and resulting grade. We define two random variables X = grade ($X = A$ or B or C) and H = hours per day devoted to homework ($H = 3$ or 4 or 5). Overall, 30% of students get an A, and 40% work 5 hours a day. A survey has found that among A students, 60% work 5 hours a day.
a. What is the prior possibility of getting an A?
b. Given that a student works 5 hours a day, what is the probability that he gets an A?

REFERENCES

Efron, B., 2005. Bayesians, Frequentists, and Scientists. Journal of the American Statistical Association 100, 1–5.

Lindquist, J.L., Arkebauer, T.J., Walters, D.T., Cassman, K.G., Dobermann, A., 2005. Maize radiation use efficiency under optimal growth conditions. Agronomy Journal 97, 72–78.

## Chapter 3

The R Programming Language and Software

1 INTRODUCTION

1.1 What Is R?

R is a programming language and a software environment (R Development Core Team, 2013). It is mainly used for data analysis, statistical computing, and graphics and has become the reference working software tool in many fields of research and development.

R is an implementation of the S programming language (Becker, Chambers, and Wilks; S-PLUS software) combined with lexical scoping semantics inspired by Scheme (Sussman). (Lexical scoping determines the rules used to search for a variable name. This is important when the same name is used with different values in different places.) R was initially created in 1993 by Ross Ihaka and Robert Gentleman (University of Auckland, New Zealand), but is now developed by the R Development Core Team since 1997. The name R is from the first names of Robert Gentleman and Ross Ihaka, but is also partly a play on the name of the earlier S language.

R is a simple and very high level programming language:

- The syntax is simple. For example, you do not need to specify the type of variable (integer, real, logical, etc.).
- R has many powerful functions, which allow you to do graphics, statistical analyses, matrix calculations, and much more with only a few lines of code.

R is an interpreted language (not compiled). This makes it easy to debug—you can execute one line at a time to verify the calculations. But it also makes execution somewhat slower than compiled languages.

R is free software, available under the GNU General Public License. Pre-compiled binary versions are provided for various operating systems (UNIX platforms, Windows, and MacOS). The source code is written in different various programming languages: C, Fortran, and R.

Full documentation is provided on the website http://www.r-project.org (R Project for Statistical Computing), which can be accessed directly from

Working with Dynamic Crop Models. DOI: http://dx.doi.org/10.1016/B978-0-12-397008-4.00003-4

within the R software. There is also a plethora of books and online material for learning to work with R. Many of these sources can be found on the R project site (http://cran.r-project.org/other-docs.html).

1.2 Why R?

Throughout this book, we show how the methods that are presented can be implemented using R. In principle one could do the calculations with any general programming language, but R has some very convincing advantages. First of all, it is a powerful language for doing mathematics and particularly statistics, because it works on vectors and matrices and has many built-in functions for statistics. Secondly, it is an open language, and anyone can contribute new packages that are then available from the R download sites. Most of the methods discussed in this book are now available as functions in such packages, which means that the methods are very easily applied. Also, we have developed a package that is a companion to this book, and, it contains all the data and code you will need to do the examples in the book for yourself, or to do the exercises.

Another major motivation for using R is its simplicity. It is really quite easy to start programming in R. Even if you are an R novice, you can quickly begin using R to run the methods in this book.

1.3 What's in This Chapter?

This chapter is meant as a rather solid introduction to R. Even if you are a complete novice, you should be able to start programming in R, and actually be able to do some rather advanced manipulations if you read this chapter carefully.

The chapter begins with instructions on how to obtain R, open the interface, and start using it. Then the notion of R objects is presented. Vectors are discussed in detail, and then other types of objects are explained (matrices, data frames, and lists). Input/output is discussed, along with methods of program control (loops and conditional execution). The notion of a function is explained, and we explain how to write your own functions. The following section is an introduction to doing graphics with R. We take a brief look at the statistical capabilities of R. We then explain how to download and use packages in R, which provide functions in specialized areas, for example the R package ZeBook associated with this book (see also Appendix 2). Finally we discuss two topics that are particularly important for systems models. The first is how to use R when the model itself is programmed in a different language. The second problem is that of execution time. We propose methods for reducing computing time. This includes an explanation of how to parallelize code with R.

2 GETTING STARTED

2.1 How to Install the R Software

- Go to the R Project for Statistical Computing website http://www.r-project.org.
- Click on CRAN in the section *Download, Packages in* the left hand column of the page.
- Choose a CRAN site near you. You can choose any site, but it may be faster if you use a nearby site.
- In the *download and install R* box, click on *Download R for Windows* if you are a Windows user. If not, you also have the choice of R for Linux or Mac OS X.
- On the next page, under subdirectories, choose *base*.
- On the next page, click on *Download R 3.0.1 for Windows* (52 megabytes, 32/64 bit). The number of the newest version will change over time.
- Execute the downloaded file to install it, and choose the interface language. You can keep the default setting for the installation.

2.2 The R Interface

When you open the R software, you will see the graphic user interface for R (RGui) (Figure 3.1). On this interface, you will see a toolbar at the top and below that the R console window with blue text providing some basic information about R. Just below that is the prompt symbol, $>$. When you type a command, it will automatically appear after the prompt.

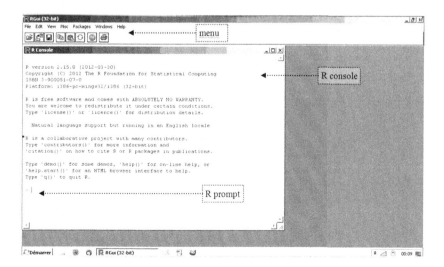

FIGURE 3.1   User interface when R is opened under windows (Rgui.exe).

2.3 Notation for R Code

We will use the following style for R instructions in this chapter.

# this is a comment	a comment begins with #
1 + 2	the instruction to type in the R prompt
[1] 3	the result printed in the R consol

2.4 Using R as a Simple Calculator

You can simply type mathematical expressions in the console window. R immediately does the calculation and outputs the results. Results are printed on consol just after the instruction on a line beginning with [1].

```
7+12
[1] 19
6-8
[1] -2
3*5
[1] 15
4/2
[1] 2
2^6
[1] 64
3^2
[1] 9
3>2
[1] TRUE
3>2&3<5
[1] TRUE
```

Some basic arithmetic and logical operations in R are shown below.

Arithmetic Operations	Description
+	Add
−	Subtract
*	Multiply
/	Divide
^	Exponentiation

The arithmetic operations work on numbers, but also on logical values, in which case TRUE is treated as 1 and FALSE is treated as 0. Thus TRUE + TRUE results in 2.

```
TRUE+TRUE
[1] 2
```

The logical operations can be applied to numbers or character strings. The result is a logical vector whose elements are either TRUE or FALSE (see below for examples).

Logical Operations	Description
> =	greater than or equal to
< =	less than or equal to
= =	exactly equal to
&	logical and
|	logical or
%in%	x %in% y is TRUE if x is contained in y

Another basic operation is outputting results. An object or an expression without an assignment statement causes the result to be printed on the monitor. There is also a print instruction, which has the form `print(x)`.

All the standard mathematical functions are available: `sqrt(x)`, `sin(x)`, `cos(x)`, `tan(x)`, `asin(x)`, `acos(x)`, `atan(x)`, `atan2(x)`, `log(x)`, `log10(x)`, `exp(x)`,... where x is a number.

```
sqrt(10)
[1] 3.162278
```

2.5 Using a Script Editor

The other basic window of R is a script window. You can create a script window by selecting *File* in the menu bar and then *New script*. Any commands that you can type in the console window you can also type in the script window. The advantage is that you can save the contents of the script window in a file. To do so, make the script window the active window (by clicking on it). Then choose *File* in the menu bar then *Save as*. When you come back to R, you can retrieve that script by selecting *Open Script* and then browsing to find the script file you saved (Figure 3.2).

The R script window in Windows has the basic text editing functions of Windows (copy, cut, paste, select, etc.).

To execute code that is in the script window, there are several possibilities. You can use the edit pull-down menu in the menu bar when the script window is active. There you have the choice between executing all the instructions in the script window, or executing just the code you have selected. This can be one or several lines of code, or even part of a line. Or you can copy and paste parts of the code into the command window.

For Windows users, you can also use a different text editor rather than the default R script window, and we recommend this choice. There are several editors specifically designed for the R language. They use different colors to help distinguish comment from code, to indicate R keywords, to help balance parentheses, etc. Two popular editors for R are Tinn-R (http://www.sciviews.org/Tinn-R) and Rstudio (http://www.rstudio.com). One can also use a generic editor like Vim (http://www.vim.org).

FIGURE 3.2   User interface of R with the default R script editor (right side) and the R console window (left side).

2.6 The Notion of an R Program

A program in R is a collection of commands which are written to accomplish some function. There are no obligatory opening or closing commands that define a program. Pieces of the program can be run separately. In fact, you will often be doing this, in order to check parts of the program.

Within the program you will have commands and comments. A comment begins with the symbol #. It is not executed.

2.7 Debugging an R Program

Most programs will require stages of testing and debugging. If there is a syntax mistake in a program, R will output an error message explaining where the error occurred and giving information about the problem. R will usually then ignore the erroneous instruction and continue with the execution of the rest of the program. Common syntax errors are unbalanced parentheses or brackets or quotes, using a vector before it has been defined (possibly due to a mistake in the name in the definition or use statement; for example, X is different from x) or calling a function with an incorrect number of arguments. These errors are quite easy to find and correct. Using an adapted text editor with highlighting and other checking capabilities may be helpful too.

To verify that the program is doing the calculations intended, one can step through the program piece by piece and check the values of the variables created, because R is an interpreted language. This is very simple, since R allows you to highlight a piece of code and then execute it.

Another useful debugging tactic is to print variables using `print(name_variable)`, especially within functions, because those variables are not available outside the function.

One trick for reducing errors is to add the instruction `rm(list = ls())` at the start of the script to clear all variables and so check that the script is self-sufficient. However, don't do this if you have variables that take time to compute and that you want to keep.

2.8 Need Help?

For any R function, you can access a help document by simply typing a question mark, followed by the name of the function or the help function. You can also search for a word using the `help.search` function.

```
# open the help document for function ls
?ls
help(ls)
# search among the help pages for a word
help.search("trigo")
```

A typical help page is structured as shown below:

name_function {base or name_package} R Documentation

Title of the section	Description
Usage	The syntax of the function with its arguments
Arguments	List and description of the arguments
Details	
Value	What is the result of calling the function?
References	
See Also	Other related functions.
Examples	Simple example you can run.

3 OBJECTS IN R

3.1 Creating Objects

In most cases one wants to work with named objects rather than directly with numbers. Either the equal sign = or the R specific <− sign can be used to store a value in a named object. The object is created when it appears in an assignment statement. An object must have an assigned value before it can be used.

Names are case sensitive, so *x* and X are different objects. Also, you probably want to avoid redefining names of pre-existing R functions or constants (for example pi).

```
x = 1
x
[1] 1
y <- 3*5.5
y
[1] 16.5
x+y
[1] 17.5
ls()
[1] "x" "y"
# names of variables are case sensitive
X
Error: object 'X' not found
```

A line without an assignment causes the result of the line to be printed on the screen.

3.2 Types of Objects

Anything that is stored by R in memory is an object. In the above example, x and y are objects. Different types of objects are stored in different ways.

The main storage modes of R object are integer (1, 2, 3, ...), numeric (identical to double for real number: 1.5, 0.333, pi, ...), logical (TRUE or T, FALSE or F) and character ("a", "b", "yield"). Another type factor is an enumerated type to handle categorical data, values from defined levels (categories).

There are several kinds of objects in R with which we will work:

- Vectors (vector) are the basic objects in R. The vector elements can be "numeric", "logical", "character", "factor", ... Note that the first objects you have created before (x and y) as real scalar are considered to be vectors of length 1.
- Matrices (matrix) and data frames (data.frame) can contain data structured in two dimensions.
- Lists (list) consist of an ordered collection of objects, possibly of different types, known as components.
- Functions (function) are used to store procedures.

We will discuss each of these types of objects in the next sections.

There are many instructions for obtaining information about objects created: str(nameofobject), typeof(nameofobject), attributes(nameofobject).

You can see all the named objects you have created using the instruction ls(). You can eliminate an object, for example *x*, using the command rm(x). You can eliminate all the objects that you have previously created using rm(list = ls()).

4 VECTORS (NUMERICAL, LOGICAL, CHARACTER)

4.1 Creation of a Vector

One way to create a vector with multiple elements is to use the function c (a,b,c,...). This creates a vector containing values a, b, c, ... The name of the function is c for concatenate.

```
v <- c(1, 3, 5.5)
v
[1] 1.0 3.0 5.5
x = 1
u <- c(x, 3, 5.5)
u
[1] 1.0 3.0 5.5
# you can combine existing vectors to create a new one
w <- c(v,10,15,u)
w
[1] 1.0 3.0 5.5 10 15 1.0 3.0 5.5
```

Note that the lines of output start with [1] indicating that the first item on the line is element number 1 of the vector.

Often one wants to create a vector with a simple sequence of numbers, or by repeating some number multiple times. The function a:b creates a vector containing the integers a, a + 1, a + 2, ... b − 1, b. A more general way of creating sequences of numbers is seq(a,b,by) which gives a starting value a, an ending value b and the step size by.

```
x <- 1:4
x
[1] 1 2 3 4
x <- seq(2,7,2)
x
[1] 2 4 6
x <- rep(3,5)
x
[1] 3 3 3 3 3
x <- rep(c(3,4),3)
x
[1] 3 4 3 4 3 4
```

Creating vectors	Description
c(a,b,c, ...)	Creates a vector with values a, b, c, ...
a:b	Creates a sequence of integers from a to b
seq(from,to,by),	Generates a sequence between a starting and a maximal value with an increment
rep(x,n)	Replicates n times x, a single number or a vector

One can give names to the individual elements of a vector, using names (x) <- vectorofnames as shown below. Then one can refer to the elements by name as shown in the next subsection.

```
v <- c(2,3,7)
names(v) = c("first","second","third")
v
  first  second  third
     2       3      7
```

4.2 Subscripting a Vector

Subscripting in R is very important. One can do quite complex operations with very few and compact instructions by using the subscripting possibilities of R.

Subscripts are enclosed in square brackets. This x[1] is the first element of x. The subscript can itself be a vector. Thus x[c(1,5,3)] returns a new vector with 3 elements, which are the first, fifth, and third element of *x*. The subscript can have negative values, which means to remove the subscripted values from the vector. Thus x[c(-1,-3)] is a new vector which is the same as *x* with the first and third elements removed. Note that it is not allowed to mix positive and negative indices in the subscript.

The subscript can contain a vector of logical values of the same length as the vector. This selects the values in positions with TRUE. This provides a very simple way to select elements of a vector which satisfy certain conditions. In the example below we want to select the elements of tmax that are greater than or equal to 15. The command tmax >= 15 creates a logical vector that is of the same length as tmax, and that is TRUE for those elements of tmax >= 15 and false otherwise. Then the command tmax[tmax >= 15] is a new vector which only contains the x values greater than or equal to 15.

```
# maximum temperature for 8 days
day <- c(11,12,13,14,15,16,17,18)
tmax <- c(10,14,13,15.5,10,20.5,15,20)
# first temperature
tmax [1]
10
# tmax on days 1,3,5 and 7
tmax[c(1,3,5,7)]
10 13 10 15
# all tmax values except first
tmax[-1]
14.0 13.0 15.5 10.0 20.5 15.0 20.0
# selection of tmax values at or above 15 (first case)
# and then at or above 15 and below 20 (second case)
tmax>=15
[1] FALSE FALSE FALSE TRUE FALSE TRUE TRUE TRUE
tmax[tmax >= 15]
```

```
15.5 20.5 15.0 20.0
(tmax>=15)&(tmax<20)
[1] FALSE FALSE FALSE TRUE FALSE FALSE TRUE FALSE
tmax[(tmax>=15)&(tmax<20)]
[1] 15.5 15.0
# tmax values below 15 or at or above 20
tmax[(tmax<15)|(tmax>=20)]
[1] 10.0 14.0 13.0 10.0 20.5 20.0
tmax[tmax %in% c(10,15,20)]
[1] 10 10 15 20
# tmax on day 14
day==14
[1] FALSE FALSE FALSE TRUE FALSE FALSE FALSE FALSE
tmax[day==14]
[1] 15.5
# tmax values for days in the set 12,13,14,15,16,17,18
day%in%12:18
[1] FALSE TRUE TRUE TRUE TRUE TRUE TRUE TRUE
tmax[day%in%12:18]
[1] 14.0 13.0 15.5 10.0 20.5 15.0 20.0
```

Subscripting Vectors	Description
x[n]	nth element
x[-n]	All but the nth element
x[1:n]	First n elements
x[-(1:n)]	Elements from n + 1 to the end
x[c(1,4,2)]	Specific elements
x["name"]	Element named "name"
x[x > 3]	All elements greater than 3
x[x > 3 & x < 5]	All elements between 3 and 5
x[x %in% c("a","b")]	Elements in the given set

One can change specific elements of a vector by assigning new values.

```
# modify part of vector
tmax <- c(10,14,13,15.5,10,20.5,15,20)
tmax[1] <- 100
tmax
[1] 100.0 14.0 13.0 15.5 10.0 20.5 15.0 20.0
tmax[tmax>=15] = 15
tmax
[1] 15 14 13 15 10 15 15 15
```

4.3 Operations on Vectors

There are many useful functions for operating on vectors. Illustrated below are functions for calculating the minimum value among the elements of the vector, the maximum, the mean, and the cumulative sum.

```
#maximum temperature for 8 days
day <- c(11,12,13,14,15,16,17,18)
tmax <- c(10,14,13,15.5,10,20.5,15,20)
min(tmax)
[1] 10
max(tmax)
[1] 20.5
mean(tmax)
[1] 14.75
cumsum(tmax)
[1] 10.0 24.0 37.0 52.5 62.5 83.0 98.0 118.0
```

The sort function applied to a vector will return a vector with the same elements, sorted into ascending or descending order (using the argument decreasing=TRUE). The order function will return a vector of integers, giving the position of the smallest value, then the position of the second smallest etc (for ascending order).

```
tmax <- c(10,14,13,15.5,10,20.5,15,20)
sort(tmax, decreasing = FALSE)
[1] 10.0 10.0 13.0 14.0 15.0 15.5 20.0 20.5
order(tmax)
[1] 1 5 3 2 7 4 8 6
tmax[order(tmax,decreasing=TRUE)]
[1] 20.5 20.0 15.5 15.0 14.0 13.0 10.0 10.0
```

A useful function for working with character vectors is the paste function. This instruction combines multiple character strings into a single string. This can be very useful in print instructions, since the print function only accepts a single object to output. Note that numbers are converted to characters automatically as necessary. The paste instruction is illustrated below.

```
yield <- 11.6
print(paste("yield = ",yield, "t/ha", sep = " "))
[1] "yield = 11.6 t/ha"
```

Further operations are presented in the table below.

Operations on Vector	Description
length(x)	Number of elements in x
max(x)	Maximum of the elements of x
min(x)	Minimum of the elements of x
range(x)	A vector c(min(x), max(x))
sum(x)	Sum of the elements of x
cumsum(x)	A vector whose ith element is the sum from x[1] to x[i]
diff(x)	A vector of length length(x) − 1. The ith element is x[i + 1] − x[i]
prod(x)	Product of the elements of x
mean(x)	Mean of the elements of x
weighted.mean(x, w)	Mean of x with weights w

median(x)	Median of the elements of x
var(x)	Variance of the elements of x (using length(x) − 1 as a divisor)
cov(x,y)	Covariance of x with y
sd(x)	Standard deviation of x
cor(x, y)	Linear correlation between x and y
quantile(x,probs =)	Sample quantiles corresponding to the given probabilities. The default is probs = c(0,.25,.5,.75,1)
scale(x)	Centers the data and divides by the standard deviation: (x − mean(x))/sd(x).To center only use the option center = FALSE, to reduce only use scale = FALSE (by default center = TRUE, scale = TRUE)
pmin(x,y,...)	A vector whose ith element is the minimum of all elements of x [i], y[i], . . .
pmax(x,y,...)	A vector whose ith element is the maximum of all elements of x[i], y[i], . . .
rank(x)	Ranks of the elements of x
unique(x)	A vector consisting of the unique elements of x (duplicates are removed)
round(x, n)	The elements of x rounded to n decimals
sample(x, size)	Resample randomly and without replacement size elements in the vector x, the option replace = TRUE allows you to resample with replacement
which.max(x)	Returns the index of the greatest element of x
which.min(x)	Returns the index of the smallest element of x
rev(x)	Reverses the elements of x
sort(x)	Sorts the elements of x in increasing order; to sort in decreasing order: rev(sort(x))
cut(x,breaks)	Divides x into intervals (factors); breaks is the number of cut intervals or a vector of cut points
match(x, y)	Returns a vector of the same length as x with the elements of x which are in y (NA otherwise)
which(x == a)	Returns a vector of the indices of x for which the argument is TRUE
na.omit(x)	Suppresses the observations with missing data (NA) (suppresses the corresponding line if x is a matrix or a data frame)

When treating large data sets, it is fairly common to encounter missing values. This can pose problems for data treatment. Most functions in R have arguments that allow you to specify how to treat missing values. Missing values in R are assigned the value NA. Computations using NA will normally result in NA.

```
# define a vector with a missing value
x <- c(3,7,4,NA,2)
2*x
[1]   6 14   8 NA   4
mean(x)
[1] NA
```

There are several ways of dealing with NA. A first function is na.omit (x), where x is a data frame. This creates a new data frame without the lines that contain NA. In many functions, you can add information as to how to treat NA as an optional argument. The argument na.rm = TRUE means that the NA should be ignored in applying the function.

```
na.omit(x)
[1] 3 7 4 2
attr(,"na.action")
[1] 4
attr(,"class")
[1] "omit"
mean(x, na.rm = TRUE)
[1] 4
```

4.4 Combining Vectors

There are many instructions which operate on two vectors of the same length. For example, the vector sum a + b adds each element of a to the corresponding element of b, so the two vectors must have the same length. If you have two vectors of unequal length, then R will repeat the shorter vector as many times as necessary to attain the length of the longer vector. If the length of the longer vector is a multiple of the length of the shorter vector, there is no particular message. If it is not a multiple, R will do the operation successfully, with only a warning message.

This is an example of the fact that R usually tries to make things work. This can be very useful, but may also let errors go unnoticed.

```
tmax <- c(10,14,13,15.5,10,20.5,15,20)
delta <- c(1,5)
tmax + delta
[1] 11.0 19.0 14.0 20.5 11.0 25.5 16.0 25.0
delta <- c(1,5,10)
tmax + delta
[1] 11.0 19.0 23.0 16.5 15.0 30.5 16.0 25.0
Warning message:
In tmax + delta :
  longer object length is not a multiple of shorter object length
```

5 OTHER DATA STRUCTURES

5.1 Matrices

A matrix is a two-dimensional array where all elements are of a single type (number, logical, character). The simplest way to create a matrix is by using the matrix function. The first input is the list of values. Then one can specify the number of rows and/or columns. In the example below, there are 16 values and it is specified that the number of rows is 8. The function deduces that the number of columns is 2. Another argument indicates whether the matrix is to be filled up one row at a time (byrow = TRUE) or one column at a time (byrow = FALSE). Optional arguments can be used to give names to the rows and columns.

Another common way to create a matrix is by stacking vectors as rows of a matrix using the function rbind or cbind to combine vectors as columns.

One can create a diagonal matrix using the function diag(x), where x is a vector that gives the values on the diagonal. If the length of x is n, this creates an n by n matrix.

```
# create a matrix of maximum temperature with two columns
mat <- matrix(c(11,12,13,14,15,16,17,18,10,14,13,15.5,10,20.5,15, 20),
ncol=2,byrow=FALSE, dimnames = list(NULL, c("day", "tmax")))
mat
       day tmax
[1,]   11 10.0
[2,]   12 14.0
[3,]   13 13.0
[4,]   14 15.5
[5,]   15 10.0
[6,]   16 20.5
[7,]   17 15.0
[8,]   18 20.0
# You can also create matrices using cbind to combine by column
day <- c(11,12,13,14,15,16,17,18)
tmax <- c(10,14,13,15.5,10,20.5,15,20)
cbind(day,tmax)
        day tmax
[1,]    11 10.0
[2,]    12 14.0
[3,]    13 13.0
[4,]    14 15.5
[5,]    15 10.0
[6,]    16 20.5
[7,]    17 15.0
[8,]    18 20.0
# Creating a diagonal matrix
xmat <- diag(c(3,7,2.4))
xmat
      [,1] [,2] [,3]
[1,]    3    0  0.0
[2,]    0    7  0.0
[3,]    0    0  2.4
```

Subscripting matrices is very similar to subscripting vectors but of course with two indices. One new type of subscript here has a blank in place of one of the subscripts. In this case, all the elements of the missing subscript are selected. Thus x[i,] is a vector containing all the elements of row i and x[,j] is a vector containing all the elements of column j. Names can be used instead of numbers for rows or columns, if these have been defined.

```
# subscripting a vector, but in 2 dimensions (rows and columns)
mat[4,2]
tmax
15.5
# a missing index means selecting all values for that index
mat[,2]
[1] 10.0 14.0 13.0 15.5 10.0 20.5 15.0 20.0
mat[1,]
  day tmax
   11   10
mat[,"tmax"]
[1] 10.0 14.0 13.0 15.5 10.0 20.5 15.0 20.0
```

Following is a table of ways of indexing a matrix, and of several functions which work with matrices.

Operations on Matrices	Description
matrix(x,nrow = ,ncol =)	Creates a matrix with elements x, with nrow rows and ncol columns. If x is not long enough to fill the matrix, it is recycled.
x[i,j]	The element in row i, column j
x[i,]	All of row i
x[,j]	All of column j
x[,c(1,3)]	All of columns 1 and 3 (a matrix)
x["name",]	All of the row by name
t(x)	Transpose of x
diag(x)	Diagonal elements of x
%*%	Matrix multiplication
solve(a,b)	Solves a %*% x = b for x
solve(a)	Matrix inverse of a
rowSums(x)	Sums of rows
colSums(x)	Sums of columns
rowMeans(x)	Means of rows
colMeans(x)	Means of columns

5.2 Data Frames

A data frame is a table like a matrix, but the columns can be of different types. It is particularly convenient for storing experimental results, where one might have a column with location (a character variable), another with treatment (a categorical variable) and then further columns with results of various measurements (numerical variables).

All the indexing methods that work for matrices also work for data frames. In addition, if the columns are named, one can access them as dataframe$columnname as illustrated below.

```
# create a data frame with maximum temperatures for 8 days
# The three columns are id (a character), day (numeric) and tmax(numeric)
data <- data.frame(id=c("a","b","c","d","e","f","g","h"),
day=c(11,12,13,14,15,16,17,18), tmax=c(10,14,13,15.5,10,20.5,15,20))
data
  id day  tmax
1  a  11  10.0
2  b  12  14.0
3  c  13  13.0
4  d  14  15.5
5  e  15  10.0
6  f  16  20.5
7  g  17  15.0
8  h  18  20.0
# subscripting works as for matrices; One can also select columns by name
data$tmax
[1] 10.0 14.0 13.0 15.5 10.0 20.5 15.0 20.0
data$day%in%12:18
[1] FALSE  TRUE  TRUE  TRUE  TRUE  TRUE  TRUE  TRUE
data$tmax [data$day%in%12:18]
[1] 14.0 13.0 15.5 10.0 20.5 15.0 20.0
# creating a data.frame as all combinations of several factors (a full factorial
design)
expand.grid(n=1:2,p=c("a","b","c"),k=c(TRUE,FALSE) )
    n p      k
1   1 a   TRUE
2   2 a   TRUE
3   1 b   TRUE
4   2 b   TRUE
5   1 c   TRUE
6   2 c   TRUE
7   1 a  FALSE
8   2 a  FALSE
9   1 b  FALSE
10  2 b  FALSE
11  1 c  FALSE
12  2 c  FALSE
```

The function str(nameofdataframe) gives information about the variables in the data frame, as illustrated below.

```
str(data)
'data.frame':    8 obs. of 3 variables:
 $ id  : Factor w/ 8 levels "a","b","c","d",...: 1 2 3 4 5 6 7 8
 $ day : num  11 12 13 14 15 16 17 18
 $ tmax: num  10 14 13 15.5 10 20.5 15 20
```

Operations on Data Frame	Description
data.frame(...)	Creates a data frame
expand.grid()	A data frame with all combinations of the supplied vectors or factors
subset(x, ...)	Returns a selection of x with respect to criteria (..., typically comparisons: x$V1 < 10)
x[,"name"]	Column named "name"
x$name	Same
str(dataframename)	Gives information about the variables in the data frame

5.3 Lists

A list is an ordered collection of objects. The elements of a list can be vectors, matrices, data frames, other lists, or any combination of those. Many functions that have complex results return those results in the form of a list. For example, a nonlinear least squares function could return the parameter values as one element of the list, their standard deviations as another element, the model residuals as a third element, etc. If you write a function (see below) and want it to return several types of results, you will put them in a list and return the list.

A list is created using the function list. Often it is convenient to give names to each element of the list. In the example below, the list has three components (name, area, crop.rotation).

Components can be recovered by name or by number. The first element of the list *x*, for example, is x[[1]]. Note the double brackets: if a single bracket is used, the result is then a list, not the object within it. The name can be used in the double brackets instead of a number, or appended to the name after a $ sign, as illustrated below.

```
# Create a list called Field1 with three named components
Field1 <- list(name = "Field1", area = 10, crop.rotation = c("wheat","maize"))
# Select the second element in the list in three different ways
Field1[[2]]
[1] 10
Field1$area
[1] 10
Field1[["area"]]
[1] 10
# If the list element is a vector, you can subscript it
Field1[[3]][2]
[1] "maize"
# Lists can be elements of a larger list
Field2 <- list(name = "Field2", area = 15, crop.rotation = c("wheat","sugarbeet"))
AllField <- list(Field1, Field2)
```

```
# Operations on elements of a list. First addition. Second logical test.
AllField[[1]]$area+ AllField[[2]]$area
[1] 25
AllField[[1]]$crop.rotation == AllField[[2]][["crop.rotation"]]
[1] TRUE FALSE
```

Operations on Lists	Description
list(...)	Creates a list of the named or unnamed arguments; list(a = c(1,2), b = "hi",c = 3)
x[,c(1,3)]	An extraction of the list with the element 1 and 3
x[[n]]	nth element of the list
x[["name"]]	Element of the list named "name"
x$name	Element of the list selected by its name

6 READ FROM AND WRITE TO FILE SYSTEM

Data you work with are very often existing data stored in the external file system of your computer. To access these data, you need to interact with your file system.

As an example, suppose we have a file *data.dat* (be careful, Windows OS often hides the extension) in a directory *c:/ZeBook/R_intro,* where the data are separated by semicolons. The first line is just general information, and the second line contains the header.

```
data for example
day;tmax;rain;cloud
11;10.0;0;yes
12;14.0;0;no
13;13.0; − ;yes
14;15.5;10;yes
15; − ;5;yes
16;20.5;0;no
17;15.0;0;no
18;20.0;5;yes
```

To input these data into R, we will use the function read.table that reads a file containing a formatted table and creates a data frame from it. read.table requires the pathname of the file to read. Rather than putting the full path name in the read.table instruction, it is often convenient first to set the working directory to the directory where the data file is stored. Then it is sufficient to put just the name of the file into the read.table function.

The working directory is set using the function setwd. As the example below shows, the backward slashes in the Window system must be replaced by forward slashes for R. After the setwd() instruction, one can use getwd()

to print the name of the current directory, and dir() to see the list of files in the directory.

```
# Set working directory.
setwd("C:/ZeBook/R_intro")
getwd()
"C:/ZeBook/R_intro"
# Print out all files in this directory.
dir()
[1] "data.dat"
```

The table can then be read as shown below. The read.table instruction has several arguments: the path name of the file (including the extension), the separator (sep = ";" for semicolon, sep = "/t" for tabulation and sep = " " for whitespace), whether or not there is a header line (header), the character that indicates missing data (na), and the symbol used for the decimal point (dec). You can also skip several lines at the start of the file using the parameter skip.

```
# Read table
data=read.table("data.dat",sep = ";",skip = 1,header = TRUE,na = " - ",dec = ".")
data
   day tmax rain cloud
1   11 10.0    0   yes
2   12 14.0    0    no
3   13 13.0   NA   yes
4   14 15.5   10   yes
5   15   NA    5  <NA>
6   16 20.5    0    no
7   17 15.0    0    no
8   18 20.0    5   yes
```

The file is read automatically as a data frame. The structure of the data frame for the example is shown below.

```
str(data)
'data.frame':   8 obs. of  4 variables:
$ day  : int  11 12 13 14 15 16 17 18
$ tmax : num  10 14 13 15.5 NA 20.5 15 20
$ rain : int  0 0 NA 10 5 0 0 5
$ cloud: Factor w/ 2 levels "no","yes": 2 1 2 2 NA 1 1 2
```

In the example, the variable named "cloud" is treated as a factor. Suppose that you want to force R to treat it as a character string. To ensure that, you can specify the variable type in the read.table function as follows.

```
# read table, specifying the type for each column
# Now the variable named "cloud" is a character vector. (Above it was a factor)
data=read.table("data.dat",sep = ";",skip = 1,header = TRUE,na = " - ",
   colClasses=c("numeric","numeric","numeric","character"))
```

```
str(data)
'data.frame':   8 obs. of  4 variables:
 $ day  : num  11 12 13 14 15 16 17 18
 $ tmax : num  10 14 13 15.5 NA 20.5 15 20
 $ rain : num  0 0 NA 10 5 0 0 5
 $ cloud: chr  "yes" "no" "yes" "yes" ...
```

R cannot easily read data from a Microsoft Excel® file. If your data are in an Excel file, you must save each tab of the Excel file as a text file with a semi-colon as separator (*csv* format). Then read.table can read these tables using the argument sep = ";" for semicolon separators. An alternative is to use a specific package (for example, xlsx or xlsReadWrite) to read Excel files, but these packages depend on the version of Excel and the operating system.

The write.table function writes a table to a file. The arguments in the example below specify the name of the object with the data to be written, the name of the data file (in the working directory), the separator, an argument to indicate whether the names of the columns are to be written in the file, an argument to indicate whether character strings are to be enclosed in double quotes, and finally an argument that gives the symbol to be used to represent a decimal point.

```
write.table(data,file = "data2.dat",sep = ";",col.names = T,row.names = F,quote = F,dec = ".")
```

There is also a simpler export function save that requires just the name of the object with the data and the name of the file. However, the data are then in XDR format. This is not readable by a text editor, but it can be re-imported into R using the load function. It may be very useful for very large data sets. You can also save other types of R objects (list, matrix, function). Then when you load them back, they are recreated with the initial name.

```
# export R data in R format file (not readable with a text editor)
save(data,file = "data.rda")
# import data from rda (it re-creates the data.frame data from the file that was saved)
load(file = "data.rda")
```

Functions to Read from and Write to File System	Description
getwd()	Gets the filepath representing the current working directory of the R process
setwd("path")	Sets the working directory to a specific filepath
dir()	Shows files in the current directory
read.table("myfile", ...)	Reads a file in table format and creates a data frame. Use sep = "" to define the separator; use header = TRUE to read the first line as a header of column names; use

Functions to Read from and Write to File System	Description
	skip = n to skip n lines before reading data. See help(read.table) for other options
read.csv("filename",header = TRUE)	This is the same as read.table but by default reads comma-delimited files
read.delim(("filename",header = TRUE)	This is the same as read.table but by default reads tab-delimited files
write.table(x,file = " myfile ", …)	Saves the data frame x to file "myfile". See help(write.table) for options
save(objects,file,…)	Saves the specified objects to the specified file in XDR format
save.image(file)	Saves all objects of the session to the specified file in XDR format
load()	Loads the datasets in XDR format (must have extension (*.rda) in the XDR platform independent binary format)
save(file,…)	Saves the specified objects (…) in XDR format
save.image(file)	Saves all objects of the session
read.delim("clipboard"), write.table (x,"clipboard", sep = "\t",col.names = NA) read.table("myfile", …)	Reads/writes a table from/to the clipboard for Excel to read a file in table format and creates a data frame. Use sep = "" to define the separator; use header − TRUE to read the first line as a header of column names; use skip = n to skip n lines before reading data. See help(read.table) for other options
read.csv("filename",header = TRUE)	This is the same as above but with defaults set for reading comma-delimited files
write.table(x,file = " myfile ", …)	Saves x after converting to a data frame. See help(write.table) for options
read.delim("clipboard"), write.table (x,"clipboard", sep = "\t", col.names = NA)	Reads/writes a table from/to the clipboard for Excel

R can also interact with several database management systems with specific packages `RODBC`, `DBI`, `RMySQL`, `RPgSQL`.

7 CONTROL STRUCTURES

7.1 Loops

A loop is a programming structure which allows code to be executed several times. It starts with a `for` or `while` statement, and then within curly brackets {} are the instructions that will be repeated.

```
# Here the instructions in curly brackets are done four times, for k = 1,2,3,4.
for (k in 1:4) { print(2^k) }
[1] 2
[1] 4
```

```
[1] 8
[1] 16
# Here is another version which does the same thing
k = 1
while (k <= 4) {
   print(2^k)
   k<-k+1
}
# In this case, you can also get the same result without a loop
k<-1:4
print(2^k)
[1] 2 4 8 16
```

As loop executions are usually slow, a general principle is that loops should be avoided, if one can find a function (apply for example) or subscripting method that does the same thing.

7.2 Conditions

The commands within curly brackets after an if statement are only executed if the condition in the if statement is TRUE. If not they are ignored. An if statement can optionally be followed by an else statement. In that case, the commands within curly brackets after the else statement are executed if the if condition is FALSE.

```
x = 11
if (x < 10) {
   print("less than 10")
} else {
   print("more than 10")
}
[1] "more than 10"
# a more compact way to have one result if the condition is true
# and a different result if the condition is false
x = 5
ifelse((x < 10), "less than 10", "more than 10")
[1] "less than 10"
```

Programming Structure	Description
for(var in seq) expr	Loop for … do …
while(cond) expr	Loop while … do …
if(cond) expr	Condition if … then …
if(cond) cons.expr else alt.expr	Condition if … then … else …
ifelse(test, yes, no)	if test is TRUE, expression yes is evaluated. If FALSE expression no is evaluated

8 FUNCTIONS

Functions are recognized by the fact that they are followed by parentheses. Thus `ls()` is a function call, `print("less than 10")` is a function call, `c(2,3)` is a function call, etc.

The elements within parentheses are the arguments that are passed to the function. There can be zero arguments, as in `ls()`, but usually functions require one or several arguments. In addition to obligatory arguments, some functions have optional arguments, but we will limit the discussion here to the obligatory arguments. These obligatory arguments have names, have a specific order, and may or may not have default values.

Consider the function `diag`, which has the form `diag(x = 1,nrow,ncol)`. The names of the three arguments are `x`, `nrow`, and `ncol`, they are in that order, and the argument `x` has the default value 1.

If one gives all the arguments in the correct order, then they needn't be named. If however the order is changed, or some arguments are omitted because the default values are used, then the arguments must be named. This is illustrated below.

```
# The arguments are not named and are assumed to be in the default order (x,nrow,ncol)
diag(1,2,3)
       [,1] [,2] [,3]
[1,]    1    0    0
[2,]    0    1    0
# x is omitted, so it takes the default value x = 1 and the other arguments must be
named
diag(nrow=2,ncol=3)
       [,1] [,2] [,3]
[1,]    1    0    0
[2,]    0    1    0
```

The value of a function is the result of executing the function. The result of executing `ls()` is a character vector, the result of executing `c(2,3)` is a numeric vector. The result of a function can also be a list. For example the function that does analysis of variance, called `aov`, returns a list with several components (see below). A function can return no value, for example, `plot`.

8.1 Writing Your Own Functions

If you write programs in R, you will eventually want to write your own functions. This is important because functions allow you to do the same calculations several times with different inputs (different arguments) each time. Each time you simply call the function with the new arguments.

Writing your own functions in R is very easy. A function begins with a statement `name_function < − function(arguments)` (the word `function` is a keyword), and then has a body of statements within curly brackets. One of those statements will be a `return(object)` statement. The value of the function (the result of calling the function) is the object in the `return` statement. It can be a list, a vector, a data frame, etc.

```
# Definition of a function called test with two arguments (x and limit)
test <- function(x, limit){
   if (x < limit) {
      result = paste("less than", limit)
   } else {
      result = paste("more than", limit)
   }
   return(result)
}
# main program, that calls function test twice
x = 11
result=test(x,15)
print(result)
[1] "less than 15"
print(test(x,10))
[1] "more than 10"
```

Function Operators	Description
function(arglist) expr	Function definition. Use braces {} around statements
return(value)	At the end of a function, return an object

9 GRAPHICS

R allows you to create wonderful graphics very simply.

A graph is created with the `plot` function. In its simplest form, the function has just two arguments, the vector of x values and the vector of y values. The `plot` function creates the axes and the labels on the axes, and then plots the points.

```
data <- data.frame(day=c(11,12,13,14,15,16,17,18),
tmax=c(10,14,13,15.5,NA,20.5,15,20),
   tmin=c(5.5,9,9,10.5,12,14.5,11,16),rain=c(0,0,NA,10,5,0,0,5))
# simple plot
plot(data$day,data$tmax)
```

The result is shown in Figure 3.3. One can personalize the graph by adding many other arguments (graphical parameters) in the plot function. Some of these are illustrated below. The arguments `xlab` and `ylab` allow you to choose the axis labels, the argument `main` gives the graph title, the argument `type` is "l" for lines and "p" for points (points is the default for the plot function), the argument `lty` gives the type of line (1 for solid line, 2 for dashed line etc.), the argument `lwd` gives the line thickness, the argument `ylim` gives the upper and lower values for the y axis and `xlim` does the same for the x axis.

Very often one wants to add to an existing graph, for example to plot a second set of points. The instructions for adding new points or lines to an existing graph are `points` and `lines`. There can be several instructions

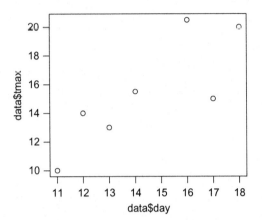

FIGURE 3.3   Graphic obtained with basic instruction `plot` with default parameters.

following a `plot` instruction and each will add new material to the graph. If you want to put several sets of data onto the same graph, you have to make sure that the axes created by `plot` cover all the values that will be plotted. If necessary, you can fix the axes ranges using the `xlim` and `ylim` parameters.

You can easily introduce a legend into a plot by using the function `legend`. The first argument is the location, then a vector of character strings, then vectors of values of plotting parameters.

```
# Plot two variables on the same graph and use graphical parameters
plot(data$day,data$tmax, xlab = "day",ylab = "T
(°C)",type = "l",lty = 1,lwd = 3,ylim=c(0,25))
lines(data$day,data$tmin, lty = 2,lwd = 3)
legend("topleft",c("Tmax","Tmin"),lty = c(1,2),lwd = c(3,3))
```

The resulting plot is shown in Figure 3.4. You can specify global graphical parameters using the `par` function, and they will apply to all subsequent graphs. To see the options available, use `help(par)`. You can draw several figures on the same graph sheet using `par(mfrow = c(nrows,ncolumns))`. This is illustrated in the instructions below and in Figure 3.5. Also illustrated are different values for the graphical parameter `type` and different line types (argument `lty`).

```
par(mfrow = c(2,2))
plot(data$day,data$tmin,lwd = 2, xlab = "day",ylab = "Tmin (°C)",main = "plot default")
plot(data$day,data$tmin,type = "l", lwd = 2, xlab = "day",ylab = "Tmin (°C)",
main = "type = l")
plot(data$day,data$tmin,type = "b", lwd = 2, xlab = "day",ylab = "Tmin (°C)",
main = "type = b")
plot(data$day,data$tmin,type = "o", lwd = 2, xlab = "day",ylab = "Tmin (°C)",
main = "type = o")
```

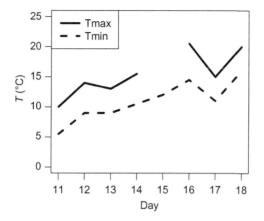

FIGURE 3.4   Graphic combining two series with plot, lines, legend function, and other graphical parameters.

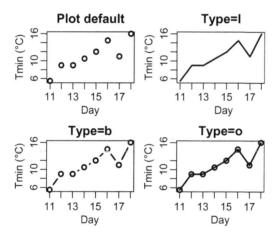

FIGURE 3.5   Graphics combined on the same plot as a 2-by-2 array on the same plot using par(mfrow = c(2,2)).

Figure 3.6 shows graphs with different plotting characters (argument pch).

```
# Plot points using a different symbol for each of the 8 days (parameter pch).
# The parameter cex determines the size of the characters
par(mfrow = c(1,1))
plot(data$day,data$tmin, pch = 1:8, cex = 2,main = "pch from 1 to 8")
```

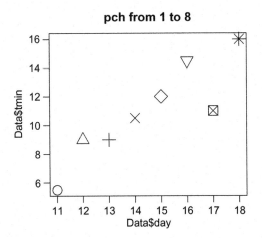

FIGURE 3.6   Plot with different types of points using argument pch.

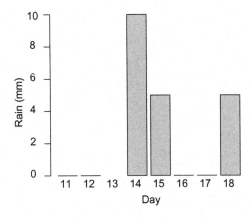

FIGURE 3.7   Barplot of rain for the different days.

By default, each new graph replaces the previous graph. So if you create several graphs in a program, at the end of the program you will only be left with the last one. Usually however you want to examine all of the graphs. To do this, put the command dev.new() before each plot. In this case, the next plot is created on a new graphics page. At the end of the program the graphics pages are stacked on top of one another.

R also has functions for creating barplots (Figure 3.7), histograms, bar and whisker graphs, and many other types of graphs. Some of these are presented in the table of commands below. Note that many of the graphics parameters are common to several types of graphs and you can check the help for specific arguments and examples.

```
# barplot. The names.arg parameter specifies the names under the bars.
barplot(data$rain,names.arg = data$day, xlab = "day", ylab = "rain (mm)")
```

Plotting Functions	Description
plot(x, y)	Creates a scatter plot of x (on the x axis) and y (on the y axis)
lines(x, y)	Adds lines connecting x,y pairs to the existing graph
par	Sets graphical parameters globally. See help(par) for full details.
(parameter = value)	Many parameters can alternatively be set within plotting functions. Here are some of the most useful parameters: type = "p" to plot points, type = "l" for lines, type = "b" for points connected by lines, type = "o" for lines and overlaid points, type = "h": for vertical lines, type = "s" for steps where the data are represented by the tops of the vertical lines, type = "S" for steps where the data are represented by the bottoms of the vertical lines, type = "n" for no plotting symbols (just axes)lty = 1 uses lines of type 1 (solid line). 2 uses dashed lines etc. (see Figure 3.4) pch = 1 uses circles as plotting characters, pch = 2 uses triangles etc. (see Figure 3.6) xlim = , ylim = specify the lower and upper limits of the axes, for example xlim = c(1, 10) or xlim = range(x) xlab = "namex", ylab = "namey" give the labels for the axes (character strings)main = main title, must be a variable of mode character sub = sub-title (written in a smaller font) mfrow = c(nrow,ncolumn) will plot nrow*ncolumn graphs on a single graph sheet. ask = TRUE will wait for confirmation before drawing a new graph
plot(y)	Plot of the values of y (on the y axis). The default x values are 1,2,...,length(y)
hist(x)	Histogram of the frequencies of x
barplot(x)	Barplot of the values of x; use horiz = TRUE for horizontal bars
boxplot(x)	box-and-whiskers plot
pairs(x)	If x is a matrix or a data frame, draws all possible bivariate plots between the columns of x
contour(z)	Contour plot (data are interpolated to draw the curves, z must be a matrix
points(x, y)	Adds points (the option type = can be used)
abline(a,b) , abline (h = y), abline(v = x), abline(lm.obj)	Draws a line of slope b and intercept a, a horizontal line at ordinate y, a vertical line at abscissa x, the regression line given by the linear regression result lm.obj
text(x, y, labels, ...)	Adds text given by labels at coordinates (x,y); a typical use is: plot (x, y, type = "n") (no ; text(x, y, names))
mtext(text, side = 3, line = 0, ...)	Adds text given by text in the margin specified by side (see axis() below); line specifies the line from the plotting area
legend(x, y, legend)	Adds the legend at the point (x,y) with the symbols given by legend. x,y can be replaced by a character string for the location such as "topright".
title()	Adds a title and optionally a sub-title
axis(side, vect)	Adds an axis at the bottom (side = 1), on the left (2), at the top (3), or on the right (4); vect (optional) gives the abcissa (or ordinates) where tick-marks are drawn
dev.new()	Draws the graph on a new graph sheet

10 STATISTICS AND PROBABILITY

10.1 Probability Distributions

A large number of probability distributions are available with R, and there are several functions for working with those distributions.

Consider first the normal distribution, which is defined by two parameters, the expectation and the standard deviation:

- The function dnorm(x,mean,sd) gives the probability density at x (a vector of values) for a normal distribution with expectation mean and standard deviation sd.
- The function pnorm(x,mean,sd) gives the cumulative probability up to x of a normal distribution with specified mean and standard deviation (sd).
- The function qnorm(q,mean,sd) gives the value of x where the cumulative distribution function equals a quantile q.
- The function rnorm(n,mean,sd) generates n random values from a normal distribution with the specified mean and standard deviation (sd).

```
# probability density
dnorm(8:12,mean=10,sd=2)
[1] 0.1209854 0.1760327 0.1994711 0.1760327 0.1209854
# cumulative probability density
pnorm(8:12,mean=10,sd=2)
[1] 0.1586553 0.3085375 0.5000000 0.6914625 0.8413447
# value where the indicated cumulative probability density is reached
qnorm(c(0.05,0.25,0.5,0.25,0.05),mean=10,sd=2)
[1] 6.710293 8.651020 10.000000 8.651020 6.710293
# generate random numbers
rnorm(5,mean=10,sd=2)
[1] 11.205795 8.720481 9.312137 10.679303 10.322486
```

There are analogous instructions for the uniform distribution (unif), the Poisson distribution (pois), the student t distribution (t), and many others.

Distributions	All These Functions Can Be Used by Replacing the Letter r with d, p, or q to Get, Respectively, the Probability Density (dfunc (x, ...)), the Cumulative Probability Density (pfunc(x, ...)), and the Value of the Quantile (qfunc(p, ...), with $0 < p < 1$)
rnorm(n, mean = 0, sd = 1)	Gaussian (normal)
rexp(n, rate = 1)	Exponential
rgamma(n, shape, scale = 1)	Gamma
rpois(n, lambda)	Poisson
rweibull(n, shape, scale = 1)	Weibull
rcauchy(n, location = 0, scale = 1)	Cauchy

rbeta(n, shape1, shape2)	Beta
rt(n, df)	Student (t)
rf(n, df1, df2)	Fisher–Snedecor (F)
rchisq(n, df)	Pearson
rbinom(n, size, prob)	Binomial
rgeom(n, prob)	Geometric
rhyper(nn, m, n, k)	Hypergeometric
rlogis(n, location = 0, scale = 1)	Logistic
rlnorm(n, meanlog = 0, sdlog = 1)	Lognormal
runif(n, min = 0, max = 1)	Uniform

10.2 Controlling the Randomness in Random Numbers

Random numbers generated by a computer are not truly random but are, rather, pseudo-random. They have some large but finite number of possible values, and they are always generated in the same order. If one starts the generator in the same place many times, the numbers that are generated will always be the same.

Often, one wants to be sure that rerunning a program many times will always generate the same random numbers. For example, suppose one is debugging a program that includes random numbers. One is verifying the calculations, and one doesn't want to deal with new random numbers each time the program is rerun. The function set.seed(seed) uses the value seed to initialize the random number generator.

```
# Run runif twice, get two different sets of random numbers
runif(3,0,1)
[1] 0.9889093 0.3977455 0.1156978
runif(3,0,1)
[1] 0.06974868 0.24374939 0.79201043
# Use set.seed before each use of runif. Get same random numbers each time
set.seed(7)
runif(3,0,1)
[1] 0.9889093 0.3977455 0.1156978
set.seed(7)
runif(3,0,1)
[1] 0.9889093 0.3977455 0.1156978
```

10.3 Basic Statistical Analyses

R has a very large number of functions for doing statistical analyses. There are books devoted just to statistical analyses with R. Here we illustrate just two classical analyses, analysis of variance and linear regression. A simple example of the analysis of variance function is aov(yield~block + N*P + K, npk). The function has two arguments; the first is a formula, the second (optional) is a data frame including columns named yield, block, N, P, and K.

A formula is an R construct that we haven't yet encountered. It has the name of a response variable on the left, then a tilde ($\sim$), then the names of explanatory variables on the right. For linear models (as here) the explanatory variables separated by + are included individually, while the notation N*P means to include each factor individually and also to include their interaction.

All of the names that appear in the formula must appear as column names in the data frame. The following code illustrates analysis of variance.

```
# example from help(aov)
data(npk, package = "MASS")
a <- aov(yield~block + N*P + K, npk)
Call:
    aov(formula = yield ~ block + N * P + K, data = npk)
Terms:
                        block           N         P         K       N:P    Residuals
Sum of Squares    343.2950    189.2817    8.4017   95.2017   21.2817     218.9033
Deg. of Freedom          5           1         1         1         1           14
Residual standard error: 3.954232
Estimated effects may be unbalanced
anova(a)
Analysis of Variance Table
Response: yield
            Df    Sum Sq   Mean Sq    F value    Pr(>F)
block        5    343.30    68.659     4.3911   0.012954  *
N            1    189.28   189.282    12.1055   0.003684  **
P            1      8.40     8.402     0.5373   0.475637
K            1     95.20    95.202     6.0886   0.027114  *
N:P          1     21.28    21.282     1.3611   0.262841
Residuals   14    218.90    15.636
---
Signif. codes:  0 '***' 0.001 '**' 0.01 '*' 0.05 '.' 0.1 ' ' 1
```

The function that does linear regression by ordinary least squares is lm. An example of the use of the function is lm(data$tmin~data$tmax)where data$tmin and data$tmax are vectors (columns of the data frame data). This function will do the linear regression of data$tmin on data$tmax. By default, an intercept is included. The formula in this case already contains the name of the data frame.

```
data <- data.frame(day = c(11,12,13,14,15,16,17,18),
tmax = c(10,14,13,15.5,NA,20.5,15,20),
tmin = c(5.5,9,9,10.5,12,14.5,11,16), rain = c(0,0,NA,10,5,0,0,5))
# evaluate the parameters of the linear regression between tmin and tmax
a <- lm(data$tmin~data$tmax)
summary(a)
Call:
lm(formula = data$tmin ~ data$tmax)
```

```
Coefficients:
(Intercept)   data$tmax
   -3.5305       0.9279
summary(a)
Call:
lm(formula = data$tmin ~ data$tmax)
Residuals:
       1        2        3        4        6        7        8
 -0.2485  -0.4601  0.4678  -0.3520  -0.9915  0.6120  0.9724
Coefficients:
              Estimate  Std. Error  t value  Pr(>|t|)
(Intercept)   -3.53053     1.31675   -2.681  0.043754 *
data$tmax      0.92791     0.08327   11.144  0.000101 ***
---
Signif. codes:  0 '***' 0.001 '**' 0.01 '*' 0.05 '.' 0.1 ' ' 1
Residual standard error: 0.7641 on 5 degrees of freedom
   (1 observation deleted due to missingness)
Multiple R-squared:  0.9613,    Adjusted R-squared:  0.9536
F-statistic:  124.2 on 1 and 5 DF,  p-value:  0.0001015
```

Functions for Statistical Analysis	Description
aov(formula)	Analysis of variance model
anova(fit,...)	Analysis of variance (or deviance) tables for one or more fitted model objects
lm(formula)	Fits a linear model
glm(formula,family =)	Fits a generalized linear model
nls(formula)	Fits a non-linear model (See Chapter 6)
predict(fit,...)	Predictions based on fitted model
coef(fit) , residuals(fit), deviance(fit), fitted(fit) df.residual(fit), df.residual(fit)	Results of model fitting. The estimated coefficients (with their standard errors), the residuals, the deviance, the fitted values, and the number of residual degrees of freedom
logLik(fit) , AIC(fit)	Logarithm of the likelihood and Akaike's information criterion of a fitted model
density(x)	Kernel density estimates of x
binom.test(), pairwise.t.test(), power.t. test(), prop.test(), t.test(), ...	Statistical tests. See help.search("test")
filter(x,filter)	Applies linear filtering to a univariate time series or to each series separately of a multivariate time series

Several functions are available for minimizing or maximizing a function, or for smoothing data. The minimization routines can be used for least squares parameter estimation for non-linear models.

Functions for Optimization and Model Fitting	Description
optim(par, fn, method = "Nelder-Mead")	General purpose optimization; par is initial values, fn is function to optimize (normally minimize) Other algorithm : method = c("Nelder-Mead", "BFGS", "CG", "L-BFGS-B", "SANN"). See Chapter 6
nlm(f,p)	Minimizes function f using a Newton-type algorithm with starting values p
approx(x,y =)	Linearly interpolates given data points; x can be an xy plotting structure
spline(x,y =)	Cubic spline interpolation
loess(formula)	Fits a polynomial surface using local fitting. Many of the formula-based modeling functions have several common arguments: data = the data frame for the formula variables, subset = a subset of variables used in the fit, na.action = action for missing values: "na.fail", "na.omit", or a function.

11 ADVANCED DATA PROCESSING

R offers powerful functions for operating on data frames (and matrices). The same operations could be handled by a programming loop, but these functions are more concise and faster.

An important instruction is apply, which applies some operation to individual lines or columns of a matrix or data frame. In the example below, we calculate the mean of each column of the data frame data. The first argument of apply is the name of the data frame. The second can be either 1 (do calculation to each row) or 2 (do calculation to each column). The third argument is the name of the function to be applied. Then there are any additional arguments to be supplied to that function.

```
data = data.frame(day = c(11,12,13,14,15,16,17,18), tmax = c(10,14,13,15.5,NA,
20.5,15,20), tmin = c(5.5,9,9,10.5,12,14.5,11,16),rain = c(0,0,NA,10,5,0,0,5))
# compute the mean of each column, removing NAs.
apply(data,2,mean, na.rm = TRUE)
       day       tmax      rain
14.500000 15.428571 2.857143
# compact way to draw plot
par(mfrow = c(1,2))
sapply(c("tmax","rain"), function(var) plot(data[,"day"],data[,var],type = "l"))
```

Another way to do operations on selected parts of a data frame is to use the by function. Here the first argument identifies the column or columns of the data frame to be treated, the second argument is a vector of the same length as the number of rows of the data base, treated as a factor, the third argument is the name of the function to be applied, and then there are any additional arguments to be supplied to that function.

```
by(data$tmax,data$rain,mean, na.rm = TRUE)
data$rain: 0
[1] 14.875
-------------------------------------------------------------
data$rain: 5
[1] 20
-------------------------------------------------------------
data$rain: 10
[1] 15.5
```

Data frames can be merged as illustrated below. This is the equivalent of a join statement in SQL query language for database to query data from two tables based on a relationship between certain columns in these tables.

```
data1 <- data.frame(day = c(11,12,13,14,15,16,17,18),
tmax = c(10,14,13,15.5,NA,20.5,15,20))
data2 <- data.frame(day = c(11,12,13,14,15,16,17,18), rain = c(0,0,NA,10,5,0,0,5))
# combine data1 and data2 using day as an index
data = merge(data1,data2,by = "day")
data
   day tmax rain
1  11 10.0    0
2  12 14.0    0
3  13 13.0   NA
4  14 15.5   10
5  15   NA    5
6  16 20.5    0
7  17 15.0    0
8  18 20.0    5
```

Advanced Data Processing	Description
apply(X,INDEX,FUN)	A vector or array or list of values obtained by applying a function FUN to margins (INDEX) of X
lapply(X,FUN)	Applies FUN to each element of the list X
tapply(X,INDEX,FUN)	Applies FUN to each cell of a ragged array defined by a vector X and a list of factors INDEX
sapply(X,FUN)	Applies FUN to each element of the list X, wrapper of lapply by default returning a vector
by(data,INDEX,FUN)	Applies FUN to data frame data subsetted by INDEX
merge(a,b)	Merges two data frames by common columns or row names
xtabs(a b,data = x)	A contingency table from cross-classifying factors

12 ADDITIONAL PACKAGES (LIBRARIES)

12.1 What Are Packages?

R packages contain functions, data, and documentation beyond what is in the basic R software. The R basic software includes about 30 packages, but there are over 4000 additional packages available. Usually, each package is dedicated to some particular type of method, for example the sensitivity package for sensitivity analysis (see Chapter 5) or the pROC package for ROC analysis (see Chapter 9). You can see the list of packages on the CRAN page (http://www.rproject.org).

You can see all the packages that you have downloaded, and all those that are available in the current session, as follows:

```
# to see all packages installed
library()
# to see packages currently loaded
search()
```

12.2 Retrieve, Install, and Load a Package

There are several ways to install a package. We illustrate using the sensitivity package (Pujol et al., 2012). If you have an Internet connection, the simplest method is to do the download and installation from within the R software. The required steps are listed below and are illustrated in Figure 3.8.

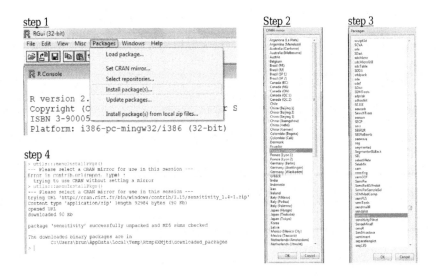

FIGURE 3.8   Screen shots of the steps for installing the sensitivity package.

Step 1: Click on *Packages* in the menu and select *Install package(s)...*
Step 2: Choose a location near you from the list of mirror sites which appear.
Step 3: Choose the `sensitivity` library from the list of available packages.
Step 4: Verify in the command window that the package has been successfully installed.

Another more advanced way to install packages is to use the install packages function as illustrated below.

```
# specify the packages to install and choose an R repository site near you
a = sapply(list("sensitivity", "boot"), install.packages , repos =
"http://cran.rstudio.com")
```

You can also separate the download and the installation. The advantage is that then, you no longer need Internet access for the installation phase. Under Windows, in the download stage you must choose to download a zip of the compiled binary form. For the installation phase, select *Install package(s) from local zip files* in the *Packages* drop-down menu. Then browse to find the location of the zip file.

Documentation for more advanced operations is available at http://cran.r-project.org/doc/manuals/R-admin.html#Installing-packages.

In order to use a package during a session, the package has to be loaded during that session. (The fact that you used the package in a previous session is irrelevant.) The package is loaded by the command `library(name)`. The warning messages in the example below can be ignored.

```
# make the sensitivity package available during this session
library(sensitivity)
Loading required package: boot
# make the sensitivity package available and display some information about it
library(help = sensitivity)
# get more detailed information about the package
help(package = sensitivity)
```

Install.Packages("*x*",repos = getOption("repos"))	Install Package "*x*"
library(*x*)	load package *x*
data(*x*)	load specified datasets

12.3 The ZeBook Package

The ZeBook package is an R library containing material related to this book. It includes the code for many of the examples that are discussed in this book (so that you can run them yourself), code for the main models that are used as examples, and data sets for the exercises. You may want to install it now.

More details about this package are presented in Appendix 2.

13 RUNNING AN EXTERNAL MODEL FROM R

In many cases, a model is coded in another programming language (C, C++, C#, JAVA, FORTRAN, PYTHON, etc.), and one does not want to recode the model in R (very time consuming and the model will probably run slower). However, one wants to use R for sensitivity analysis, calibration, evaluation, or data assimilation, since R allows one to do all that quite simply, using powerful functions.

One can in fact have the best of both worlds. One can leave the model in its original language and nonetheless use R for working with the model. This is possible because one can execute a compiled program from R. There are, however, some limitations. The model must read parameters and input variables from external files, and must also write output to an external file. Also, the other language must be a language that can produce an executable binary file.

Running the model from an R program is then quite simple. One first writes the parameter and input variable files using R. Then one uses R to launch the execution of the model. Finally, one reads the files with the model outputs into R.

This is illustrated in the example below, using the simple maize model. The model was written in Python and then compiled to give the executable file "*maize.exe*". (Available in the R package ZeBook but for Windows OS only.) In the R program, first one writes the weather data to the file "*weather.dat*". This is the input needed by the model. The next two lines create a data frame with parameter names in column one and values in column two and then the next line writes this to a file that will be read by the model. The function shell("maize.exe") causes the execution of the model program. The program writes the results in a file called "*result.dat*". The last line of the program reads this file into R.

```
library(ZeBook)
# define the working directory where maize.exe resides
setwd("c:/ZeBook/R_intro")
# write one year of weather data into the file weather.dat
write.table(maize.weather(working.year=2010,
working.site=1),file="weather.dat",sep=";",col.names=T,row.names=F,quote=F,dec="."
)
# prepare the parameter values, and write into file "param.ini"
param <- maize.define.param()["nominal",]
param <- rbind(data.frame(param=names(param),value=param,stringsAsFactors=F),
c("sdate",100),c("ldate",250))
write.table(param,file="param.ini",sep=";",col.names=T,row.names=F,quote=F,dec="."
)
# launch the execution of the maize model
```

```
shell("maize.exe")
# retrieve the results which are in the file result.dat
result <- read.table(file="result.dat",sep=";",header=T,dec="." )
```

Alternatively, if the model is written as a Dynamic Link Library (dll), it can be run from R (see R function dyn.load). Also, if the model is encoded using Virtual Laboratory Environment (VLE) software (http://www.vle-project.org) there is a special R package rvle that permits easy interfacing with R. In these cases information is transferred directly without accessing the system file, which will increase the speed of execution.

14 REDUCING COMPUTING TIME

Several of the methods presented in this book (for parameter estimation, uncertainty and sensitivity analysis, data assimilation) require many simulations with the model; the number of simulations can easily be in the thousands or even larger. If the model itself is large, the overall computing time may be quite long, taking perhaps days or weeks. It is therefore important to consider ways of reducing computation time.

14.1 Good Programming Practices

As already mentioned, it is worthwhile avoiding the use of loops in R programs as far as possible. Where the same result can be obtained using R functions or R subscripting features, this will almost always be much faster than programming with loops (see apply and other advanced data processing functions).

Writing data to system files is a relatively slow operation. Reducing the number of write operations may substantially reduce computing time.

14.2 Measuring the Computation Time for Each Piece of Code

R provides tools for measuring the time required for each piece of code. This can be a valuable tool for understanding exactly where savings in execution time are most important. If you think that execution time will be a problem, you can use these tools on a simplified version of the program, in order to predict if there will be problems and what pieces of the code will be mostly responsible.

The function system.time(expression) gives the time required (in seconds) to execute the expression in parentheses. The function returns the CPU time required to execute the user instructions (user), the CPU time required by the system on behalf of the user (system), and finally the total time (elapsed). The example below measures the time required for running the maize model for one field in one year.

```
library(ZeBook)
```

```
# maize.model2 is a function that simulates the simple maize model (Appendix A1).
system.time(r <-
maize.model2(maize.define.param()[1,],maize.weather(2010,1),100,250))
    user  system  elapsed
   0.58   0.11    0.69
```

The Rprof function (for profiling) measures the time taken by each of the instructions in a program. In the illustration below, the first instruction is Rprof(tmp < − tempfile(), interval = 0.01, memory.profiling = TRUE). This starts the profiling. The first argument is the name of the file where the profiling information will be written. The second argument is the sampling time, in seconds. In this example, the program is sampled 100 times every second (in order to identify the instruction being executed). The last argument is memory.profiling. Setting this to TRUE means that the profiling information is to be written to the file.

After this instruction one places the program to be analyzed, and then at the end the instruction Rprof(), which ends the profiling. Then the instruction summaryRprof(tmp, memory = "both") prints out the results of the profiling. The first argument here is the name of the file with the profiling information. The second argument memory = "both" instructs the program to print out both timing information and memory allocation information.

```
# beginning of the profile
Rprof(tmp <- tempfile(),interval = 0.01, memory.profiling = TRUE)
weather <- maize.weather(2010,1)
param <- maize.define.param()[1,]
r <- maize.model2(param,weather,100,250)
write.table(r,file = "res.tmp")
# end of the profile
Rprof()
# summary of R functions used in maize.model2
summaryRprof(tmp, memory = "both")
$by.self
```

	self.time	self.pct	total.time	total.pct	mem.total
names <−	0.19	40.43	0.19	40.43	13.7
maize.weather	0.13	27.66	0.44	93.62	34.6
structure	0.06	12.77	0.06	12.77	4.3
[.data.frame	0.03	6.38	0.12	25.53	8.2
==	0.02	4.26	0.02	4.26	1.6
&	0.01	2.13	0.01	2.13	1.0
maize.model2	0.01	2.13	0.01	2.13	0.7
Rprof	0.01	2.13	0.01	2.13	0.0
write.table	0.01	2.13	0.01	2.13	0.4

`$by.total`

	total.time	total.pct	mem.total	self.time	self.pct
maize.weather	0.44	93.62	34.6	0.13	27.66
names <−	0.19	40.43	13.7	0.19	40.43

[.data.frame	0.12	25.53	8.2	0.03	6.38
[	0.12	25.53	8.2	0.00	0.00
structure	0.06	12.77	4.3	0.06	12.77
==	0.02	4.26	1.6	0.02	4.26
&	0.01	2.13	1.0	0.01	2.13
maize.model2	0.01	2.13	0.7	0.01	2.13
Rprof	0.01	2.13	0.0	0.01	2.13
write.table	0.01	2.13	0.4	0.01	2.13

```
$sample.interval
[1] 0.01
$sampling.time
[1] 0.47
unlink(tmp)
```

14.3 Recoding a Portion of the R Code in an Externally Compiled Language

Because R is interpreted rather than compiled, it is relatively slow. If execution time is a problem, then it would probably be advantageous to recode portions of the R program in a language that can be compiled. The most time-consuming portions of the program can be identified as explained in the previous subsection.

We have already shown how R can interface with a model written in another language. Keeping the model in another language, or translating from R into a compiled language, is one way to make the program faster.

Another approach is to convert your code into R-C hybrid code. As an example, we could create a file r_c.c in a directory c:/ZeBook/R_intro, containing the following C function.

```
void Csuitearith2(double *data,double *suite)
  {
  int i ;
  suite[0] = 1 ;
  for (i = 1 ; i < = *data ; i++){suite[0] = suite[0] + 2;}
  }
```

You will need Rtools and the appropriate system path for R, MinGW, and Perl. Then, you can compile the R-C code in a DOS console window using the command "*R CMD SHLIB r_c.c*"

Under R, you will need to define the function that calls the complied C function.

```
setwd("c:/ZeBook/R_intro")
# R version
suitearith2R <- function(n) {
  suite <- 1
  if (n > 0){for (i in 1:n) {suite <- suite + 2}}
return(suite)
```

```
}
# C version
suitearith2C <- function(data) {
 if (!(is.loaded("Csuitearith2")))
 dyn.load("r_c.dll")
 .C("Csuitearith2", as.double(data), suite = double(1))$suite
}
# comparative test of performance
system.time(a<-suitearith2R(1000000))
   user  system  elapsed
   1.12    0.02     1.14
system.time(b<-suitearith2C(1000000))
   user  system  elapsed
      0       0        0
```

Using The R package *RCPP* (Eddelbuettel and Francois, 2012) can also be a good option to optimize part of your R code.

14.4 Parallelization of R Code

The use of parallel computing with R is an advanced R programming feature that can reduce execution time substantially in some cases. The concepts of parallel programming and the application to the maize model will be presented in detail.

Most computers today are equipped with multi-core processors or multiple processors on the mother board; they typically have 2−6 independent or core processing units (PUs). Increasing processing speed with a single processor runs into problems of heat generation. These problems can be mitigated by using multiple PUs. However, the extra computational capacity brought by multiple PUs is underutilized by the standard R program.

The principle of parallel programming is simple. It is based on the existence of multiple PUs which can be mobilized and which are able to communicate with the main process (Figure 3.9). The main process will define a sequence of independent processes, divide this sequence into N sub-sequences, and send each sub-sequence to a different PU. Then, each PU treats its sub-sequence and returns the corresponding results. Finally, the main process takes over again to recover the N sub-sequences and aggregate the results.

In R, there are several libraries that can handle this type of parallel com putation. Since the R version 2.14.0, you can use the `parallel` package, based on part of the older `snow` package (Simple Network Of Workstations, Tierney et al., 2012). With this package, you can improve the performance by executing advanced parallel programming. The `parallel` package is used to declare a set of N available PUs. The portion of the code will be distributed among the N PUs. Each of the N units will independently process the assigned sub-sequence and return the results.

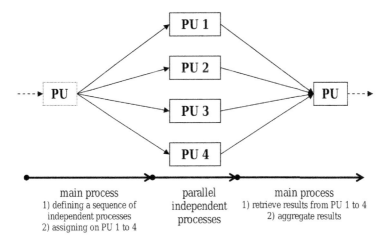

FIGURE 3.9   Illustration of parallel computing with four processing units with the `parallel` package.

```
library(ZeBook)
library(parallel)
# example with a simulation design with random generation with a uniform
distribution
n <- 10000
mat.param <- sapply(1:7,function(k)runif(n,min = maize.define.param()["binf",k],
    max = maize.define.param()["bsup",k]))
colnames(mat.param) <- colnames(maize.define.param())
mat.param = cbind(id = 1:nrow(mat.param),mat.param)
# for one single weather, sdate and ldate
weather <- maize.weather(working.year = 2010, working.site = 1)
sdate <- 100
ldate <- 250
# 1) define the function that is run in each parallel PU
# id is used to construct arguments and the input variables required
# sdate,ldate, ncore are arguments to be passed to the parallel process
maize_cluster <- function (id, sdate,ldate,ncore)
{
Bmax = max(maize.model2(mat.param[id,],weather,sdate = sdate,ldate = ldate)$B,na.rm = TRUE
)
# return output variables for process id
return (c(id = id,Bmax = Bmax))
}
# test of maize_cluster for id = 1
maize_cluster(1,sdate,ldate,ncore)
# 2) creating the Cluster: definition of computational units mobilized.
```

```
# Example for the definition of four units.
ncore <- 4
cl <- makeCluster (ncore)
# 3) define a calculation sequence.
sequence <- mat.param[,"id"]
# 4) export global variables to parallel process units.
# Each calculation unit needs access to global variables
# including the function and the data from the main process
clusterExport (cl, list ("maize.model2","mat.param", "weather"))
# 5) Allocation of the calculation sequence to calculation units
# The same R function maize_cluster is applied to the cluster "cl"
# where id is picked from sequence
# supplementary arguments (sdate,ldate,ncore) are passed to maize_cluster too
list.result <- clusterApply (cl, sequence, maize_cluster, sdate,ldate,ncore)
# 6) Closing cluster
stopCluster(cl)
# 7) Aggregate results from the parallel computations (a list) to the appropriate
structure
mat.result_paral <- matrix(unlist(list.result),byrow = TRUE,ncol = 2)
# compare to a classical version
mat.result_nonparal <- maize.simule(mat.param[,2:8], weather, sdate, ldate,
all = FALSE)
all(mat.result_nonparal == mat.result_paral[,2])
```

Figure 3.10 shows an example of comparison of performance of a parallelized program to a basic program.

If you want to achieve a similar parallelization of your code, you need to write a very similar script with the seven steps. In some cases, when working with simulation functions that read and write to the file system (for example, running a simulator from an external program, see above), you will need to manage it. So, you need to know in which PUs you are operating from and to have multiple ncore paths for reading/writing files. In the clustered function, you can add the instruction core <- id%ncore% and use it to explore the right path. To see how the sequence is assigned to the cores, you can try clusterSplit(cl,sequence). Depending on the computer (especially the CPU) being operated, you need to adapt the size of the cluster to the optimal size.

Parallel computing has its peculiarities and difficulties. Debugging capabilities are relatively limited, so it is important to verify that the results obtained by the parallelized functions are consistent with those obtained from the initial functions. Consequently, parallelization should probably be used only when computation time becomes a serious problem (for example, a frequent and long computation taking several hours or days) and after exploring other means of reducing computing time.

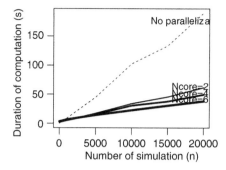

FIGURE 3.10 Test of performance on a Windows computer with a four core CPU with hyper threading technology (eight threads) (Intel Xeon CPU W3550 3.07 GHz and 4.00Go of Ram). The parallelized program can run up to five times faster than the basic program!

EXERCISES

Easy

1. Encoding data as a vector.

Table. Observed and Simulated Yield for Two Levels of Nitrogen Input

Nit (Nitrogen Input)	Yobs (Observed Yield)	Ysim (Simulated Yield)
100	6.0	6.2
100	6.5	6.0
100	6.1	5.8
150	8.0	8.3
150	7.8	8.0
150	7.9	8.3
150	-	8.0

Encode these data as three separate vectors, Nit, Yobs, Ysim.

2. Selection of data.

Select the fourth value of Yobs.

Select Yobs for Nit = 100.

Select Yobs value greater than Ysim.

3. Graphical representation of data.

Plot Yobs against Ysim with appropriate legend and title.

Add the y = x line.

4. Some simple calculations.

Calculate the vector D = Yobs − Ysim and print D in the console window. Then do the same for D^2 and the absolute value of D.

Calculate the sum of the elements of D^2 (SSE) and the average (MSE).

Moderate

1. Combining data.

From the previous vector Nit, Yobs, Ysim, create a data.frame with Nit defined as a factor.

Sort the table by increasing Yobs.

2. Calculation on a table.

Calculate the mean and sum of Ysim for each level of nitrogen.

3. Creation of a function.

Create a function taking as arguments Yobs and Ysim and returning as results the sum and mean of D^2 (SSE and MSE).

4. Graphical representation of data.

Plot Yobs against Ysim with appropriate legend and title, with different symbols and colors for each level of nitrogen. Use a larger font than the default for text and symbols.

Add the y = x line, using `abline`.

High

1. Programming.

Table. Daily Mean Temperature

Day	Tmean (°C)
1	25.5
2	22.0
3	23.5
4	8.0
5	6.5
6	5.5
7	6.0
8	8.0
9	12.0
10	15.0

Encode these data as a data.frame. Build a function that takes as arguments this data.frame and two parameters, Tbase and Tmax (with default values Tbase $-$ 7 and Tmax $=$ 25). The function should calculate the temperature sum with a base of Tbase and a maximum of Tmax, by using a loop (for) and condition (if/else).

2. Analysis of variance.

Analyze the effect of nitrogen input on Yobs with an analysis of variance. Plot mean yield for each level of nitrogen as a bar plot, with the mean yield value on it.

3. Linear model.

Estimate the parameters of a linear model between Yobs and Ysim. Add the linear relation on the previous graph.

4. Reading external data.

Copy-paste the following lines in a text editor such as Notepad and save it to a file called *yield.csv*. NS is for not available data.

```
Wheat yield data
site;year;yield
Grignon;2007;6.5
Grignon;2008;7.5
Grignon;2009;NS
Grignon;2010;5.0
Grignon;2011;7.0
Toulouse;2007;7
Toulouse;2008;8
Toulouse;2009;7.5
Toulouse;2010;6.5
Toulouse;2011;7.5
```

5. Write the R script to correctly read the file created, defining `site` as character, `year` as factor, and `yield` as numeric.

6. Compute the mean and variance by site and by year. Plot the evolution of yield for both sites on one single graph with a line for the mean for each site. Add a legend.

REFERENCES

Crawley, M.J., 2007. The R Book. Wiley, Chichester, England, p. 950.

Eddelbuettel, D., Francois, R., 2012. R package Rcpp: Seamless R and C++ Integration. http://cran.r-project.org/web/packages/Rcpp/.

Pujol, G., 2012. R package sensitivity: Sensitivity Analysis. http://cran.r-project.org/web/packages/sensitivity/.

Tierney, L., Rossini, A.J., Li, N., Sevcikova, H. 2012. R package snow: Simple Network of WorkStations for R. http://cran.r-project.org/web/packages/snow/.

R Development Core Team, 2013. List of R documentation. http://cran.r-project.org/other-docs.html.

R Development Core Team, 2013. R: A language and environment for statistical computing. R Foundation for Statistical Computing, Vienna, Austria, http://www.r-project.org.

R Development Core Team, 2013. R package parallel. http://stat.ethz.ch/R-manual/R-devel/library/parallel/doc/parallel.pdf.

R Development Core Team, 2013. The R Manuals—An Introduction to R. Notes on R: A Programming Environment for Data Analysis and Graphics. http://cran.r-project.org/manuals.html.

## Chapter 4

Simulation with Dynamic System Models

1. INTRODUCTION

In this chapter, we present concepts for simulating dynamic system models using digital computers, which are referred to as 'computer simulation'. Our use of simulation is consistent with a paraphrased dictionary definition: 'the imitative representation of the functioning of one system or process by means of the functioning of another, such as a computer simulation of a process' (http://www.merriam-webster.com/dictionary). Thus, we refer to simulation as the numerical solution of a system model to produce values of the variables that represent the real system components over time. Interestingly, another definition of simulation is that it is a counterfeit or a sham. Although this definition is not helpful as we attempt to learn how to simulate dynamic models, it does remind us that the results from computer simulation models are simplifications of reality; they do not reproduce exactly the behavior of the real system that the model represents.

The simulation (or computer simulation) process is different from that of modeling. In modeling, mathematical models are developed based on understanding of the system being studied and assumptions about system boundaries, components, state variables, etc., as explained in Chapter 1. The result of a modeling process is a set of mathematical relationships that are intended to represent the system. The result of a simulation process is a set of numerical results that represent the dynamic behavior of the system, given the particular model, explanatory variables, parameters, and time period used. In particular, the simulation produces values of all state variables at each point in time over a time duration that is set by the analyst. For example, simulated results of the crop model described in Chapter 1 would be the age of the crop (in degree days, $TT(t)$) on each day t, and daily values of leaf area index ($LAI(t)$) and above-ground biomass of the crop, ($B(t)$), starting at a simulated time $t = 0$ at plant emergence and continuing until a time t when the crop reaches maturity (defined by a specific degree day age). The simulation program may also produce additional output variables, such as auxiliary variables and various transformations of state variables. Example results will be given later in this chapter. Some refer to this process as numerical solution of dynamic models, particularly when the systems are

Working with Dynamic Crop Models. DOI: http://dx.doi.org/10.1016/B978-0-12-397008-4.00004-6

represented by differential equations, such as the model describing the volume of water in each tank from Chapter 1, or partial differential equations. The numerical results are produced by a computer program developed specifically to simulate the model. Thus, when one wants to simulate a system, computer code is developed based on the mathematical model of the system.

In this chapter, we present basic methods that are widely used to simulate crop and other dynamic system models. Many books have been written on techniques for simulating dynamic systems (e.g., MATLAB, http://www.mathworks.com/products/matlab/; Press et al., 1996). Also, software packages have been developed to help one create computer code to simulate a system being studied, some of which allow software users to manipulate Forrester diagram symbols on the screen to first of all develop the system model, and then to simulate it after specific explanatory variables and parameters are quantified using the software interface (e.g., Stella, http://www.iseesystems.com/softwares/Education/StellaSoftware.aspx). Although these modeling and simulation software packages can be very useful, each may have limitations when one attempts to use it to model complex systems, and in any case it is important for one to understand the methods being used to simulate a system to avoid misuse or misinterpretation of results. Most crop models are written in more general programming languages, including Fortran and C. Although the methods described here can be implemented in any program language, we will demonstrate their use in the R Software.

There are several essential elements for computer simulation programs: (1) initialization, (2) control of simulated time, (3) computation of state variables at each simulated time step, and (4) a final section. In the initialization element, the initial values of all state variables are defined, a starting time and a time step are quantified, and values are given to each parameter and explanatory variable. A stopping time may also be specified to end the computations after the passage of a given amount of time or the crossing of a threshold by a particular variable, such as when the crop exceeds a certain degree day age. If there are time-varying inputs, such as environmental variables that change over time, like temperature and solar radiation, vectors of these variables may be read from files and stored for use during simulated time steps when state variables are updated.

Control of simulated time begins with the program setting time equal to the starting time specified in the initialization section. Time then loops through all dynamic model calculations until the ending condition is met. When the ending condition is met, summary calculations are made, files are written to store results, and tables and graphical results are displayed in the final section. A new set of simulated results could be obtained by looping back through the entire set of elements if desired, and in fact this is frequently done in many applications of dynamic system models. This process produces values of state variables at discrete time intervals depending on the chosen value of the time step, approximating the continuous-time solution of the model equations.

What happens during each time step is central to the dynamic simulation process. In this section, there are three essential steps. First is the calculation of changes in each of the state variables. This is where the differential or difference forms of the model equations are used in the simulation computer code. The second step is that all state variables are updated to their values at the next time step. And finally, time is updated by incrementing the current time by the time step that was set in the initialization section. Figure 4.1 shows these steps that are included in computer simulation programs. There are sometimes other important considerations that need to be taken into account when writing the programs and an example of such considerations is discussed later.

In Chapter 1, we showed that dynamic system models can be written either as differential equations or as difference equations. The basic steps for simulating each of these forms of models are similar, but there are important differences that must be considered when developing the computer code for one vs. the other. In continuous model simulations, written as a set of differential equations (Equation 1 in Chapter 1), the goal is to mimic as closely as possible an exact mathematical solution of the differential equations. This has important implications on the choice of the time step (Δt) as will be

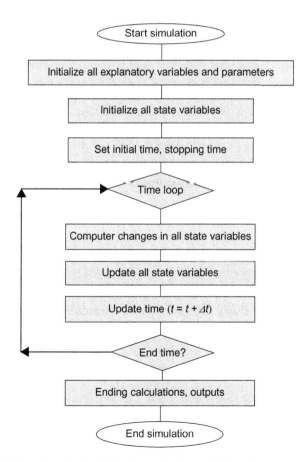

FIGURE 4.1   Flow chart showing the basic steps and control of time for computer simulation.

shown later in this chapter. For models written as difference equations, the time step is decided upon when the functions describing changes in state variables and model equations are written. So, in both cases, discrete time steps are used to compute and update state variables, but in the simulation of difference equations, functional relationships usually depend strongly on the time step chosen for the model. This will be discussed later.

In the remaining sections of this chapter, we provide additional guidelines about the implementation of the basic set of steps in Figure 4.1 for simulating dynamic system models that are expressed either as differential equations in continuous time or as difference equations using discrete time steps. Examples will be given for simulating different types of dynamic systems.

2. SIMULATING CONTINUOUS TIME MODELS (DIFFERENTIAL EQUATION FORM)

In simulating systems modeled as continuous differential or partial differential equations, the goal is to closely approximate the actual solution of the equations in continuous time. Numerical methods are used since the system models cannot usually be solved to produce analytical functions that express the state variables as functions of time. This happens either because system models are much more complex than those for which analytical solutions can be derived and/or because environmental conditions that influence system components vary in complex patterns over time and space. Thus, not only are system models simplifications of reality, simulations of those system models also provide only approximations to the exact solutions of the formulated models. In other words, there are numerical errors introduced in the simulation process for continuous system models, which can add to the errors associated with the assumptions and simplifications made when developing the model (e.g., choice of components to include in the system, state variables chosen to describe the dynamic behavior of the system, processes included and their functional forms, and parameters of the model). Although these numerical errors can cause considerable problems if the analyst is unaware of them, there are ways to evaluate them and to minimize their influence on simulated results.

2.1 Mathematical Basis for Continuous Simulation

In Chapter 1, we presented the general form of a dynamic system model as a set of ordinary differential equations:

$$\frac{dU_1(t)}{dt} = g_1(U(t), X(t), \theta) \tag{1}$$

$$\vdots$$

$$\frac{dU_S(t)}{dt} = g_S(U(t), X(t), \theta)$$

The $U_i(t)$ are state variables at time t, $X(t)$ is a vector of explanatory variables at time t, θ is the vector of parameters, and the g_i are functions. The basic mathematical basis for simulating continuous models written as a system of first order differential equations is that the continuous form of those equations can be approximated by using the definition of a derivative.

$$\frac{dU(t)}{dt} \sim \lim_{\Delta t \to 0} \left\{ \frac{U(t + \Delta t) - U(t)}{\Delta t} \right\} \qquad (2)$$

Equation (2) indicates that as Δt approaches 0, the ratio in brackets approaches the derivative of $U(t)$. It shows only one equation for describing the rate of change of one state variable, but this same relationship can be written for each of the S equations represented in Equation (1).

The Euler method for simulating continuous models is based on Equation (2) by removing the limit, and rewriting the equation in its so-called 'finite difference' form. Rearranging the equation, one obtains the Euler form for a single state variable model:

$$U_i(t + \Delta t) = U_i(t) + g_i(U(t), X(t), \theta) \cdot \Delta t \qquad (3)$$

This shows how one can compute the value of a state variable at time $t + \Delta t$ by computing the derivative of the state variable at time t, multiplying it by the time step, and adding the result to the original value of the state variable at time t. When one writes this approximation for each of the state variables, there are S equations,

$$U_1(t + \Delta t) = U_1(t) + g_1(U(t), X(t), \theta) \cdot \Delta t \qquad (4)$$

$$\vdots$$

$$U_S(t + \Delta t) = U_S(t) + g_S(U(t), X(t), \theta) \cdot \Delta t$$

Note that this is similar to Equation (2) in Chapter 1 for the general form of a system model in discrete time. The difference is that, in Equation (4) above, the g function is the derivative of the state variable computed at time t, whereas the g functions for discrete time models are not usually represented by differential equations. Instead, those g functions include various approximations to estimate the overall change in a state variable over a pre-selected time interval to represent the dynamic behavior of the system, sometimes referred to as 'functional models' (Addiscot and Wagenet, 1985). There is more about these differences in a later section of this chapter.

2.2 The Euler Method

Implementation of the Euler method for simulating continuous models is done simply by writing a program to compute values of each state variable at time $t + \Delta t$ by computing the derivatives (or rate of change) of each continuous model state variable at time t (e.g., see Equation (1)), multiplying each of these values by Δt, then adding these values to values of the state variable at time t (which are already known), then repeating this set of calculations as depicted in Figure 4.1. When this is done, one may have to include additional information to prevent simulated values of state variables from falling outside of their physical, chemical, or biological ranges. A good example is the model of water volume dynamics in Chapter 1. When this model (Equation (6) in Chapter 1) is implemented using the Euler method, the numerical value of one of the tank volumes may fall below 0.0 by a small amount. Although this may not seem important if it is a very small negative number, it will result in program failure when the program attempts to compute the square root of a negative number. Code is needed to set lower limits of 0.0 to avoid this problem.

The choice of Δt is very important for continuous system simulation, primarily because of errors that can accumulate. Smaller values of Δt typically result in lower errors. This can be shown by using the Taylor Series approximation of a continuous variable ($U(t)$) for values of time that are near time t (small Δt in the equation below).

$$U(t + \Delta t) = U(t) + \Delta t \cdot U(t)^{(1)} + \frac{(\Delta t)^2}{2!} \cdot U(t)^{(2)} + \cdots + \frac{(\Delta t)^n}{n!} \cdot U(t)^{(n)} \quad (5)$$

$$\text{where} \quad U(t)^{(1)} = \frac{dU(t)}{dt}; \cdots U(t)^{(n)} = \frac{dU^{(n)}(t)}{dt^{(n)}}$$

and $n! = 1 \cdot 2 \cdot 3 \cdots n$

Thus, by computing first, second, third, $\cdots$ through the nth derivatives of $U(t)$ and using them in the above equation, we can very closely approximate the exact value of U at time $t + \Delta t$. Although there are numerical methods based on the approach of computing higher derivatives (Press et al., 1996), this is not practical for our dynamic system models because we seldom have expressions for computing those higher derivatives. Note that the first two terms on the right hand side of Equation (5) are exactly equal to Equation (3) for computing $U(t + \Delta t)$. The 'local error' in simulating continuous systems, or the error associated with updating the state variables from time t to time $t + \Delta t$, can be written using all of the terms in Equation (5) that are missing from Equation (3).

$$error = \frac{(\Delta t)^2}{2!} \cdot U(t)^{(2)} + \frac{(\Delta t)^3}{3!} \cdot U(t)^{(3)} + \cdots + \frac{(\Delta t)^n}{n!} \cdot U(t)^{(n)} \quad (6)$$

This local error can be thought of as an error due to truncating Equation (5) after considering only the first derivative term. This local error is proportional to $(\Delta t)^2$ and higher powers of (Δt), which can be approximated by $(\Delta t)^2$ when Δt is very small. In that case, the $(\Delta t)^2$ dominates and local error is said to be proportional to $(\Delta t)^2$. The global error associated with using this method to simulate from time $t = 0$ until time $t = M$ can be expressed by multiplying the local error (Equation (6)) by the number of steps needed to simulate over a time interval of M (computed by $\dfrac{M}{\Delta t}$); thus global error for the Euler method is said to be proportional to (Δt). Although one cannot compute ahead of time what the errors will be when simulating a model, this general relationship means that global errors may be proportional to the size of Δt used in the Euler method.

This effect is easily seen by comparing an exact solution of a simple dynamic model with that computed by simulating the model using different values of Δt in the Euler method. Consider the exponential growth model, expressed as:

$$\frac{dY(t)}{dt} = a \cdot Y(t) \tag{7}$$

with the exponential growth constant a and the state variable $Y(t)$. The analytical (exact) solution to this equation is:

$$Y(t) = Y(0) \cdot e^{a \cdot t} \tag{8}$$

An R program was written to simulate this simple model for 40 time units using different values of Δt, assuming that $a = 0.10$, and the initial value $Y(0) = 1.0$. The program also compares the analytical solution shown in Equation (8) with the simulated solution using $\Delta t = 2.0$. The program listing is given below, showing the model and its main program in R.

```
######Euler Method#############################################
exponential.model < - function(a,Y0,duration = 40,dt = 1)
{
     # Initialize variables
     # K : number of time steps
  K = duration/dt + 1
     # 1 state variable, as 1 vector initialized to NA
     # Y: Size of the population
  Y = rep(NA,K)
     # Initialize state variable Y[1] = Y0
     # time variable
time = 0
     # Integration loop
k = 1
while (time < duration){
     # Calculate change of state variable
dY = a*Y[k]*dt
     # Update state variable
Y[k + 1] = Y[k] + dY
     # Update time
time = time + dt
```

```
k = k + 1
  }
   # End loop
return(round(as.data.frame(cbind(time  = seq(0,duration,by = dt)  ,
Y)),10))
}
#### Main program #################################################
a = 0.1
Y0 = 1.0
duration = 40
dt = 2
   # comparison between analytical and Euler solutions
x = seq(0,duration,by = 0.1)
y = Y0*exp(a*x)
plot(x,y,type = "l",lwd = 3)
lines(exponential.model(a,Y0,duration,dt),lty = 2,lwd = 3,
col = "black")
legend("topleft",legend = c("Analytical Solution","Numerical
Solution"),lwd = c(2,2),lty = c(1,2),col = c("black","black"),
cex = 0.75)
#End of file
```

The program can also be used to simulate Y using different values of Δt to show how global error increases as Δt is varied. Figure 4.2 shows a graph of simulated values of global error $(Y_{exact} - Y_{Euler})$ at time $t = 40$ days obtained by repeating the Euler solution for values of Δt varying from 0.01 to 4.0 days. This graph shows that the global error increases strongly with Δt, and that global error is proportional to Δt, although not exactly linearly. The exact relationship between global error and Δt depends on the model

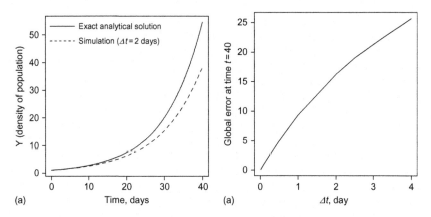

FIGURE 4.2   (a) Comparison of exact solution of the exponential growth equation for 40 time units (using values of $a = 0.1$ and $Y(0) = 1.0$) with the solution created using the Euler method to simulate exponential growth using a Δt of 2.0. (b) Error of Euler integration $(Y_{exact} - Y_{Euler})$ at time = 40, dependent on Δt.

and all of the explanatory variables used to solve it. This type of analysis is recommended when one intends to obtain an accurate solution of a continuous model. However, seldom is an analytical solution possible and one cannot compute global error. Instead, one can vary Δt and compare successive values of the solution at various simulation times. A graph of Y_{Euler} at a specific time vs. Δt will show small differences for low Δt values, but will diverge as Δt increases beyond a particular value. Such an exercise will help one select time steps that result in acceptable errors. Simulated solutions of system models using Euler and other numerical methods will always be approximations to an exact solution.

Can one determine the value of Δt that will produce reliable approximations to the exact solution for system models? The simple answer to this question is no, however, there are guidelines that one can use to obtain an initial value to try. Simulated results using this initial guess value should then be compared with simulated results using smaller values of Δt, for example, one-tenth of the original guess. If simulated results are approximately the same for both values, then either value can be used. If differences are too large relative to one's objectives, then Δt should be further reduced in iterations until successive simulated results are acceptably close to each other.

Acceptable values of the time step depend on the dynamic behavior of the system. For system models with rapidly changing state variables, Δt should be small. The time constant of a system is the amount of time that is required for a system variable to reach 63.2% of its final value after a step change in an input is forced onto the system. For linear models, the time constant (τ) is the inverse of the coefficient that is multiplied with the state variable. In Equation (7), for example, τ would be equal to $(1/a)$. One should select a Δt as an initial guess that is less than one-tenth of this value. Note, however, that different state variables in a system model will have different response times, so one would need to estimate τ values for each state variable and select an initial guess of Δt that is less than one-tenth of the smallest time constant. Note that most system models are highly complex and one cannot really compute time constants using this approach. However, this process may provide reasonable values. In all cases when the objective is to approximate the real continuous dynamic solution, one should try different time steps to determine an acceptable value of Δt for the system being modeled.

2.3 Improved Euler Method

Although the most widely used method is the Euler method, there are many other ways to obtain approximate solutions to first order, non-linear differential equation formulations of dynamic system models. Two of these general methods are summarized in this section, and they are typically available in software developed to simulate dynamic system models (such as Stella, http://www.iseesystems.com/softwares/Education/StellaSoftware.aspx). The motivation for these methods is to increase the accuracy of approximate solutions to continuous system models, thus to improve the solution for a specified value of Δt.

One method is referred to as the Improved Euler method, but is also known by other names including the second-order Runge-Kutta method and the predictor−corrector method. The term 'second order' is based on the fact that global error in this method is proportional to $(\Delta t)^2$ assuming that higher order terms are negligible for small Δt values. This error is smaller in magnitude than global error in the Euler method, which is proportional to Δt. There are also higher order methods, such as the popular fourth-order Runge-Kutta method (Press et al., 1996) with global error proportional to $(\Delta t)^4$.

The basis for many of these improved methods is easily demonstrated using the Improved Euler method. Note that in the simple example above, the model equation is used to compute the rate of change of the state variable at time t. This rate is assumed to be constant over the time step duration and the value at time $= t + \Delta t$ is computed using Equation (3). If we could instead compute the average rate of change of the state variable over the time interval, the updated state variable would more closely approximate the real value. One way to approximate the average rate of change is to compute the rate at time t and at time $t + \Delta t$, and then average them. Note, however, that the model function that computes rate of change depends on the state variable at a point in time, and we do not know the state variable value at time $t + \Delta t$. So, in the Improved Euler method, the state variable at time $t + \Delta t$ is first estimated using the basic Euler method, that estimate is then used to estimate the rate of change for $t + \Delta t$, the average rate is estimated using the rates at both time t and $t + \Delta t$, and this average is used to update the state variable. Mathematically, this is shown as follows for a single state variable model:

$$U_1(t + \Delta t) = U_1(t) + \frac{1}{2}\left[\frac{dU_1(t)}{dt} + \frac{dU_1^P(t + \Delta t)}{dt}\right] \cdot \Delta t \qquad (9)$$

where

$$U_1^P(t + \Delta t) = U_1(t) + g_1(U_1(t), X(t), \boldsymbol{\theta}) \cdot \Delta t \qquad (10)$$

$$\frac{dU_1^P(t + \Delta t)}{dt} = g_1\left(U_1^P(t + \Delta t), X(t + \Delta t), \boldsymbol{\theta}\right) \qquad (11)$$

and the superscript P refers to predicted value at time $t + \Delta t$. The R program for simulating the simple exponential growth model is shown below, comparing analytical and simulated solutions using the improved Euler method and $\Delta t = 2.0$. Note that this program is similar to the one shown above, and it includes the use of the Euler method for predicting *YP*, the value of the state variable estimated for time $t + \Delta t$. However, the integration loop in the program below contains several lines of code to compute each of the terms in Equations (9) to (11) for updating the single state variable. As will be seen when one runs both methods using Δt of 2.0, the global error is much smaller for the Improved Euler method.

```
#####Improved Euler Method########################################
exponential.model2 < - function(a,Y0,duration = 40,dt = 1)
{
     # Initialize variables
     # K : number of time steps
  K = duration/dt + 1
     # 1 state variable, as 1 vectors initialized to NA
     # Y : Size of the population
  Y = rep(NA,K)
     # Initialize state variable at start of simulation, year 0
Y[1] = Y0
     # time variable
time = 0
     # Integration loop
k = 1
while (time < duration){
     # Calculate change of state variable at time t
dY = a*Y[k]*dt
     # Predict value of state variable at time t + Δt
YP = Y[k] + dY
     # Estimate change of state variable at time t + Δt
dYP = a*YP*dt
     # Predict value of state variable at time t + Δt
Y[k + 1] = Y[k] + (1/2)*(dY + dYP)
     # Update time
time = time + dt
k = k + 1
  }
     # End loop
return(round(as.data.frame(cbind(time = seq(0,duration,by = dt),
Y)),10))
}
######### Main program ###########################################
a = 0.1
Y0 = 1.0
duration = 40
dt = 2
     # comparison between analytical and Euler solutions
x = seq(0,duration,by = 0.1)
```

```
y = Y0*exp(a*x)
plot(x,y,type = "l",lwd = 3)
lines(exponential.model2(a,Y0,duration,dt),lty = 2,lwd = 3,
col = "black")
legend("topleft",legend = c("Analytical Solution","Numerical
Solution"),lwd = c(2,2),lty = c(1,2),col = c("black","black"),
cex = 0.75)
#End of file
```

2.4 Simulating Continuous System Models with Multiple State Variables

The examples given in the R code above are for only one state variable. As noted in the flow diagram in Figure 4.1, one needs to compute changes that occur during one time step for all of the state variables before updating any of them and before updating time. This is important for continuous models, where one is trying to obtain an accurate solution. As we will see later when we discuss simulation of difference models, there are other practical considerations that should be taken into account. An example is given here of a model of the population dynamics of insects, Equation (26) from Chapter 1, which is rewritten below to compute the changes in the state variables over the given time step.

$$dE(t) = [r_b \cdot A(t) \quad r_E \cdot E(t) - m_E \cdot E(t)] \cdot \Delta t$$

$$dL_1(t) = [r_E \cdot E(t) - r_{12} \cdot L_1(t) - m_1 \cdot L_1(t)] \cdot \Delta t$$

$$dL_2(t) = [r_{12} \cdot L_1(t) - r_{23} \cdot L_2(t) - m_2 \cdot L_2(t)] \cdot \Delta t$$

$$dL_3(t) = [r_{23} \cdot L_2(t) - r_{34} \cdot L_3(t) - m_3 \cdot L_3(t)] \cdot \Delta t \qquad (12)$$

$$dL_4(t) = [r_{34} \cdot L_1(t) - r_{4P} \cdot L_4(t) - m_4 \cdot L_4(t)] \cdot \Delta t$$

$$dP(t) = [r_{4P} \cdot L_4(t) - r_{PA} \cdot P(t) - m_P \cdot P(t)] \cdot \Delta t$$

$$dA(t) = [r_{PA} \cdot P(t) - m_A \cdot A(t) + i_A(t)] \cdot \Delta t$$

There are seven state variables $(E(t), L_1(T), L_2(t), L_3(t), L_4(t), P(t),$ and $A(t))$ for population densities of eggs, larva in each of four stages, pupae, and adults, respectively. The R program for this model is given below:

```
# Population Dynamics Model with Age Classes #################
# Simulated using the Euler Method
population.age.model =
function(rb = 3.5,mE = 0.017,rE = 0.172,m1 = 0.060,r12 = 0.217,
m2 = 0.032,r23 = 0.313,m3 = 0.022,r34 = 0.222,m4 = 0.020,r4P = 0.135,
mP = 0.020,rPA = 0.099,mA = 0.027,iA = 0,duration = 100,dt = 1){
     # state variables
     # E : egg stage. population density (number per ha)
     # L1 : larvae1 stage. population density (number per ha)
```

```
# L2 : larvae2 stage. population density (number per ha)
# L3 : larvae3 stage. population density (number per ha)
# L4 : larvae4 stage. population density (number per ha)
# P : pupae stage. population density (number per ha)
# A : adult stage. population density (number per ha)
# V : matrix of state variables (one per column)
V = matrix(NA,ncol = 7,nrow = duration/dt + 1,
dimnames = list(NULL,c("E","L1","L2","L3","L4","P","A")))
    # initialization of state variables
V[1,] < − c(5,0,0,0,0,0,0)
    # Simulation loop; k is the number of the time step
for (k in 1:(duration/dt)){
    # Calculate rates of change of state variables
dE = (rb*V[k,"A"] − rE*V[k,"E"] − mE*V[k,"E"])*dt
dL1 = (rE*V[k,"E"] − r12*V[k,"L1"] − m1*V[k,"L1"])*dt
dL2 = (r12*V[k,"L1"] − r23*V[k,"L2"] − m2*V[k,"L2"])*dt
dL3 = (r23*V[k,"L2"] − r34*V[k,"L3"] − m3*V[k,"L3"])*dt
dL4 = (r34*V[k,"L3"] − r4P*V[k,"L4"] − m4*V[k,"L4"])*dt
dP = (r4P*V[k,"L4"] − rPA*V[k,"P"] − mP*V[k,"P"])*dt
dA = (rPA*V[k,"P"] − mA*V[k,"A"] + iA)*dt
    # vector of rates of change
dV = c(dE,dL1,dL2,dL3,dL4,dP,dA)
    # Update state variables
V[k + 1,] < − V[k,] + dV
  }
      # End simulation loop
return(round(as.data.frame(cbind(time = (1:(duration/dt + 1))
*dt − dt, V)),10))
}
######################### MAIN program #########################
    # 1) simulation
sim <−
population.age.model(rb = 3.5,mE = 0.017,rE = 0.172,m1 = 0.060,
r12 = 0.217,m2 = 0.032,r23 = 0.313,m3 = 0.022,r34 = 0.222,m4 = 0.020,
r4P = 0.135,mP = 0.020,rPA = 0.099,mA = 0.027,iA = 0,duration = 40,
dt = 0.01)
    # 2) values simulated at different time:
sim[sim$time = = 0,]
sim[sim$time = = 10,]
sim[sim$time = = 40,]
    # 3) Graphical representation
graph.param = data.frame("V" = c("E","L1","L2","L3","L4","P","A"),
"lty" = c(2,2,2,2,3,4,1), "lwd" = c(2,1,1,1,1,2,3))
plot(c(0,max(sim$time)), c(0,max(sim[, −1])),type = "n",xlab = "time
(day)",ylab = "population density")
null = sapply(c("E","L1","L2","L3","L4","P","A"),function(v)
lines(sim$time,sim[,v],
lty = graph.param[graph.param$V = = v,"lty"],lwd = graph.param
[graph.param$V = = v,"lwd"]))
```

```
legend("topright", legend = graph.param$V, lty = graph.param$lty,
lwd = graph.param$lwd, cex = 0./5)
#End of file
```

Figure 4.3 shows simulated results for this model over the first 40-day period of time. Initial conditions for this example were an egg population density of 5.0 and population densities of 0.0 for all other age classes at time $t = 0$. The parameters used to obtain the results in Figure 4.3 are shown in the R program code given above.

There are several features of this program that are important to note. First, the amount of change is calculated for each state variable (lines 23–31) before any state variable is updated (line 33). This follows the recommended steps in Figure 4.1 and is very important when simulating continuous system models. All changes of state variables must be computed for time t, using values of all state variables at the same time t. Then, all state variables should be updated to simulate their values at time $t + \Delta t$ and then time should be updated to time $= t + \Delta t$.

Next, note that the program defines a vector, dV, where the rates of change of all state variables are contained, in order (line 31). Also note that the state variables are all copied into a matrix, with the columns being the state variables and each row containing all values of the state variables for a specific time (the matrix is defined in lines 16–17 and the row for each day is filled in on line 33). All of the state variables are updated using matrix addition, $V[k + 1,] = V[k,] + dV$ (k is the counter for the number of time steps). This is equivalent to

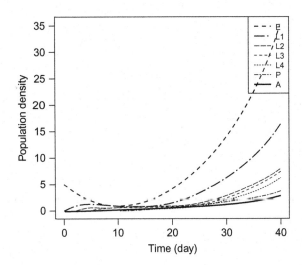

FIGURE 4.3   Example of simulated results for the population dynamics model with age structure using the R code and the parameter and initial condition values shown in Section 2.4. The model has seven state variables, one for each of the age classes, which are defined according to physiological stages of the insect.

updating each state variable using a separate equation for each, but this use of a vector of state variables and a matrix to hold all of the simulated results over time is an efficient way to allow for simplification of the program. This approach is used in many continuous system models, such as models with state variables that vary over space in two or three dimensions (e.g., diffusion models and hydrology models in addition to population dynamics models). The reason for doing this is that as the number of state variables becomes large, the code for many of the equations needed for calculations of rates of change of state variables and/or updating each state variable are similar. In the R code for the population model with age classes, for example, the equations for updating all state variables are the same, so it is much more compact to write this using vector and matrix manipulations. Note that we could also write this entire model using vector and matrix equations to further simplify the code. We chose not to do so for this model, however, because it would have made the code less readable. This population dynamics model with age classes is also given in Appendix A showing how the simulation code looks when vector and matrix notation and algebra are used for the entire model.

A final point about this program is that it does not have a statement similar to those contained in the code in Sections 2.2 and 2.3 where time is updated using an explicit statement such as time = time + dt or time $[k + 1]$ = time$[k]$ + dt. The latter formulation allows for each time step that was used for simulating the system model to be stored in a vector so that it can be printed or used to plot state variables vs. time. Instead, the times at which the calculations are done are calculated within the #return statement (line 36):

$$return(round(as.data.frame(cbind(time = (1:(duration/dt + 1)) *dt - dt, V)), 10)) \quad (13)$$

The total time for simulating the model is given by the argument dura-tion. In the call to the model in the R code for this problem, we have duration = 40. From this total simulation time and the integration time step, we can compute the number of integration steps needed as (duration/dt + 1), which is shown in Equation (13). This statement also computes simulated time for each step by multiplying the number of steps by dt, then subtracting dt since initial simulated time = 0. Equation (13) creates a vector of time values with a value for each time step in the simulation and combines this vector with the matrix V, which contains the simulated values of the insect population state variables. This approach is considered better than adding a value of dt to the previous time at each time step as the model is being simulated. The reason is that numerical errors accumulate when thousands or millions of additions are performed using very small numbers of dt that are stored in the computer using numerical approximations of continuous numbers.

2.5 Simulating Continuous System Models Using the Runge-Kutta Fourth Order Method

The R program above for the population model makes use of the Euler method. Could a more complex continuous model like this be simulated using the Improved Euler method or a higher order method? The answer is yes, and those more accurate methods are frequently used. One can easily program the additional steps needed to predict the state variables at time $t + \Delta t$, using them to estimate the rates of change at time $t + \Delta t$, then computing the average rates and using them to update all of the state variables. However, many software packages have several methods to use, including Euler, Improved Euler, and Runge-Kutta Fourth Order (e.g., MATLAB, http://www.mathworks.com/products/matlab/; Stella, http://www.iseesystems.com/softwares/Education/StellaSoftware.aspx; and the R programming language, http://www.r-project.org/).

The Runge-Kutta Fourth Order method is widely used for simulating differential equation system models because of its relative simplicity and greater accuracy relative to the Euler and Improved Euler methods described earlier. Thus, we give the equations for implementing this method and then show how this method can be implemented for simulating continuous system models in R.

The approach is similar to the Improved Euler method. For each time when the program updates the state variables, four intermediate rates are used to compute an average rate during the specified time interval. These equations are summarized below (Press et al., 1996), where $g(y(t))$ represents the right hand side of the differential equation for the model. Note that $y(t)$ can be a single state or a vector of state variables and g represents the functional form of a dynamic model.

$$k_1 = g(y(t)) \cdot \Delta t$$

$$k_2 = g(y(t) + \frac{1}{2}k_1) \cdot \Delta t$$

$$k_3 = g(y(t) + \frac{1}{2}k_2) \cdot \Delta t \qquad (14)$$

$$k_4 = g(y(t) + k_3) \cdot \Delta t$$

$$y(t + \Delta t) = y(t) + \frac{1}{6} \cdot k_1 + \frac{1}{3} \cdot k_2 + \frac{1}{3} \cdot k_3 + \frac{1}{6} \cdot k_4$$

All that is needed to solve these equations, and thus obtain a value for the state variable y at time $t + \Delta t$, is the initial value of y at time t, Δt, and the model equation (or equations if there are multiple state variables). Note, however, that if there are multiple state variables, the incremental change in each state variable must be computed before updating any of them, as noted earlier. Note also that programming these intermediate steps requires several lines of computer code for each state variable, thus it is much more efficient to make use of pre-programmed functions in R or other software. Shown below is the R program for the homogenous insect population dynamics model with age classes, using the R function for solving a system of ordinary differential equations using the Runge-Kutta Fourth Order method (rkMethod ('rk4')). Note in the Main program below that the R library for solving differential equations (deSolve) is required.

```
###Population Age Model Using Runge−Kutta 4th Order Method####
population.age.model.ode =
function(rb = 3.5,mE = 0.017,rE = 0.172,m1 = 0.060,r12 = 0.217,
m2 = 0.032,r23 = 0.313,m3 = 0.022,r34 = 0.222,m4 = 0.020,r4P = 0.135,
mP = 0.020,rPA = 0.099,mA = 0.027,iA = 0,duration, dt, method){
       # states variables − same as in Euler program listing
       # Egg, Larva1, Larva2, Larva3, Larva4, Pupae, and Adult
  E0 = 5
  L10 = 0
  L20 = 0
  L30 = 0
  L40 = 0
  P0 = 0
  A0 = 0
       # defining the model as ordinary differential equations
  predator.prey.ode <− function(Time, State, Pars) {
    with(as.list(c(State, Pars)), {
  dE = (rb*A − rE*E − mE*E)*dt
  dL1 = (rE*E − r12*L1 − m1*L1)*dt
  dL2 = (r12*L1 − r23*L2 − m2*L2)*dt
  dL3 = (r23*L2 − r34*L3 − m3*L3)*dt
  dL4 = (r34*L3 − r4P*L4 − m4*L4)*dt
  dP = (r4P*L4 − rPA*P − mP*P)*dt
  dA = (rPA*P − mA*A + iA)*dt
  return(list(c(dE, dL1, dL2, dL3, dL4, dP, dA)))
    })
    }
  sim = ode(y = c(E = E0,L1 = L10,L2 = L20,L3 = L30,L4 = L40,P = P0,
  A = A0), times = seq(0,duration,by = dt), func = predator.prey.ode,
  parms = c(rb,mE,rE,m1, r12,m2,r23,m3,r34,m4,r4P,mP,rPA,mA,iA),
  method = rkMethod(method))
  return(as.data.frame(sim))
  }
################### MAIN program #########################
library(deSolve)
       # 1) simulation with RK4 integration
       # classical 4th order Runge − Kutta, fixed time step
```

```
sim <-
population.age.model.ode(rb = 3.5,mE = 0.017,rC - 0.172,m1 = 0.060,
r12 = 0.217,m2 = 0.032,r23 = 0.313,m3 = 0.022,r34 = 0.222,m4 = 0.020,
r4P = 0.135,mP = 0.020,rPA = 0.099,mA = 0.027,iA = 0,duration = 40,
dt - 1,method = "rk4")
    # 2) values simulated at different time:
sim[sim$time = = 0,]
sim[sim$time = = 10,]
sim[sim$time = = 40,]
    # 3) Graphical representation
graph.param = data.frame("V" = c("E","L1","L2","L3","L4","P","A"),
"lty" = c(2,2,2,2,3,4,1), "lwd" = c(2,1,1,1,1,2,3))
plot(c(0,max(sim$time)), c(0,max(sim[, - 1])),type = "n",xlab = "time
(day)",ylab = "population density")
null = sapply(c("E","L1","L2","L3","L4","P","A"),function(v)
lines(sim$time,sim[,v],
lty = graph.param[graph.param$V = = v,"lty"],lwd = graph.param
[graph.param$V = = v,"lwd"]))
legend("topright",  legend = graph.param$V,  lty = graph.param$lty,
lwd = graph.param$lwd, cex = 0.75)
#end of file
```

The R function to solve ordinary differential equations (ODE) in this example uses the method 'rk4'. The method is specified as an argument in the call to the model on line 41. It is then used as the last argument of the ode function, line 31. One who implements a different continuous system model in R using the Runge-Kutta Fourth Order method would make use of these functions (available in the R library deSolve). This amounts to the coding of the specific equations for the rates of change of each state variable in continuous time, similar to the seven equations (lines 18−24) and lines of code above, then the solution is obtained for those new model equations. This makes the task of coding much simpler than if one has to code the specific solution method as was shown earlier.

2.6 Which Method Should Be Used?

There is no single answer to this question. Many people who develop their own programs to simulate continuous dynamic system models, in R or other programming languages, use the Euler method because it is simple to implement and to modify. However, numerical errors may be large, and it is important for the modeler to evaluate the model performance over the range of conditions being studied to confirm that solutions are accurate. Numerical accuracy is particularly problematic for models that are highly nonlinear and have discrete components such as thresholds. Because many computer programming environments have functions that implement higher order methods, such as the Runge-Kutta

Fourth Order method, it is relatively easy to select any of the available methods and compare results across methods to ensure that the method being used and the time step are adequate and produce accurate results.

3. SIMULATION OF SYSTEM MODELS IN DIFFERENCE EQUATION FORM

System models may be written using difference equations instead of continuous differential equations. Although the form of these difference equation system models is similar to that of the Euler equation that approximates continuous system solutions, there are several very important differences. Equation (15) shows the general form of these equations.

$$U_1(t + \Delta t) = U_1(t) + g_1[U(t),X(t), \theta] \tag{15}$$

$$\vdots$$

$$U_S(t + \Delta t) = U_S(t) + g_S[U(t),X(t),\theta]$$

System models developed as systems of difference equations are relatively easy to simulate using the basic steps shown in the flow chart in Figure 4.1. We show examples of R code to simulate the soil water balance model and the simple maize growth model in Chapter 1.

Note that the g_i functions are not multiplied by Δt in Equation (15) on the right hand side as is the case for the Euler method as shown in Equation (3). This is because most difference equation system models are developed with the assumption that the time step Δt is equal to 1. Furthermore, the functional relationships of the model (e.g., the g_i in Equation (15)) are developed to compute the net changes of state variables over a unit time step at time t. The g_i functions in Equation (15) may be very different from those used in continuous models of the same system. These equations are usually simplified functions developed by the modeler to describe complex behavior of systems; Addiscot and Wagenet (1985) referred to such models as functional models to distinguish them from continuous system models that include processes that interact in real time to affect the dynamic behavior of the system. Difference model functions may not be continuous equations; they may contain various discontinuities, non-linearities, and empirical relationships.

An important implication of this formulation is that one cannot change the time step in difference models. The functions that describe the rates of change of each state variable are written explicitly for a unit time step. This also means that one cannot use the concepts summarized above for selecting an integration method or estimating error associated with numerical approximation of the system. The integration method is defined by the difference model equations selected by the modeler as approximations to the system.

3.1 Example Soil Water Balance Model

A simple soil water balance model was presented in Chapter 1 in Equations (17) to (22). This model, used to compute a reference drought index (ARID, see Woli et al., 2012), was developed as a difference model with a time step of one day. Here, we use this model as an example to demonstrate simulation of models written with all of the functions dependent on the time step. Below is the R program listing for this model.

```
############Soil Water Balance and Drought Index Model#########
# Model described in Chapter 1
watbal.model.arid = function(WHC, MUF, DC, z, CN, weather, WP, FC,
WAT0 = NA)
{
    #WHC :Water Holding Capacity of the soil (cm3.cm^-3)
    #MUF :Water Uptake coefficient (mm^3 mm^-3)
    #DC :Drainage coefficient (mm3.mm^-3)
    #z :root zone depth (mm)
    #CN :Runoff curve number
    # Maximum abstraction (a parameter used to compute runoff)
S = 25400/CN^-254
    # Initial Abstraction (a parameter used to compute runoff)
IA = 0.2*S
    # WATfc : Maximum Water content at field capacity (mm)
WATfc = FC*z
    # WATwp : Water content at wilting Point (mm)
WATwp = WP*z
    # input variable describing the soil
    # WP : Water content at wilting Point (cm^3.cm^-3)
    # FC : Water content at field capacity (cm^3.cm^-3)
    # WAT0 : Initial Water content (mm)
if (is.na(WAT0)) {WAT0 = z*FC}
    # Initialize variable
    # WAT : Water at the beginning of the day (mm) : State variable
WAT = rep(NA, nrow(weather))
    # supplementary variable ARID drought index.
    # computed as the ratio of transpiration to potential
    # transpiration. (See Woli, 2010)
    # A value of ARID = 0 means no crop water stress;
    # a value of ARID = 1 means maximum stress & no growth
ARID = rep(NA, nrow(weather))
    # initialize state variable to amount of water at time 0
WAT[1] = WAT0
ARID[1] = NA
    # integration loops
for (day in 1:(nrow(weather)^-1))
{
    # Calculate rate of change of state variable WAT
    # Compute maximum daily water uptake by plant roots, RWUM
```

```
RWUM = MUF*(WAT[day]—WATwp)
    # Calculate the amount of water lost by transpiration (TR)—
prior to RAIN, RO, and DR
TR = min(RWUM, weather$ETr[day])
    # Compute Surface Runoff (RO)
if (weather$RAIN[day] > IA){RO = (weather$RAIN[day] —
0.2*S)^2/(weather$RAIN[day] + 0.8*S)}else{RO = 0}
    # Calculate the amount of deep drainage (DR)
if (WAT[day] + weather$RAIN[day]—RO > WATfc){DR =
DC*(WAT[day] + weather$RAIN[day]—RO—WATfc)}else{DR = 0}
    # Update state variables
dWAT = weather$RAIN[day]—RO—DR—TR
WAT[day + 1] = WAT[day] + dWAT
    # compute the ARID index. Note that it is an auxiliary
variable, not a "state variable" as is WAT[day]
if (TR < weather$ETr[day]) {ARID[day + 1] = 1 —
TR/weather$ETr[day]} else {ARID[day + 1] = 0.0}
    }
    # Volumetric Soil Water content (fraction : mm.mm^-1)
WATp = WAT/z
return(data.frame(day = weather
################### MAIN PROGRAM #######################
# TO DO : change path to your working directory (with '/')
setwd("C:/RWork/")
weather_all = read.table("weather_all.csv",sep = ";",header = TRUE,
na = c(—99))
    # choose weather for a site and a year
    # change to weather year you want (2007 or 2008)
    # select site in France (1 to 10)
weather = subset(weather_all,(idsite = = 1)&(year = = 2007))
    # model parameters
WHC  = 0.15
MUF = 0.096
DC = 0.55
z = 400
CN = 65
    # input variables describing soil
    # WP : Water content at wilting Point (cm^3.cm^-3)
soil.WP = 0.06
    # FC : Water content at field capacity (cm^3.cm^-3)
soil.FC = soil.WP + WHC
    # WAT0 : Initial Water content (cm^3.cm^-3)
soil.WAT0 = soil.FC*z
sim = watbal.model.arid(WHC, MUF, DC, z, CN, weather,
soil.WP,soil.FC, soil.WAT0)
    # Write output to a file (for use in notepad or excel)
options(digits = 3)
write.table(format(sim), file = "sim.WAT.csv", quote = FALSE,
sep = "\t", dec = ".", row.names = FALSE)
    # Produce graphical output of the state variables
```

```
par(mfrow = c(3,1), mar = c(4.1,4.1,1.1,0.2))
barplot(sim$RAIN, xlab = "day", ylab = "Rain (mm)" )
barplot(sim$ETr, xlab = "day", ylab = "ETr (mm)")
plot(sim$day,sim$WATp*100, xlab = "day", ylab = "WATp
(%)",type
= "l",lwd = 2,ylim = c(0,soil.FC*110))
abline(h = soil.FC*100, lty = 2)
abline(h = soil.WP*100, lty = 2)
dev.new()
plot(sim$day,sim$ARID, xlab = "day", ylab = "ARID
index",type = "l",lwd = 2,ylim = c(0,1))
# End of file
```

Note that this program has basically the same structure used to simu-late continuous dynamic system models, with a main program (lines 63−102) and a function that contains all of the model equations to com-pute changes in state variables and to update them (lines 3−62). However, note the absence of a Δt in the above R program; the state vari-able in this model is total soil water (mm), $WAT[day]$, and this variable is updated to produce $WAT[t + 1]$ each day by adding all net changes com-puted in day t to the value of WAT on day t (lines 53, 54). Note that the program computes volumetric soil water content ($w(t)$ in Chapter 1, and $WATp$ in this program), by dividing $WAT[day]$ by the root zone depth, z (line 61). We do not have the flexibility to change Δt for this difference equation model. In this case, model functions were written specifically for a time step of one day.

We point out a few other important features of this program. First, the model requires daily weather data (rainfall, and reference evapotranspiration, ETr). These daily records are in a file named 'weather_all.csv' (read in lines 66−67). The R program must specify a working directory from which it opens and reads these daily values, putting them in a table named 'weather.' In this example, the weather file is in the directory 'C:/RWork' (line 65).

Simulated results were produced by the above code using the 2007 weather year, shown in Figure 4.4. (The year of weather and the site are specified in line 71.) This figure shows graphs of rain, ETr, and volumetric soil water content vs. day of year. The $ARID$ drought index (bottom graph in Figure 4.4) shows an estimate of the degree of drought stress of a grass refer-ence crop vs. time. Note that $ARID$ is not a state variable in this model. Instead, it is an auxiliary variable that is computed as a function of state vari-ables, here as a function of $WAT[day]$ (lines 57−58). There is a sequence of calculations that computes the net change in soil water on the current day (lines 42−53) before computing soil water for $day = day + 1$ (line 54). Figure 4.5 shows these same variables simulated for the year 2008. Note the different patterns of volumetric soil water content and $ARID$ during the time course of each year.

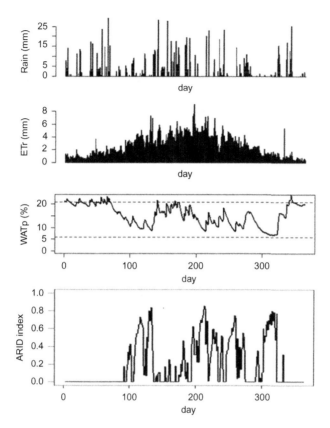

FIGURE 4.4 Simulated results from the simple soil water balance and ARID drought index model using weather data for the year 2007. Daily graphs of rain (top figure) and potential evapotranspiration (ETr, second graph), daily simulated values of volumetric soil water content (third graph), and daily simulated values of the ARID drought index.

Another important point is that the model simulates state variables only for discrete points in time, even though the real processes happen continuously over time. The program for this difference model uses a particular sequence of calculations of processes that take place during each day, and the order of these calculations may be very important. For example, note that *TR* (transpiration rate, mm/d) is computed strictly using the value of soil water of the current day and the *ETr* (potential evapotranspiration rate as read from an input file, mm/d) of that same day (lines 42–45). Then, surface runoff is computed based only on *RAIN* (rainfall amount, mm/d) for that same time = day, and neither *TR* nor *RO* (surface runoff, mm/d) depend on other processes that are taking place during that same day. However, drainage (*DR*, mm/d) is computed by adding the day's rainfall to the soil water at time = day and subtracting runoff (lines 50–51).

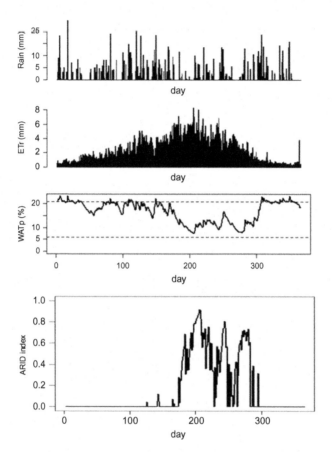

FIGURE 4.5   Simulated results from the simple soil water balance and ARID drought index model using weather data for the year 2008. Daily graphs of rain (top figure) and potential evapotranspiration (ETr, second graph), daily simulated values of volumetric soil water content (third graph), and daily simulated values of the ARID drought index.

This assumes that rainfall and runoff both take place before drainage is computed in the day's sequence of calculations.

This sequence of calculations can be modified by computing drainage using the current day's soil water content before accounting for rainfall and runoff, then computing runoff. The R code for this modification in the sequence of calculations is shown here.

```
# Calculate rate of change of state variable WAT
# Compute maximum water uptake by plants for day, RWUM
RWUM = MUF*(WAT[day]-WATwp)
# Calculate the amount of water lost by transpiration
# (TR) - prior to RAIN, RO, and DR
TR = min(RWUM, weather$ETr[day])
```

```
     # Calculate the amount of deep drainage (DR)
if (WAT[day] > WATfc){DR = DC*(WAT[day] − WATfc)}else{DR = 0}
        # Compute Surface Runoff (RO)
if (weather$RAIN[day] > IA){RO = (weather$RAIN[day] −
0.2*S)^2/(weather$RAIN[day] + 0.8*S)}else{RO = 0}
          # Update state variables
dWAT = weather$RAIN[day] − RO − DR − TR
WAT[day + 1] = WAT[day] + dWAT
```

Compare these statements with those in the earlier code listing (lines 40−54) to see that in the above code, DR on a day depends only on the current day's soil water content in excess of field capacity (line 8) whereas in lines 50−51 of the original code, DR is computed by adding the current day's rainfall and subtracting the current day's runoff from the current day's soil water. The results are different; there is more soil water after days of rainfall in this latter case. An exercise is given to demonstrate this. In fact, depending on the amount of rainfall, one could obtain unrealistic values of soil water content. This would not happen in continuous models if processes were modeled using differential or partial differential equations and Δt was small enough to produce accurate results.

What order should one use for calculating net changes in state variables during a discrete time step in difference models? To be consistent, one should compute all changes in state variables based on the current day's state variables and explanatory variables before computing updated state variable values for the next day. However, the order of calculations during a fixed time step can have large effects in difference models. If all processes in a model have time constants that are much longer than the preselected time step, difference models may be good approximations of Euler solutions of continuous time differential equation models. Generally, however, some processes may cause rapid changes in one or more state variables (i.e., they may have time constants less than the time step), whereas others occur more slowly. Model developers need to choose the order of calculations during one time step such that they produce good approximations for the changes in all state variables during that time step. This is a major reason that functional relationships and the order of calculations in difference models are different from those in models that are represented by differential or partial differential equations. In the soil water model, rainfall may start and end within one hour, causing very rapid changes in soil water content. In contrast, drainage and root water uptake rates occur at slower rates. Generally, one should develop the relationships and order of calculations within a time step to account for the processes that cause relatively large changes in state variables within the time step. Although there is no universal answer to the question posed above, one could also consider when the different processes tend to occur during a time step. In the soil water example, if we assume that rainfall generally occurs late in the day and drainage occurs after rainfall occurs, then the original calculations (lines 42−53) would best mimic that sequence with transpiration computed first, then runoff, and then drainage. Rainfall can be thought of as a discrete event in this model where changes to state variables occur immediately within a single time step.

3.2 Simple Maize Crop Model

A simple maize crop growth model was presented in Chapter 1 in Equations (11) to (16). This model was written in a difference equation format using a daily time step and weather inputs of maximum and minimum daily temperatures and daily solar radiation. This is a potential yield model in that we assumed no water or nutrient stresses and no damage by pests and diseases. An R program to simulate this model is shown below. The code has a function that performs all of the calculations to compute thermal time (lines 12−13, 26), leaf area index (*LAI*) (lines 20−25, 28), and biomass (*B*) (lines 14−19, 27) each day. In this program, one can select any particular year of weather data to use (between 2007 and 2011) using the file 'weather_all.csv' (read in lines 38−40). The main program is structured to first simulate results for the year 2010 (select 'weather' in line 41, 'call model' in lines 65−67), produce graphical outputs of each of the daily weather variables from the file, and also graph thermal time, *LAI*, and *B* (lines 75−83, see outputs in Figure 4.6). Then, the program simulates two years (2010 and 2011 are selected in the code in lines 86−88, and the model is called twice in lines 90−93) and graphs are produced to compare *LAI* and *B* simulated results vs. day of year for the two years (lines 95−103). This demonstrates an important feature associated with using a function for the model. This function can be called using any combination of years, values of parameters, and initial conditions to produce simulated values over time. This feature is very important for many types of model uses, such as those presented in later chapters (e.g., for sensitivity analysis, parameter estimation, and other applications).

```
############## Simple maize model Described in Chapter 1 #########
maize.model = function (Tbase, RUE, K, alpha, LAImax, TTM, TTL,
weather, sdate, ldate)
{
TT <- rep(NA, ldate)
B <- rep(NA, ldate)
LAI <- rep(NA, ldate)
TT[sdate] <- 0
B[sdate] <- 1
LAI[sdate] <- 0.01
for (day in sdate:(ldate-1)) {
dTT <- max((weather$Tmin[day] + weather$Tmax[day])/2-
Tbase, 0)
if (TT[day] <= TTM) {
dB <- RUE * (1 - exp(-K * LAI[day])) * weather$I[day]
}
```

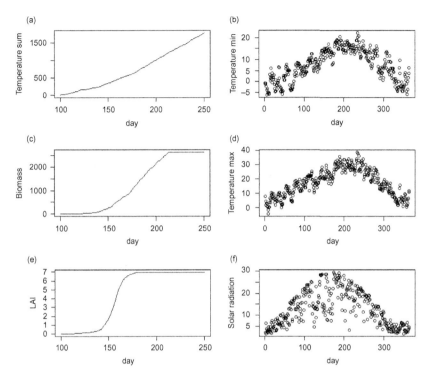

FIGURE 4.6 Simulated results of the simple maize crop model using R code in Section 3.2 for (a) computed temperature sum (TT[day]), (b) minimum daily temperature (Tmin[day]), (c) simulated biomass (B[day]), (d) maximum daily temperature (Tmax[day]), (e) simulated leaf area index (LAI[day]), and (f) daily solar radiation values (I[day]).

```
else {
dB <- 0
}
if (TT[day] <= TTL) {
dLAI <- alpha * dTT * LAI[day] * max(LAImax - LAI[day], 0)
}
else {
dLAI <- 0
}
TT[day + 1] <- TT[day] + dTT
B[day + 1] <- B[day] + dB
LAI[day + 1] <- LAI[day] + dLAI
}
return(data.frame(day = sdate:ldate, TT = TT[sdate:ldate],
LAI = LAI[sdate:ldate], B = B[sdate:ldate]))
}
########### MAIN PROGRAM ####################################
# TO DO: set directory for location of weather data
# in the file "weather_all.csv"
```

```
set.wd("C:/RWork")
    # Part 1: working with the model
weather_all = read.table("weather_all.csv",sep = ";",header = TRUE)
    # change the year of the weather you want (1984 to 2011)
    # the site in France (1 to 40)
weather = subset(weather_all,(idsite = =1)&(year = =2010))
head(weather)
    # Define parameter values of the model function
    # Tbase: the baseline temperature for growth (°C)
Tbase <- 7.0
    # RUE: radiation use efficiency (g.MJ⁻¹)
RUEmax <- 1.85
    # K : extinction coefficient (−)
K <- 0.7
    #alpha: relative rate of leaf area index increase for
    # small values of leaf area index ((°C.day)−1)
alpha <- 0.00243
    #LAImax: maximum leaf area index (m2 leaf/m2 soil)
LAImax <- 7.0
    #TTM: temperature sum for crop maturity (°C.day)
TTM <- 1200
    #TTL: temperature sum at the end of leaf area increase (°C.day)
TTL <- 700
    # sdate: sowing date
sdate <- 100
    # ldate: last date
ldate <- 250
    # Running the model
output <-
maize.model(Tbase,RUEmax,K,alpha,LAImax,TTM,TTL,weather,sdate,
ldate)
    # Write output to a file (to open with notepad or excel)
options(digits = 3)
write.table(format(output), file = "output.csv", quote = FALSE,
sep = "\t", dec = ".", row.names = FALSE)
    # Produce graphical output of state variables
dev.new()
par(mfcol = c(3,2))
plot(output$day,output$TT, xlab = "day", ylab = "Temperature
sum",type = "l")
plot(output$day,output$B, xlab = "day", ylab =
"Biomass",type = "l")
plot(output$day,output$LAI, xlab = "day", ylab = "LAI",type = "l")
    # Produce graphical output of the input variables
plot(1:365,weather$Tmin, xlab = "day", ylab = "Temperature min")
plot(1:365,weather$Tmax, xlab = "day", ylab = "Temperature max")
plot(1:365,weather$I, xlab = "day", ylab = "Solar Radiation")
```

```
# Part 2-run the model for two years, 2010 and 2011
weather1 = subset(weather_all,(idsite = = 1)&(year = = 2010))
head(weather1)
weather2 = subset(weather_all,(idsite = = 1)&(year = = 2011))
head(weather2)
output1 = maize.model(Tbase,RUEmax,K,alpha,LAImax,TTM,TTL,
weather1,sdate,ldate)
output2 = maize.model(Tbase,RUEmax,K,alpha,LAImax,TTM,TTL,
weather2,sdate,ldate)
        #plot Biomass and LAI vs. day of year for the 2 years
dev.new();plot(output1$day,output1$B, xlab = "day", ylab =
"Biomass", type = "l")
lines(output1$day,output2$B, xlab = "day", ylab =
"Biomass",type
= "l",col = "red")
dev.new();plot(output1$day,output1$LAI, xlab = "day", ylab = "LAI",
type = "l")
lines(output1$day,output2$LAI, xlab = "day", ylab =
"LAI",type
= "l",col = "red")
        #compute simulation day of maturity for the 2 years
print(min(which(output1$TT > TTM)))
print(min(which(output2$TT > TTM)))
# End of file
```

Note that the net rates of change of both *LAI* and *B* are based on the values of these state variables on the current day (e.g., *LAI*[*day*] and *B*[*day*]), and the weather data from that same day (lines 14−29) before updating the state variables for *day* + 1. The order of these calculations is not critical, except for the fact that the value of *dTT* (thermal time at time = day) is needed to compute *dLAI*, the net change in *LAI* for that day. This is because the changes to *LAI*

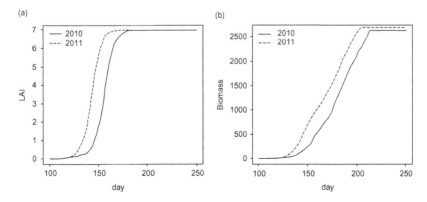

FIGURE 4.7   Simulated results of the simple maize crop model using R code in Section 3.2 for two years (2010 and 2011). (a) Leaf area index (*LAI*) and (b) Biomass (*B*).

[*day*] and *B*[*day*] occur slowly relative to the time step; there are no processes in this model similar to rainfall in the soil water balance model.

Figure 4.7 shows the graphs of *LAI* and *B* created by the above code for two years (2010 and 2011), demonstrating that there can be large effects of temperature and solar radiation variations over time on model performance. Although this is a very simple model, these effects are similar to those that are obtained by more comprehensive crop models when water and nutrients are not limiting. If one wants to include water limitations, either daily values of soil water conditions would need to be provided as external environmental inputs, or a soil water model component would need to be added, as shown in Chapter 1.

By comparing the R program for this simple maize model with the model description in Chapter 1 (Equations (11) to (16)), one should appreciate the importance of developing the model equations, including all of the functions to be included, before writing the computer program to simulate results. This helps ensure that the model description is complete and that it accurately documents what is coded in the model. This is true for both continuous time differential equation models and models developed as difference equations.

It is also very useful to develop functions for different components of a system model. There are good examples of crop models that have been programmed using modular approaches such that model code for system components is written using different functions. There are many benefits of modular model programming, including the ability to 'unplug' a model component and 'plug in' another one, making it easier to maintain and modify the computer code (Jones et al., 2001). For example, the APSIM and DSSAT cropping system models were developed using modular approaches (Keating et al., 2003; Jones et al., 2003).

3.3 Additional Comments on Difference and Continuous Model Simulation

One reason that many agricultural models (e.g., crop, soil, disease) are developed using difference equations and a daily time step is that these models depend strongly on weather conditions, and usually weather data are available on a daily basis. Some difference models, however, use an hourly time step in an attempt to improve the calculations of runoff and infiltration as well as photosynthesis and development rates that are nonlinearly related to temperature and solar radiation. However, because weather data are typically only available on a daily basis, model developers have to estimate hourly values of weather (e.g., temperature, solar radiation, rainfall, humidity, and wind speed) to compute hourly changes in the crop and soil state variables.

To further contrast the difference vs. differential equation model forms, consider the single soil volume model discussed in Section 3.1 above. To create a continuous form model of this system, one would write a differential equation, similar to Equation (17) in Chapter 1, as follows:

$$\frac{dW(t)}{dt} = P(t) - RO(t) - T(t) - D(t) \qquad (16)$$

This states that the rate of change in soil water volume is equal to the rates of precipitation minus runoff, transpiration, and drainage. Of course, one could also include other processes, such as lateral flow rate into/out of the soil and soil evaporation rate. One could also model the vertical distribution of soil water and the rates of redistribution in the soil.

There are several important implications if one uses this continuous model formulation. First, each of the processes on the right hand side of Equation (16) is considered to be continuous in time, thus affecting soil water on an instantaneous basis. Furthermore, the rates of these processes may vary by more than an order of magnitude, with $P(t)$ varying considerably over time within each day and among days; rainfall rates could easily exceed 100 mm h^{-1} whereas transpiration rates would be closer to 1 mm h^{-1} on relatively hot days. All rainfall and runoff on a particular day may occur within less than one hour. Thus, the time step needed for accurate simulations of water input to the soil via rainfall and runoff processes in continuous time differential equation models would need to be small, perhaps on the order of minutes or even seconds in some cases. The second important implication of using this continuous formulation is that one would need rainfall rate (also potential evapotranspiration rate) information at sub-daily time steps. A third implication is that the process equations for modeling runoff, transpiration, and drainage would be different. Additionally, accurate modeling of transpiration and drainage in this simple model at sub-daily time steps may require that one models the variability of soil water vs. soil depth, which would require the use of the Richards equation for soil water transport in soils. There are examples in the literature of such continuous models, some of which are contained in cropping system models (e.g., the RZWQM model, Ahuja et al., 1993: Hanson et al., 1999), however, there are many more examples of difference soil water balance models that use daily time steps (e.g., in DSSAT (Jones et al., 2003), STICS (Brisson et al., 2003), and APSIM (Keating et al., 2003) cropping system models).

Earlier in this chapter, we discussed models of dynamic systems with a focus on continuous processes, with an implicit assumption that state variables change smoothly vs. time as a result of these processes. However, many or perhaps most agricultural system models contain discrete events that are assumed to occur at an instant in time. These events may change values of state variables in the dynamic models instantaneously. An example of an event would be an irrigation or nitrogen fertilizer application. These events may be scheduled based on information that is input from a file and known about beforehand, or they may be 'triggered' based on the system state variables.

For example, if one is simulating a real experiment that included an irrigation event on day of year 95, totaling 5 cm, then this would be read into the R program on day 95 and the model would need to add 5 cm of water to the soil water volume on that day. Of course, one could model the infiltration and run-off of an irrigation event similar to rainfall in continuous time models by specifying a start and end time of the irrigation event and the rate of irrigation during that time period. However, in difference soil water balance models with a daily time step, this irrigation event occurs on a particular day during which time the water is added to the soil much like rainfall is added to the soil in the example shown in Section 3.1 above.

Events can be triggered internally in a simulation. For example, if one wants to simulate the irrigation demand for a particular season, irrigation events can be modeled to occur on a day when the soil water content drops below a critical threshold, with a specified daily amount of irrigation to apply. These amounts can be summed during a season to estimate the total irrigation requirement for the particular year and crop being simulated. Other examples of discrete events are pesticide applications in insect or disease models and events that remove grain or biomass from the simulated fields.

EXERCISES

Easy

1. Modify the R program shown in Section 2.2 to simulate the following dynamic model with one state variable (using the Euler method).

$$\frac{dY(t)}{dt} = r \cdot [K - Y(t)]$$

where
$Y(t)$ = state variable
t = time in days, d
r = relative rate of change of the state variable, d^{-1}, and
K = maximum sustained value of the state variable

Assume that $r = 0.2$ and $K = 40$, both constant values. The initial Y, $Y(0)$, is equal to 0.0.

a. What is the time constant (τ) for this model? What value of Δt would you select as a first try at simulating the model? Why?
b. Simulate values of $Y(t)$, starting at $t = 0$ and continuing until $t = 20$ d, using the Δt selected above. Plot $Y(t)$ vs. time.
c. Simulate the system again, but this time, use a Δt that is only half as large as used above. Plot results. Repeat this solution a third time, reducing your original Δt by a factor of 10. Compare the graphs. What can you conclude about an appropriate value of Δt for this model?

2. Develop an R program to simulate the following homogenous population continuous dynamic model (logistic equation) from Chapter 1:

$$\frac{dA(t)}{dt} = r \cdot A(t)\left[1 - \frac{A(t)}{K}\right]$$

where
$A(t)$ = population density, number ha^{-1}
t = time in days, d
r = rate of reproduction per individual in the population, d^{-1}, and
K = carrying capacity of the environment, number ha^{-1}
For this solution, use values of 0.05 and 20.0 for r and K, respectively, and use a Δt of 1.0 d. The initial population $A(0)$ is 1.5 individuals. Simulate the population for 100 days using the Euler method.
   a. Show the graph of $A(t)$ vs. t.
   b. Repeat this simulation using the Improved Euler method and the same Δt. How different are the results? What value of Δt would you need to use with the Euler method to obtain results with the Improved Euler method that differ at most by 0.01 individuals?

3. Develop an R program to simulate the homogenous population model above using the 4th Order Runge-Kutta method. Use a Δt value of 2.0 for this problem. You can use the equations in this chapter for this or you can use the 'rk4' method that is a function in R.
   a. Compare results with those of problem 2. Which method is more accurate?
   b. Add statements to your R program to compute the exact solution for this homogenous population model, and to compute the error between the simulated and exact values for each time step. Vary Δt between values of 0.01 and 4.0, and plot the maximum error for each Δt vs. Δt.

Moderate

4. Develop an R program to simulate the volume and height of water in each of the two tanks for the tank problem shown in Figure 3 of Chapter 1. The equations for this model are (from Equation (6) in Chapter 1):

$$\frac{d(V_1(t))}{dt} = i(t) - C_1\left[\left(\frac{2gV_1(t)}{A_1}\right)\right]^{1/2}$$

$$\frac{d(V_2(t))}{dt} = C_1\left[\left(\frac{2gV_1(t)}{A_1}\right)\right]^{1/2} - C_2\left[\left(\frac{2gV_2(t)}{A_2}\right)\right]^{1/2}$$

   a. Simulate values of $V_1(t)$, $V_2(t)$, $H_1(t)$, and $H_2(t)$ for 1000 seconds. The water flow rate into the tank ($i(t)$) is 0.2 m^3 s^{-1} for the first 50 s of the simulation, then $i(t)$ drops to 0.0 for the duration of the 1000 s. Values of the constants and parameters are:

$$g = 9.8 \text{ m s}^{-2}, A_1 = 10 \text{m}^2, A_2 = 30 \text{m}^2, C_1 = 0.8 \text{ m}^2, C_2 = 0.4 \text{ m}^2$$

b. Plot $H_1(t)$ and $H_2(t)$vs. time between $t = 0$ and 1000 s. Also, plot $V_1(t)$ and $V_2(t)$ vs. time between $t = 0$ and 1000 s.

5. You are asked to evaluate the differences in simulated soil water and ARID that is caused by using different sequences of calculations in the discrete time difference equation model discussed in this chapter. Section 3.1 shows the original R program with one sequence; and a different sequence of calculations is given later in Section 3.1. Compute the differences in both WAT[*day*] and ARID[*day*] for two different years of weather data (2010 and 2011). Graph the differences and discuss results. Which sequence do you think is more reasonable?

Difficult

6. You are asked to develop an R program to simulate a continuous flow biological reactor described in Chapter 1, homework problem 5. Time units are in hours and you are to simulate the concentration of substrate [$S(t)$], concentration of product [$B(t)$] inside the reactor, AND the cumulative amount of product B that comes out of the reactor. Note that the variables without brackets are masses ($S(t)$ and $B(t)$), whereas these same symbols with brackets represent the concentration of those masses inside the reactor (e.g., [$S(t)$] and [$B(t)$]), computed by dividing the masses by V, the volume of the reactor vessel. The diagram and equations for this system are given below:

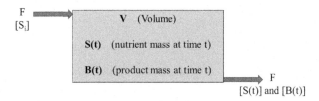

Diagram for a continuous flow biological reactor

Equations for Tank Reactor Problem

$$\frac{dS(t)}{dt} = F[S_i] - F[S(t)] - \mu \cdot (aB(t))$$

$$\frac{dB(t)}{dt} = \mu B(t) - F[B(t)]$$

with
$$\mu = \frac{\mu_{max}[S(t)]}{K_S + [S(t)]}$$

Furthermore, you are to compare these values for different flow rates of liquid into the reactor. Assume the following values for the problem:

$V = 20$ liters (volume of the reactor vessel)
$F = 1, 5, 10,$ and 14 liter h-1 (compare results for each flow rate)
$[S_i] = 10$ mg liter^{-1} (concentration of substrate flowing into the reactor)
$B(0) = 10$ mg (initial condition mass of B)
$[S(0)] = 0.0$ mg l^{-1}, initial concentration of S in the reactor, or $[S]$ at $t = 0$
$a = 0.013$
$K_S = 7$ mg liter-1
$\mu_{max} = 0.2$ h-1
$T_F = 70$ h (length of time to simulate the reactor dynamics)

Use the Euler method. Specific questions are:
a. Try a Δt of 0.10 h, but determine if this is an adequate time step; if it is not acceptable, set Δt to an acceptable value.
b. Show graphs of results of $[S((t)]$ and $[B(t)]$ vs. time for $t = 0$ until $t = 50$ h, using a suitable Δt.
c. Show a plot of cumulative product B that flows from the reactor during the 70 h for each inflow (F) rate. How does cumulative B produced depend on F?
d. For the case when $F = 10$ liter h-1, simulate the dynamics to determine if steady state is reached. If it is, what are values of $[S(t)]$ and $[B(t)]$ at steady state?
e. Modify the solution using the improved Euler method and compare results using different values of Δt.

7. Develop an R program to simulate a rabies epidemic in a population of foxes, based on the model equations given in Anderson et al. (1981) as shown in Keen and Spain (1992) on pages 386-388.

$$\frac{dS(t)}{dt} = k_B S(t) - k_D S(t) - k_1 C(t)S(t) - k_G S(t)\left(\frac{N(t)}{K}\right)$$

$$\frac{dI(t)}{dt} = k_1 C(t)S(t) - k_C I(t) - k_D I(t) - k_G I(t)\left(\frac{N(t)}{K}\right)$$

$$\frac{dC(t)}{dt} = k_C I(t) - k_D C(t) - k_M C(t) - k_G C(t)\left(\frac{N(t)}{K}\right)$$

Where:
$S(t) = $ number of susceptible foxes
$I(t) = $ number of foxes infected with rabies
$C(t) = $ number of carrier (rabid, diseased) foxes

$N(t)$ = total number of foxes $(S + I + C)$
K = carrying capacity of environment
k_x = rate constants for births (k_B), deaths (k_D), infections (k_I), and environment effects (k_G). The k_G term is actually $(k_G = k_B - k_D)$.

Also the rate of change of the total fox population is the sum of the above rates for each category of foxes:

$$\frac{dN(t)}{dt} = k_G S(t) - k_D N(t) - k_M C(t) - k_G N(t) \left(\frac{N(t)}{K} \right)$$

Use the Euler method to simulate the population dynamics with $\Delta t = 1$ day.

The rate constants are:

$$k_B = 0.00274 \quad k_D = 0.00137 \quad k_G = 0.00137$$
$$k_I = 0.21833 \quad k_C = 0.033562 \quad k_M = 0.200$$

The carrying capacity of foxes is $K = 2.0$ foxes km^{-2}. Initial conditions at t = 0 are:

$$N(0) = 2 \quad I(0) = 0.1 \quad C(0) = 0. \quad \text{Note that } N(t) = S(t) + I(t) + C(t)$$

a. Simulate this epidemic for 40 years and plot $N(t)$, $I(t)$, and $C(t)$ vs. time over this time period. This should be a pretty realistic description of a rabies epidemic as it occurs in European foxes as found by Anderson et al., 1981 (Nature 289:765-771).

b. Does the population of foxes reach steady state in your simulation? If not, why not?

8. Temperature has a major effect on the dynamic behavior of plants, diseases, and other biological organisms. To demonstrate how temperature effect can be included to influence population processes of a simple homogenous predator-prey population model, you are asked to develop an R program to simulate the predator-prey model given below.

$$\frac{dH(t)}{dt} = r_H \cdot H(t) \cdot \left[1 - \frac{H(t)}{K_H} \right] - m_H \cdot L(t) \cdot H(t)$$

$$\frac{dL(t)}{dt} = m_H \cdot L(t) \cdot H(t) \cdot eff - m_L \cdot L(t)$$

where $H(t)$ is the population density of aphids (number per ha), the host, $L(t)$ is the population density of ladybeetles (number per ha), r_H is the relative reproductive rate for aphids, K_H is the environment carrying capacity for aphids (number per ha), m_H is the relative rate of predation (proportional to both aphid and ladybeetle population densities), eff is the efficiency with which the rate of reproduction of ladybeetles occurs due to feeding on aphids, and m_L is the relative rate of mortality of ladybeetles.

a. First, you are asked to simulate these dynamics assuming that temperature does not affect the rate of reproduction of either the host (aphids) or prey (ladybeetles). Use the Euler method to simulate the model and a Δt value of 0.2 days. Simulate the populations for a total of 200 days. The parameter values to use for this solution are assumed to be constant values:

$$r_H = 1 \quad K_H = 10 \quad m_H = 0.2 \quad eff = 0.5 \quad m_L = 0.2$$

Plot $L(t)$ and $A(t)$ vs. time for the 200 days. Assume that the initial populations are $L(0) = 4.0$ and $H(0) = 2.0$.

b. Now, the model is modified to cause a temperature effect on the reproductive rate (r_H) of aphids. Although temperature would also in reality affect the other processes in the model, you should only change this rate now. To do this, you will need to read in daily weather data from the weather_all.csv file used for the simple maize and soil water models. The starting day for the populations in this case will be on day 120 so that temperatures are reasonable values for the aphids and ladybugs. So, initial values of $H(0)$ and $L(0)$ should be the same as above, and use the same time step for the Euler method, and read year 2010. Thus, one new feature of this part of the problem is that daily maximum and minimum temperatures must be read in. The second modification to your program is that you will need to modify aphid reproductive rate daily as temperature varies each day. We assume that there is a maximum rate of reproduction at an average daily temperature of 26°C ($r_{H\text{-}max}$), which is assumed to be 1.0 (the same value as r_H above). Then the following equation is used to compute daily values of r_H:

$$r_H = r_{H\text{-}max} \cdot \frac{\left(T_{\text{avg}}(t) - 8.0\right)}{(25.0 - 8.0)} \quad for\ T_{\text{avg}}(t) \quad between\ 8\ and\ 25C$$

$$r_H = r_{H\text{-}max}\ for\ T_{\text{avg}}(t) \quad greater\ than\ 25C$$

$$r_H = 0.0\ for\ T_{\text{avg}}(t) \quad less\ than\ or\ equal\ to\ 8C$$

and

$$T_{\text{avg}}(t) = \frac{(T_{\text{max}}(t) + T_{\text{min}}(t))}{2}$$

These equations restrict r_H to be between values of 0.0 and r_H-max as daily temperature average varies. How are the population dynamics of aphids and ladybeetles affected by this change?

a. Repeat the simulation with temperature effects, this time using the year 2011. How do the results vary between the two years?

b. Discuss how you would modify the other functions in the model.

9. Refer to homework assignment number 7.d in Chapter 1. You were asked
to develop the finite difference equations (or Euler method equations) to
approximate the solution to a continuous diffusion problem where gas diffusion occurs along a pipe. Here you are asked to develop an R program
to simulate this distributed system.

Assume that you will divide the diffusion pathway into 20 compartments. Number the compartments starting using numbers (1,2,3, etc.) but
note that distance along the flow path is (compartment number $*\Delta x$).
Simulate the diffusion along a linear gradient (one dimension) of 20 compartments. Let $D = 15 \text{ cm}^2 \text{ min}^{-1}$, $\Delta x = 10$ cm. (Note units are in cm vs.
mm as indicated in the PowerPoint. The equation is the same, however,
so do not convert cm to mm to solve this problem.) Use a Δt value of
0.1 min. Initial conditions for the 20 compartments are $C_x(t) = 50$ for the
first 10 compartments and $C_x(t) = 0.0$ for linear compartments 11 to 20.
Note also that the boundary condition concentrations are no flows! There
is no flow from the outside into the first compartment nor flow from
compartment 20 to the outside. Thus, you have to modify the above equation for both ends of the diffusion path.

a. Plot the concentration vs. time (minutes) in the last compartment for
the full simulation period of 200 minutes.

b. For time $= 20$ minutes, plot the concentration vs. distance along the
diffusion pathway. Plot concentration vs. distance for times 50, 100,
150, and 200 minutes also.

c. Repeat the solution, but using different boundary conditions. For this
case, assume that the pipe is open on both ends and outside concentration is 0.0 at both ends.

REFERENCES

Addiscott, T.M., Wagenet, R.J., 1985. Concepts of solute leaching in soils: A review of modeling approaches. J. Soil Sci. 36, 411–424.

Ahuja, L.R., DeCoursey, D.G., Barnes, B.B., Rojas, K.W., 1993. Characteristics of macropore transport studied with the ARS root zone water quality model. Trans. ASAE. 36, 369–380.

Anderson, R.M., Jackson, H.C., May, R.M., Smith, A.M., 1981. Population dynamics of fox
rabies in Europe. Nature 289, 765–771.

Brisson, N., Gary, C., Justes, E., Roche, R., Mary, B., Ripoche, D., et al., 2003. An overview of
the crop model STICS. Eur. J. Agron. 18 (3–4), 309–332.

Hanson, J.D., Rojas, K.W., Shaffer, M.J., 1999. Calibration and evaluation of the root zone water
quality model. Agron. J. 91, 171–177.

Jones, J.W., Luyten, J.C., 1998. Simulation of biological processes. In: Peart, R.M., Curry, R.B.
(Eds.), Agricultural Systems Modeling and Simulation. Marcel Dekker, Inc., pp. 19–62.

Jones, J.W., Keating, B.A., Porter, C.H., 2001. Approaches to modular model development.
Agric. Syst. 70, 421–443.

Jones, J.W.G., Hoogenboom, C.H., Porter, K.J., Boote, W.D., Batchelor, L.A., Hunt, P.W., et al.,
2003. The DSSAT cropping system model. Eur. J. Agron. 18 (3–4), 235–265.

Keating, B.A., Carberry, P.S., Hammer, G.L., Probert, M.E., Robertson, M.J., Holzworth, D., et al., 2003. An overview of APSIM, a model designed for farming systems simulation. Eur. J. Agron. 18, 267–288.

Keen, R.E., Spain, J.D., 1992. Computer Simulation in Biology: A BASIC Introduction. Wiley-Liss, John Wiley & Sons, Inc., New York.

Press, W.H., Teukolsky, S.A., Vetterling, W.T., Flannery, B.P (Eds.), 1996. Numerical Recipes in FORTRAN 90: The Art of Parallel Scientific Computing. 2nd edition Cambridge University Press.

Woli, P., Jones, J.W., Ingram, K.T., Fraisse, C.W., 2012. Agricultural Reference Index for Drought (ARID). Agron. J. 104, 287–300.

Methods

## Chapter 5

Uncertainty and Sensitivity Analysis

1. INTRODUCTION

1.1 Why Model Outputs Are Uncertain

Models used in agricultural and environmental sciences include three sources of uncertainty, namely input variables, parameter values, and equations. Input variables correspond to variables whose values vary between sites and/or year and can be measured. Climatic variables, such as daily mean temperature or radiation, are typical examples of input variables. Climatic variables can be measured from weather stations, but their values are often imperfectly known due to error of measurement or due to the absence of a weather station in the sites of interest.

Parameters correspond to model components whose values cannot be directly measured but which need to be estimated from expert knowledge, from data, and from both expert knowledge and data. When parameters are estimated from expert knowledge, the accuracy of the estimates depends on expert bias and the method used to interview the experts (Low Choy et al., 2009). When parameters are estimated from data, the accuracy of the parameter estimates depends on the estimation technique and on the quality of the data set (see Chapter 6).

Model equation is another source of uncertainty. Several alternative models may be available for a given practical problem. In such cases, the traditional approach is to take a model selection process to find the best model from which one makes practical applications (see Chapter 9 for examples of model selection techniques). Several criteria have been proposed for selecting models but potential problems have been recognized by statisticians (Draper, 1995; Yuan and Yang, 2005). Uncertainty in model equations will only be briefly considered in this chapter. We will ignore uncertainty in model structure and also model residual error.

1.2 How to Describe the Uncertainty of a Quantity of Interest

The uncertainty about a quantity of interest is usually described by defining this quantity as a random variable. Fuzzy logic is another approach that

Working with Dynamic Crop Models. DOI: http://dx.doi.org/10.1016/B978-0-12-397008-4.00005-8
161

won't be considered in this chapter. A random variable can be either discrete or continuous depending on its definition.

Example

Weed Model

Consider the weed model described in Appendix 1. This dynamic model includes several output variables such as weed plant number, weed seed production, weed seed bank, and wheat yield as a function of crop succession and weed management-related variables. The initial weed plant number is one of the input variables of this model and this quantity is typically uncertain. As this variable is discrete, its uncertainty can be described using a discrete random variable, for example, a variable following a Poisson distribution (see Chapter 2). Assume that for an agricultural field of interest the modeler considers that the initial weed plant number should be equal to 3 plants per m^2, but they are not entirely sure that this value is correct. They may describe the uncertainty about this quantity using a Poisson probability distribution with a mean set equal to 3 plants per m^2 (Figure 5.1A). According to this distribution, the probability that the actual weed density is lower than 5 plants per m^2 is higher than 0.91, but there is almost a 9% chance of getting a weed density higher than this value. This example shows how a probability distribution can be used to describe the range of values that can be taken by a model input variable of interest and to describe the lack of knowledge about a phenomenon.

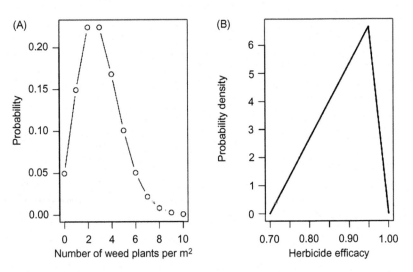

FIGURE 5.1 Probabilities of weed number per m^2 calculated from a Poisson probability distribution of mean set equal to 3 plants per m^2 (A). Triangle probability distribution representing uncertainty about herbicide efficacy (min. efficacy = 70%, max. efficacy = 100%, mode = 95% of killed weed plants) (B).

Probability distributions can also be used to describe uncertainty about parameter values, as illustrated in Figure 5.1B for a parameter of the weed model representing herbicide efficacy (the fraction of the weed population killed by the herbicide). The uncertainty about the value of this parameter can be described using a continuous probability distribution. As the parameter value cannot be higher than 1 or lower than zero, a bounded probability distribution must be used to describe the uncertainty about this quantity. Here, we used a triangle distribution with lower bound, mode, and upper bound equal to 0.7, 0.95, and 1, respectively. The R code is presented in Section 6.

1.3 Objectives of Uncertainty and Sensitivity Analysis

Uncertainty analysis consists of quantitatively evaluating uncertainty in model components (input variables, parameters, equations) for a given situation, and deducing an uncertainty distribution for each output variable rather than a single value (Vose, 1996). The objective of this analysis is to determine how the uncertainties on some of the model components (inputs, parameters, equations) translate into uncertainties in model outputs.

It can be used, for instance, to compute the probability of an output variable of interest (e.g., crop grain protein content, number of spores of a fungus) to exceed some threshold. Uncertainty analysis is a key component of model-based risk analysis and decision making because it provides risk assessors and decision makers with information about the accuracy of model outputs. For example, uncertainty analysis was used to estimate the probability of reaching the European Union target of water nitrate concentration under various scenarios characterized by different sets of farmers' practices (Lacroix et al., 2005). In pest risk analysis, uncertainty analysis was used by several authors to estimate probability of entry and establishment, spread of invasive species, and to assess efficiency of management options (Stansbury et al., 2002).

As in uncertainty and sensitivity analysis, input variables and parameters have the same role, uncertain input variables and parameters will be termed uncertain 'factors'. Formally, let us consider that a given model includes K uncertain factors, i.e., uncertain parameters and/or input variables, noted $z_1, \ldots, z_K$. The purpose of the uncertainty analysis is to define some probability distributions for $z_1, \ldots, z_K$ and to calculate the probability distribution of one or several model outputs $f(z_1, \ldots, z_K)$. As the function f is usually complex, it is not possible to calculate the probability distribution of $f(z_1, \ldots, z_K)$ analytically and specific methods have to be used (see Section 2 of this chapter).

The aim of sensitivity analysis is to determine how sensitive the output of a model is with respect to elements of the model which are subject to uncertainty. For dynamic models, sensitivity analysis is closely related to the study of error propagation. Two types of sensitivity analyses are usually distinguished, *local* sensitivity analysis and *global* sensitivity analysis (Saltelli et al., 2000). Local sensitivity analysis focuses on the local impact of uncertain factors on model outputs and is carried out by computing partial derivatives of the output variables with respect to the input factors. With this kind of method, the uncertain factors are allowed to vary within small intervals around nominal values, but these intervals are not related to the uncertainty in the factor values. Contrary to local sensitivity analysis, global sensitivity analysis considers the full domain of uncertainty of the uncertain model factors. In global sensitivity analysis, the uncertain factors are allowed to vary within their whole range of variation.

Sensitivity analysis may have various objectives, such as:

- to study relationships between model outputs and model inputs
- to identify which input factors have a small or a large influence on the output
- to identify which input factors need to be estimated or measured more accurately
- to detect and quantify interaction effects between input factors
- to determine possible simplification of the model

For example, sensitivity analysis was used to identify the parameters of the wheat dynamic crop model Azodyn that need to be estimated more accurately in order to predict wheat yield and grain protein content (Makowski et al., 2006). Sensitivity analysis was also used to identify the most important factors influencing the predicted efficiencies of different pest risk management options (Stansbury et al., 2002).

2. A SIMPLE EXAMPLE USING UNCERTAINTY AND SENSITIVITY ANALYSIS

In this section, we present a simple example to show how uncertainty and sensitivity analysis can be used in practice. We consider the simple generic infection model for foliar fungal plant pathogens defined by Magarey et al. (2005):

$$W = \min\left\{ W_{\max}, \frac{W_{\min}}{g(T)} \right\}$$

and

$$g(T) = \left(\frac{T_{\max} - T}{T_{\max} - T_{\mathrm{opt}}} \right) \left(\frac{T - T_{\min}}{T_{\mathrm{opt}} - T_{\min}} \right)^{(T_{\mathrm{opt}} - T_{\min})/(T_{\max} - T_{\mathrm{opt}})}$$

if $T_{\min} \le T \le T_{\max}$ and zero otherwise.

where T is the mean temperature during wetness period (°C), W is the wetness duration required to achieve a critical disease intensity (5% disease

severity or 20% disease incidence) at temperature T. T_{min}, T_{opt}, T_{max} are the
minimum, optimal, and maximum temperature for infection, respectively,
W_{min} and W_{max} are the minimum and maximum possible wetness duration
requirements for critical disease intensity, respectively. This model was used
to compute the wetness duration requirement as a function of temperature
for many species and was included in a disease forecast system (Magarey
et al., 2005, 2007).

T_{min}, T_{opt}, T_{max}, W_{min}, and W_{max} are five species-dependent parameters
whose values have been estimated from experimental data and expert knowledge for different foliar pathogens (e.g., Magarey et al., 2005; EFSA 2008).
However, for some species these parameters are uncertain, due to the limited
availability of data (Magarey et al., 2005) and, in such cases, it is important
to perform uncertainty and sensitivity analysis.

Here, we consider a fungus species for which no specific data are available and we define a probability distribution for this fungus from the values
estimated for 51 other fungal species by Magarey et al. (2005). In order to
describe the uncertainty about the model parameters, we assume that the
probability distribution of the five model parameters $\theta = (T_{min}, T_{max}, T_{opt}, W_{min}, W_{max})'$ follows a log normal distribution, i.e., that $log(\theta)$ follows
a normal distribution. This probability distribution was selected in order
to ensure that the model parameters are positive. The expected values,
variances, and covariances of normal distribution of $log(\theta)$ were estimated
from the 51 values of $\theta = (T_{min}, T_{max}, T_{opt}, W_{min}, W_{max})$ estimated for the
51 fungi (Magarey et al., 2005). The expected value of $log(\theta)$ was set
equal to the mean of the log of the 51 parameter values, i.e.,
$E[log(\theta)] = (1.13, 3.47, 3.05, 1.84, 3.1)'$ and the correlation matrix of $log(\theta)$
was set equal to the correlation matrix computed from the 51 estimated
values, i.e.,

$$R = \begin{pmatrix} 1.00000000 & 0.07622954 & 0.1356981 & 0.23517795 & 0.01197128 \\ 0.07622954 & 1.00000000 & 0.6640263 & 0.07781161 & 0.22259201 \\ 0.13569809 & 0.66402633 & 1.0000000 & 0.11148523 & 0.22516422 \\ 0.23517795 & 0.07781161 & 0.1114852 & 1.00000000 & 0.71877132 \\ 0.01197128 & 0.22259201 & 0.2251642 & 0.71877132 & 1.00000000 \end{pmatrix}.$$

Note that a possible alternative would consist of using an empirical distribution, with the 51 sets of parameters. The response curve of W vs. T obtained
with the mean parameter values is presented in Figure 5.2A.

Ten thousand parameter values were randomly generated by Monte Carlo
sampling using the R function `rmvnorm(Num,mean,SIG)` where Num is the
number of Monte Carlo samples, mean is the expected value of $log(\theta)$, and
SIG is its covariance matrix. The 10,000 corresponding responses of W vs. T
were computed and a sample of 20 out of the 10,000 response curves is displayed in Figure 5.2B for illustration. The distribution of the 10,000 response

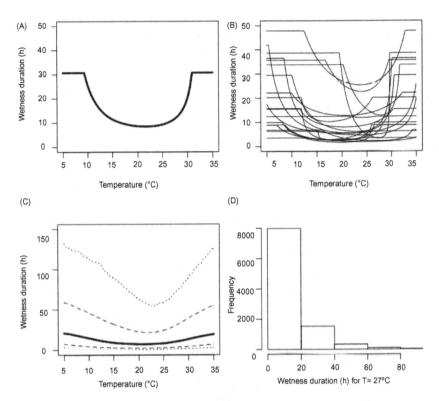

FIGURE 5.2 Predicted wetness duration requirements for infection. A: Predictions obtained with the mean parameter values reported by Magarey et al. (2005). B: Sample of 20 response curves generated by Monte Carlo simulation. C: Percentiles 1%, 10%, 50%, 90%, and 99% of the 10,000 simulated wetness duration requirements as a function of temperature. D: Distribution of the 10,000 simulated wetness duration requirements for $T = 27°C$.

curves is summarized in Figure 5.2C by the percentiles 1%, 10%, 50% (median), 90%, and 99% using the R function quantile. The R code is presented in Section 6.

The results show that the uncertainty is less important when the temperature during the wetness period T is close to 25°C, i.e., the estimated optimal temperature for the fungus (Figure 5.2C). The distribution of W obtained for $T = 27°C$ is skewed (Figure 5.2D); the median is equal to 14h, the 10% percentile is equal to 3.3h, and the 90% percentile is equal to 28.9h. The difference between the 10% and 90% percentiles is larger for lower or higher temperatures. For example, for $T = 10°C$, the 10% percentile was equal to 4.4h and the 90% percentile was equal to 46h.

In order to identify the main sources of uncertainty, sensitivity indices were computed for the five model parameters for several temperatures T. Sensitivity of the model output W to parameter values was quantified by

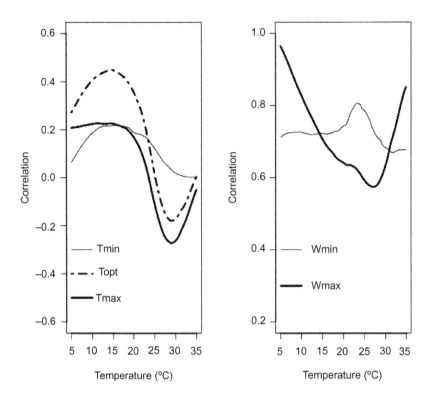

FIGURE 5.3   Sensitivity indices for the five model parameters as a function of temperature. Sensitivity indices correspond to correlations between parameter values and wetness duration requirements estimated from 10,000 Monte Carlo simulations.

calculating correlations between W and parameter values using the 10,000 Monte Carlo simulations. Results are shown in Figure 5.3 for all parameters as a function of T. A correlation close to $+1$ or -1 indicates a strong influence of the parameter on the model output. A correlation close to zero indicates that the parameter is not influential. More sophisticated sensitivity indices could have been computed (Saltelli et al., 2000, 2004) as shown in Section 4 of this chapter.

Figure 5.3 shows that the correlation between W and the parameter T_{min} is close to zero for temperatures higher than 30°C. This result shows that the model output is not very sensitive to the values of these two parameters. The parameter T_{opt} had a strong and positive effect on W for temperatures in the range 10–15°C, and a strong and negative effect for temperatures in the range 27–32°C. Its effect is negligible for extreme temperatures, i.e., when T is close to 5°C or to 35°C and when T is close to 20°C. The parameter T_{max} had a negative effect on W, but its effect is negligible for extreme temperatures. When T is close to 5°C or to 35°C, the model output is sensitive to only one parameter: W_{max}. This sensitivity analysis thus reveals that the model output is sensitive to four parameters, T_{min}, T_{opt}, T_{max}, and W_{max}, and that the effect of these parameters is strongly dependent on the temperature.

3. UNCERTAINTY ANALYSIS

The main steps of an uncertainty analysis are:

i. Definition of the objective of the analysis
ii. Definition of the sources of uncertainty
iii. Generation of values of the input factors and computation of the model outputs
iv. Presentation of the results

Each step is presented below and illustrated with one or two examples. We also briefly address the ensemble modeling approach.

3.1 Step i. Definition of the Objective of the Analysis

At this step, the modeler must define as precisely as possible the objective of the uncertainty analysis. In practice, it consists of identifying one or several model outputs of interest or, alternatively, a function of some of the model outputs.

Example

Weed Model

We illustrate this idea with the weed model presented in Appendix 1. The modeler can perform a large variety of analyses with this model. For example, they may choose to analyze the yield loss due to the non-application of herbicide in one year, e.g., in the third year of the simulation period. In this case, the analysis will concern the difference between the simulated yield in the case of herbicide application in year 3 and the simulated yield in case of non-application of herbicide in year 3. The yield loss in year 3 will be used below to illustrate the principles of the uncertainty analysis.

3.2 Step ii. Definition of the Sources of Uncertainty

At this step, the modeler needs to identify all the sources of uncertainty. The number and type of uncertainties will be different depending on the context. For example, if the modeler wants to predict yield one month before harvest, they could consider that the climatic variables are known from sowing until the date of prediction and unknown the last month before harvest. On the other hand, if yield is predicted before sowing, the climatic variables should be considered as unknown between sowing and harvest.

The modeler must list the uncertain parameters and input variables and, when relevant, list different variants for some of the equations of the model. Possible values for uncertain inputs and parameters must then be described using probability distributions as explained above. These probability distributions can be specified using different sources of information:

- Expert knowledge
- Direct measurements (e.g., temperatures, rainfall, water balance)
- Variances and confidence intervals of estimated parameter values
- Use of coefficients of variation (e.g., ±10%, ±20%) defined from expert knowledge

Expert knowledge elicitation consists of defining probability distributions for one or several uncertain variables from the results of expert interviews. Various methods have been proposed for conducting expert interviews in a rigorous manner in order to avoid different kinds of bias (Low Choy et al., 2009). The use of expert knowledge raises some issues as each expert may have their own personal uncertainty. In order to deal with this issue, it is usually recommended to interview several experts and to ask them to provide data supporting their views.

In some cases, it is possible to define probability distributions of uncertain quantities from direct measurements. This is generally possible for some of the model inputs, especially climatic input variables (e.g., temperatures, rainfall) for which large series of measurements are usually available. Climate variability is frequently described using several years of climatic data, for example, 30 years of daily temperatures.

For parameters, it is often relevant to define probability distributions from the results of the statistical method used for estimating parameter values. If a frequentist statistical method (see Chapter 6) is used for parameter estimation, probability distributions can be defined from the variances and/or confidence intervals of the parameter estimators. If model parameters are estimated using a Bayesian method (see Chapter 7), the outcome of the estimation procedure is a posterior probability distribution that can directly be used for the uncertainty analysis.

When the model includes several uncertain factors, the question of the independence/correlation of these factors is important. In the example considered in Section 2, the model parameters were not considered independent and the correlations were estimated from published data. However, in many cases, the information about the correlations of the uncertain factors is limited. In such cases, it is recommended to assess the sensitivity of the results to the assumptions made on the probability distributions.

When the data and expert knowledge about the uncertain factors is very limited, it is possible to define lower and upper bounds using some coefficient of variation (e.g., ±10%, ±20%) as illustrated below. The modeler should keep in mind that, with this approach, the results may be highly dependent on the chosen coefficient of variation.

Example

Weed Model

We consider here that the 16 model parameters are uncertain. They are defined as random variables with independent uniform probability distributions. The lower and upper bounds of these distributions were defined as the nominal values ±10% with two exceptions:
- as the two parameters mh and beta.1 cannot exceed one, their upper bounds are set equal to one
- as the nominal value of mc is equal to zero, the lower and upper bounds of this parameter are set equal to zero and 0.1, respectively.

TABLE 5.1 Nominal Values, Lower and Upper Bounds of the 16 Parameters of the Weed Model

Parameter	Lower Bound	Nominal Value	Upper Bound
Mu	0.756	0.84	0.924
V	0.54	0.6	0.66
phi	0.495	0.55	0.605
beta.1	0.855	0.95	1
beta.0	0.18	0.2	0.22
chsi.1	0.27	0.3	0.33
chsi.0	0.045	0.05	0.055
delta.new	0.135	0.15	0.165
delta.old	0.27	0.3	0.33
Mh	0.882	0.98	1
Mc	0	0	0.1
Smax.1	400.5	445	489.5
Smax.0	266.4	296	325.6
Ymax	7.2	8	8.8
rmax	0.0018	0.002	0.0022
Gamma	0.0045	0.005	0.0055

The nominal values, lower bounds, and upper bounds of the 16 model parameters are presented in Table 5.1. See Appendix 1 for further details about the model.

3.3 Step iii. Generation of Values of the Uncertain Factors and Computation of the Model Outputs

For some simple models, it is possible to calculate the exact probability distribution of the model output $f(z_1, \ldots, z_K)$ from the probability distributions of the uncertain input variables and parameters $z_1, \ldots, z_K$. For example, if the model is defined by $f(z_1, z_2) = z_1 + z_2$ and if z_1 and z_2 are two independent normally distributed random variables, $f(z_1, z_2)$ is normally distributed and its mean and variance can be easily calculated from the means and variances of z_1 and z_2; $f(z_1, z_2) \sim N(\mu_1 + \mu_2, \sigma_1^2 + \sigma_2^2)$, where μ_1 and μ_2 are the expected values of z_1 and z_2, and σ_1^2 and σ_2^2 are their variances.

However, in most cases it is not possible to calculate the probability distribution analytically and other approaches should be used. A first approach consists of linearizing the model using its derivatives. If the uncertain factors are all normally distributed, it is then possible to calculate the probability distribution of the linearized model analytically; it is a normal distribution whose mean and variance are functions of the means and variances of the uncertain factors (de Rocquilly, 2006). A limit of this approach is that its application is restricted to the cases where the uncertain factors are normally distributed. As discussed in Section 1.2 of this chapter, it is sometimes more appropriate to use other distributions, especially when the random variables are discrete or when they are bounded (Figure 5.1). Another issue is that the result of this approach can be unreliable when the linear approximation is not accurate. A linear approximation for a complex system model is likely to be very poor. This does not seem like a real possibility for our models.

For these reasons, it is usually recommended to use another approach, based on Monte Carlo simulations. This approach consists of randomly generating a sample of N values of the uncertain factors from their probability distributions, and then computing the N corresponding values of the model outputs of interest. The result is N sets of uncertain factors $z_i = (z_{1i}, \ldots, z_{Ki})$ and the corresponding N values of the model outputs $Y_i = f(z_{1i}, \ldots, z_{Ki})$, $I = 1, \ldots, N$. This approach can be implemented with the original model, it does not require any linearization of the model equations, and can be applied with a variety of probability distributions (not only with normal distributions). On the other hand, this approach relies on a large number of model simulations (N) and its implementation can thus be computationally demanding.

The N sets of values of the uncertain factors can be easily generated with R. Values can be sampled from normal, uniform, Poisson, and binomial probability distributions using the rnorm, runif, rpois, and rbinom R functions, respectively. The user needs to specify the parameters of the probability distributions (e.g., lower and upper bounds for uniform distribution), and the sample size (N) when using these functions.

Example

Weed Model

$N = 1000$ values were generated for each of the 16 parameters of the weed model. For example, 1000 values of Ymax were generated using the following R code:

```
runif(1000, 7.2, 8.8)
```

where 7.2 and 8.8 are the lower and upper bounds defined in Table 5.1.

These values are displayed in Figure 5.4 for two parameters, namely Ymax and mh. A sub-sample of size 100 is also presented in order to show that the parameter space is not well covered by a sample of small size.

The choice of the value of N is critical when implementing the Monte Carlo method. The use of a too small N value may lead to inaccurate results; some part of the space defined by the uncertain factors may be unexplored in this case, and the approximation of the probability distribution of the model output may be inaccurate. On the other hand, the use of a very high N value will lead to a large number of model simulations that may be both time consuming and useless. The choice of the value of N is a compromise between computation time and accuracy. This choice can be made by calculating some of the characteristics of the distribution of the model output values, and plotting these characteristics as a function of N. One option is to plot the mean of the N model output values versus N ($N = 1, 2, \ldots 100, 1000\ldots$), and then to

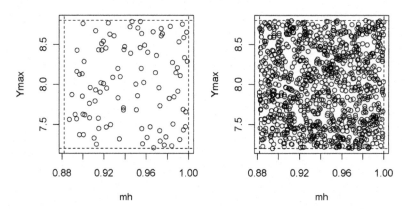

FIGURE 5.4 Values of the parameters mh (efficacy of the herbicide) and Ymax (potential crop yield) randomly generated from their uniform distributions. The dashed lines indicate the lower and upper bounds of the uniform distributions of the two parameters. N was set to 100 (left) and to 1000 (right).

use this plot to identify the N value above which the mean of the N output values changes only marginally.

Example

Weed Model

A sample of size 10,000 was generated for the 16 model parameters. The model was then run with each one of the 10,000 sets of parameter values leading to a distribution of 10,000 simulated yield loss values. The mean value of yield loss distribution was then computed using the first $N = 1, 2, 3...10,000$ model output values. The exercise was repeated for several yield loss percentiles and the results were plotted (Figure 5.5). The results show that the estimated yield loss mean and percentiles vary widely with N when N is small, i.e., when N is lower than 100. Then, the mean and percentile values tend to stabilize and become

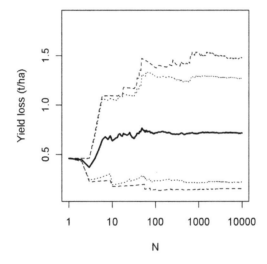

FIGURE 5.5   Mean yield loss (thick line), 5- and 95-yield loss percentiles (dotted lines), 1- and 99-yield loss percentiles (dashed lines) calculated for $N = 1, 2, 3, ...10,000$ Monte Carlo simulations.

very stable when N is equal to or higher than 1000. In this example, it is thus acceptable to use $N = 1000$ Monte Carlo simulations to perform the analysis. However, as the weed model runs very quickly, it is also possible to consider larger N values.

3.4 Step iv. Presentation of the Results

The distribution of the model output values generated by the Monte Carlo method can be described and summarized in a number of ways. It is possible to present the distribution graphically using scatter plots, histograms, density plots, etc. It is also useful to summarize the distribution of the model output values by its mean, median, standard deviation, and quantile values. When several outputs are considered, it is often useful to study the relationship between different outputs using scatter plots and correlation coefficients.

Example

Weed Model

The histograms of the distributions of the yields with and without herbicide and of the yield loss due to the non-application of herbicide are presented graphically in Figure 5.6. Yield loss ranges from almost zero to more than 1.5 t ha^{-1} and the relative yield loss reaches more than 20% in some of the model simulations. Yield and yield loss percentiles are presented in Table 5.2. The results of the uncertainty analysis showed that there is a 99% chance that the simulated yield loss exceeds 0.15 t ha^{-1} and a 1% chance that it exceeds 1.48 t ha^{-1}.

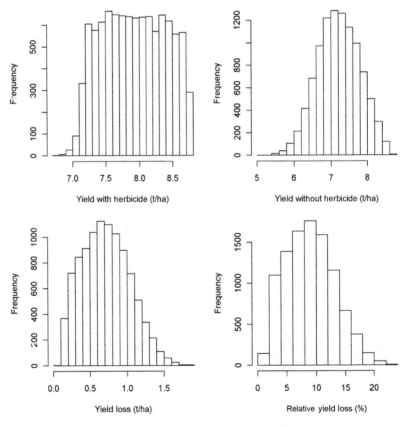

FIGURE 5.6 Histograms of wheat yield with and without herbicides (year 3), of the yield loss (yield with herbicide − yield without herbicide) and of the relative yield loss (100 yield loss − yield with herbicide) obtained with N = 10,000 Monte Carlo simulations.

TABLE 5.2 Yield and Yield Loss Percentiles of the Distributions ($N = 10,000$)

	1%	5%	95%	99%
Yield with herbicide (t/ha)	7.08	7.21	8.66	8.75
Yield without herbicide (t/ha)	5.91	6.28	8.17	8.42
Yield loss (t/ha)	0.15	0.22	1.27	1.48

The possible range of yield loss values is thus very large showing that the relatively low uncertainty considered in the model parameter values (±10%) induces a large uncertainty about yield loss values.

4. SENSITIVITY ANALYSIS

4.1 Local Sensitivity Analysis vs. Global Sensitivity Analysis

Local sensitivity analysis is based on the local derivatives of output with respect to one uncertain factor z, which indicate how fast the output increases or decreases *locally* around given values of the factor z. The derivatives can sometimes be calculated analytically, but they are usually calculated numerically for complex models. Problems may arise if the derivative of the model does not exist at some points. In addition, the derivatives may depend strongly on the z-value. This problem is illustrated in Figure 5.7a where three derivatives are reported.

Let's define $Z_i = (z_{1i}, \ldots, z_{ki}, \ldots, z_{Ki})$ as a set of values of uncertain factors. The local (first-order) sensitivity coefficient $S_k^{local}(Z_i)$ for the kth factor is defined as the partial derivative of the output variable with respect to the kth factor, calculated at the value $Z_i = (z_{1i}, \ldots, z_{ki}, \ldots, z_{Ki})$:

$$S_k^{local}(Z_i) = \frac{\partial f(z_1, \ldots, z_k, \ldots z_K)}{\partial z_k}\Big|_{Z_i}$$

This criterion is equivalent to the slope of the calculated model output in the parameter space. The $S_k^{local}(Z_i)$ criterion is an absolute measure of sensitivity, which depends on the scales of model output and of the factor. A standardized version, called the relative sensitivity, is defined by:

$$S_k^{local}(Z_i) = \frac{\partial f(z_1, \ldots, z_k, \ldots z_K)}{\partial z_k}\Big|_{Z_j} \times \frac{z_{ki}}{f(Z_i)}$$

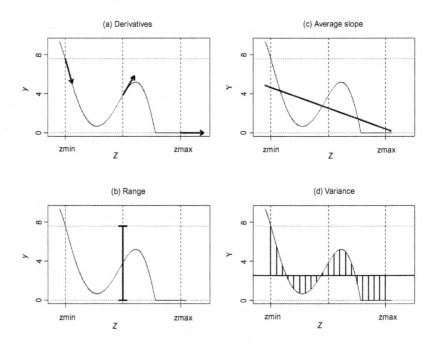

FIGURE 5.7   Four approaches for analyzing the sensitivity of a model output Y to an uncertain factor z.

Local sensitivity analysis can be used to study the role of some parameters or input variables in the model. But this method is less useful than global sensitivity analysis when the purpose of the analysis is to study the effect of uncertainty of several factors on model outputs.

In *global sensitivity analysis* (Figures 5.7 b, c, d), on the other hand, the output variability is evaluated when the input factors vary in their whole uncertainty domains. This provides a more realistic view of the model behavior when used in practice. There are several methods to perform global sensitivity analyses and the next sections of this chapter are concerned with their description, while the book edited by Saltelli, Chan, and Scott (2000) is a comprehensive reference.

The global degree of association between the factor z and the model output Y over the interval $[Z_{min}, Z_{max}]$ can first be measured through a model approximation. For instance, if the crop model is approximated by a linear relationship between z and Y (Figure 5.7c), sensitivity can be measured by the squared regression coefficient or by the linear correlation between z and Y. This approach was used to analyze the sensitivity of the output of the Magarey model to its parameters in Section 2 of this chapter. It is a simple and often efficient way to measure sensitivity, provided the model approximation is adequate.

The approaches illustrated in Figures 5.7b and 5.7d are different since they do not rely on a model approximation, in principle at least. The

sensitivity criterion illustrated in Figure 5.7b is simply based on the range of
the model output when z runs within $[Z_{min}, Z_{max}]$. In the approach illustrated
in Figure 5.7d, sensitivity is measured by the variance of Y over $[Z_{min}, Z_{max}]$.
Several methods are presented below to illustrate the approaches described
in Figures 5.7b and 5.7d.

4.2 One-at-a-Time Methods and Morris

The most intuitive method to conduct a sensitivity analysis is to vary one
factor at a time, while the other factors are fixed at their nominal values. The
relationship between the model output $f(z_1, \ldots, z_{k-1}, z_k, z_{k+1}, \ldots, z_K)$ and the
value of the kth uncertain factor determines a one-at-a-time response profile.
Drawing response profiles is often useful, at least in preliminary stages.
However, we have already argued that more global methods are preferable,
because they take account of and quantify interactions between input factors.

In practice, the model responses $f(z_1, \ldots, z_{k-1}, z_k, z_{k+1}, \ldots, z_K)$ are calculated for Q equi-spaced discretized values of the factor z_k. If the number of
sensitivity factors is not too large, graphical representations are the best way
to summarize the response profiles. Alternatively, summary quantities may
be calculated for each factor's profile, and compared between factors. Bauer
and Hamby (1991), for instance, proposed using the following index for each
factor z_k

$$I_k^{BH} = \frac{\max_{z_k} f(z_1, \ldots, z_{k-1}, z_k, z_{k+1}, \ldots, z_K) - \min_{z_k} f(z_1, \ldots, z_{k-1}, z_k, z_{k+1}, \ldots, z_K)}{\max_{z_k} f(z_1, \ldots, z_{k-1}, z_k, z_{k+1}, \ldots, z_K)}$$

This index can be approximated by the difference between the maximum
and minimum simulated values. In its most restricted application, one-at-a-
time sensitivity analysis is applied at the nominal values of the uncertain factors only. In that case, it gives information on the model only in a small
neighborhood of the nominal values. However, interesting results may be
obtained by calculating one-at-a-time local sensitivity criteria for a lot of different sets of factor values.

This idea is exploited by Morris (1991). Morris defines the elementary
effect of the kth factor for a set of factor values scenario $Z_i = (z_{1i}, \ldots, z_{k-1i}, z_{ki}, z_{k+1i} \ldots, z_{Ki})$ as

$$d_k(Z_i) = \frac{f(z_{i1}, \ldots, z_{ik-1}, z_{ik} + \Delta, z_{ik+1}, \ldots, z_{iK}) - f(z_{i1}, \ldots, z_{ik-1}, z_{ik}, z_{ik+1}, \ldots, z_{iK})}{\Delta}$$

where $z_{ki} + \Delta$ is a perturbed value of z_{ki}.

Suppose that $z_{min(k)}$ and $z_{max(k)}$ are the lower and upper bounds of the kth
uncertain factor. The principle of the Morris's method is to sample a series
of sets of uncertain factors, $Z_i = (z_{1i}, \ldots, z_{k-1i}, z_{ki}, z_{k+1i} \ldots, z_{Ki})$, $I = 1, \ldots, N$,
in a K-dimensional space defined by the values $[z_{min(k)}, z_{min(k)} + \delta, z_{min(k)} + 2\delta, \ldots, z_{max(k)}]$, $k = 1, \ldots, K$, and to calculate $d_k(Z_i)$ for each sampled value.

The K-dimensional space corresponds to a grid including a finite number of values of factors comprised between $z_{\min(k)}$ and $z_{\max(k)}$, $k = 1, \ldots, K$. Examples of two-dimensional grids (two uncertain factors) including either 5^2 or 9^2 values are shown in Figure 5.8.

The resulting distribution of the elementary effects $d_k(Z_i)$, $I = 1, \ldots, N$, of the kth factor is then characterized by the mean and variance of $d_k(Z_i)$, $I = 1, \ldots, N$:

$$\mu_k^* = \frac{\sum_{i=1}^{N} |d_k(Z_i)|}{N}$$

$$\sigma_k^2 = \frac{\sum_{i=1}^{N} \left[d_k(Z_i) - \frac{1}{N} \sum_{i=1}^{N} d_k(Z_i) \right]^2}{N}$$

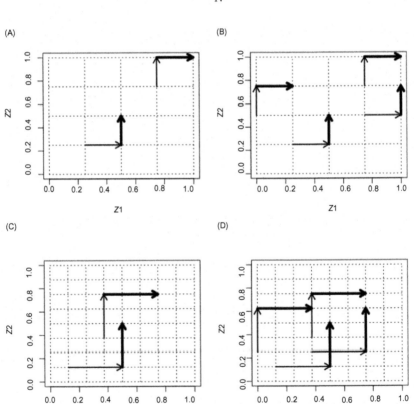

FIGURE 5.8 Grids defined for two uncertain factors z_1 and z_2 ($K = 2$). Grids A and B include 5^2 values and grids C and D include 9^2 values. Each pair of thin and thick arrows corresponds to one replicate. Two replicates ($N = 2$) are displayed in A and C, and four replicates ($N = 4$) are displayed in B and D. Each individual arrow corresponds to one jump of size $\Delta = 1$ (A, B) or of size $\Delta = 3$ (C, D).

A high mean indicates a factor with an important influence on the output. A high variance shows that the value of $d_k(Z_i)$ is highly dependent on the value of the uncertain factor Z_i and indicates either a factor interacting with another factor or a factor whose effect is non-linear. It is also possible to standardize the results in order to constrain the elementary effect means and variances to be between 0 and 1.

The implementation of the Morris method requires the definition of the size of the grid, of the value of the jump Δ, and of the number of replicates N. Figure 5.8 shows examples of values sampled in 5×5 and 9×9 two-dimensional grids. Figures 5.8 A, B show two and four replicates generated using a jump $\Delta = 1$. Figures 5.8 C, D show two and four replicates generated using a jump $\Delta = 3$.

With the package sensitivity of R, the Morris method can be applied with the Morris function. When using this function, the modeler needs to specify the number or the names of the uncertain factors, the lower and upper bounds of these factors, the jump value, the size of the grid, and the number of replicates.

Example

Weed Model

The Morris method was applied to the yield output of the weed model in order to calculate μ_k^* and σ_k^2 for the 16 uncertain parameters. The lower and upper bounds of the parameters were defined in Table 5.1. The method was implemented with the Morris function of the sensitivity R package with a grid of size 4^{16}, a jump $\Delta = 2$, and a number of replicates $N = 500$:

```
morris(model=weed.simule    ,    factors=paraNames,    r=500,
design=list(type="oat",levels=4 , grid.jump=2)
```

weed.simule corresponds to a function including the model, paraNames is a list of names of uncertain factors, r = 500 indicates the number of replicates, levels indicates the size of the grid, and grid.jump defines the size of the jump.

The results are presented in Figure 5.9. When the herbicide is applied, the most influential factor (the factor with the highest value of μ^*) is Ymax; the potential yield value and the other factors have almost no influence on the simulated yield. Ymax is also the most influential factor when no herbicide is applied in year 3, but its μ^* value is lower and three other parameters seem to have an influence on the simulated yield, namely mu (parameter defining the size of the weed seed bank before soil labor), mh (efficacy of the herbicide), and beta.1 (parameter defining the size of the weed seed bank after soil labor). The high values of σ obtained for beta.1 and mu indicate that these parameters have either a non-linear effect on the simulated yield and/or have interactions with other parameters (Figure 5.9). The R code is presented in Section 6.

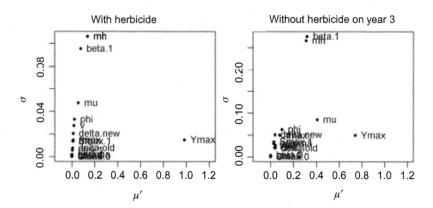

FIGURE 5.9 Results of the Morris method obtained with the weed model. The method was implemented with a grid including 4 levels per factor, a jump equal to 2, and 500 replicates.

4.3 Correlation-Based Method and Analysis of Variance

4.3.1 Correlation-Based Method

This approach consists of randomly generating factor values by Monte Carlo sampling. The principle is to randomly generate N values of the input factors $Z_i = (z_{1i}, \ldots, z_{ki}, \ldots, z_{Ki})$ $i = 1, \ldots, N$, and to compute the model output for each scenario, $f(Z_i)$, $i = 1, \ldots, N$, in a similar way to what is done for an uncertainty analysis. The statistical methods related to regression (see for example Venables and Ripley, 1999) are then used to represent and to measure the sensitivity of the output variables with respect to the input factors. These methods are presented below.

Correlation coefficients can be used to quantify the relationships between input factors and output variables. This approach was already illustrated in Section 2 with a simple epidemiological model. We define the method more formally below.

Let

$$s_Y^2 = \frac{1}{N} \sum_{i=1}^{N} \left[f(Z_i) - \overline{f} \right]^2 \quad \text{and} \quad s_{z_k}^2 = \frac{1}{N} \sum_{i=1}^{N} \left[z_{ki} - \overline{z}_k \right]^2$$

denote the empirical variances of $Y = f(Z)$ and z_k in the simulations, and let

$$\text{cov}(Y, z_k) = \frac{1}{N} \sum_{i=1}^{N} \left[f(Z_i) - \overline{f} \right] \left[z_{ki} - \overline{z}_k \right]$$

denote their covariance. Then the PEAR (Pearson Product Moment Correlation Coefficient) coefficient between z_k and Y is defined by

$$r_{Y,z_k} = \frac{\mathrm{cov}(Y, z_k)}{s_Y s_{z_k}}.$$

It varies between -1 and $+1$ and it measures the degree of linear association between the variations of z_k and those of Y. Figure 5.2 (see Section 2 of this chapter) shows how correlation coefficients can be used to detect influential parameters.

Some non-linear associations may remain undetected and underestimated by the PEAR coefficient. An alternative is the Spearman correlation coefficient, which is calculated on the ranks of z_k and Y. The Spearman correlation coefficient is more adequate in cases of strongly non-linear, but still monotonous, relationships.

With the PEAR or Spearman coefficients, no account is taken of the possible effects of input factors other than z_k. In contrast, the partial correlation coefficient (PCC) aims at measuring the association between z_k and Y after eliminating possible effects due to other input factors z_j, $j \neq k$. The PCC coefficient is similar to the PEAR correlation coefficient, but it is calculated with $f(Z_i)$ and z_{ki} replaced by the residuals of the following two regression models

$$f(Z_i) = b_0 + \sum_{j \neq k} b_j z_{ji} + \varepsilon_i, \; z_{ki} = c_0 + \sum_{j \neq k} c_j z_{ji} + \varepsilon'_{ki}$$

where b_j and c_j are regression coefficients to be estimated by ordinary least squares.

Regression models give a general framework for studying the influence of all input factors simultaneously. By approximating the crop model under study, they make it possible to evaluate the influence of each input factor. Consider for instance the regression model with first-order effects only:

$$f(Z_i) = b_0 + \sum_{k=1}^{K} b_k z_{ki} + \varepsilon''_{ki}$$

where the b_k are the regression coefficients to be estimated and ε''_{ki} is the approximation error term. The regression coefficients are estimated by least squares. The quality of the adjustment is synthesized typically by calculating the model coefficient of determination R^2, that is, the percentage of output variability explained by the model.

The estimated regression coefficients $\hat{b}_k$ can be considered as sensitivity measures associated with the factors z_k, provided they are standardized with respect to the variability in Y and in z_k. The standardized regression coefficients (SRC) are defined as the quantities $\hat{b}_k \dfrac{s_{z_k}}{s_Y}$.

Many more principles and techniques of regression are useful for sensitivity or uncertainty analysis, but it is out of the scope of the present chapter to present them all. However, a few remarks can be made:

- the regression model can be extended in order to incorporate interactions between input variables, qualitative as well as quantitative factors, quadratic as well as linear effects
- when the number of terms in the model is large, model selection techniques (stepwise regression for instance) may become a precious aid to interpretation, since they can eliminate factors with negligible influence
- the regression techniques presented here are good essentially at capturing linear effects between z_k and the Y

4.3.2 ANOVA Based on Complete Factorial Designs

The sensitivity analysis of a model is similar to an experiment where nature is being replaced by the model. It follows that the classical theory of experimental design provides very useful tools for sensitivity analysis. In particular, factorial designs make it possible to evaluate simultaneously the influence of many factors, with possibly a very limited number of runs. An additional practical advantage is that the methods of analyses are available in general statistical packages.

Despite the analogy between natural experiments and sensitivity analyses, some differences must be pointed out. First, there is nothing like measurement error in simulated experiments, at least when the model is deterministic. As a consequence, there is no residual variance and it is unnecessary to replicate the same scenarios and to introduce blocking, whereas replication and blocking are key components of designed experiments. The second difference is that the number of runs may quite often be much larger in simulation studies than in real experiments.

Many books are dedicated to the design of experiments. A very good reference on factorial designs and response surface methods is Box and Draper (1987).

With K uncertain factors and m modalities per factor, there are m^K distinct input scenarios. The (unreplicated) complete m^K factorial design consists of running simulations for each of these scenarios exactly once. For example, let us consider two uncertain factors z_1 and z_2. A complete factorial design including 3 modalities per factor is defined by

z_1	z_2
1	1
2	1
3	1
1	2
2	2
3	2
1	3
2	3
3	3

where 1, 2, 3 are the first, second, third modality of the uncertain factors z_1 and z_2. This design includes 9 combinations of factors.

The common point between the complete factorial design and the one-at-a-time profiles is that each factor is studied at a restricted number of levels. However, the major difference is that the emphasis in factorial designs is on making all combinations of the m modalities of the factors.

A less favorable consequence is that the complete factorial design requires many runs when the number of factors under study is large. For this reason, the 2^K and 3^K factorial designs are the most frequently used complete factorial designs when the number of factors is large. These designs are very useful to quantify interactions between factors.

The analysis of variance is based on the decomposition of the response variability between contributions from each factor and from interactions between factors. This decomposition is related to the statistical theory of the linear model.

For the sensitivity analysis of a deterministic model, the main interest lies in comparing the contributions of the factorial terms to the total variability, while formal testing of hypotheses has no real meaning since there is no residual variability. It follows that the most useful information lies in the sums of squares. The sum of squares associated with the main effect of the kth uncertain factor z_k is defined as

$$SS_k = m \sum_{j=1}^{m} \left(\bar{f}_{z_{kj}} - \mu \right)^2$$

where m is the number of modalities considered for the factor z_k, $\bar{f}_{z_{kj}}$ is the mean of the model simulations obtained for the jth modality of the factor z_k, and μ is the mean of all the model simulations.

By dividing the sums of squares by the total variability, the following 'anova' sensitivity indices can be easily calculated:

- main effects sensitivity indices

$$S_1 = \frac{SS_1}{SS_T}, \; S_2 = \frac{SS_2}{SS_T};$$

- interaction sensitivity indices

$$S_{12} = \frac{SS_{12}}{SS_T};$$

- total sensitivity indices such as

$$TS_1 = \frac{SS_1 + SS_{12}}{SS_T} \quad \text{or} \quad TS_2 = \frac{SS_2 + SS_{12}}{SS_T},$$

which summarize all factorial terms related to a particular factor. The above formulae are valid for two factors.

The main effect SS_k is the sum of squares associated with the factor k and $SS_{1...s}$ is the sum of squares associated with interactions between up to s factors.

Example

Weed Model

The ANOVA method was used to analyze the sensitivity of yield with and without herbicide to five model parameters, namely Ymax, mh, mu, beta.0, and beta.1. Main effects and interaction sensitivity indices were calculated using a 3^5 experimental design based on the minimum and maximum parameter values reported in Table 5.1 and the means of these values. The design was defined using the expand.grid R function. With this design, the number of model simulations used to calculate the indices was equal to $3^5 = 243$. The indices were calculated using the aov R function. The R code is shown in Section 6. The results (Figure 5.10) showed that the most important factor was Ymax, although the effect of this factor on yield tended to be less important without herbicide. The interactions between parameters were small in general with one exception; the interaction between mh and beta.1 when no herbicide is applied on year 3.

4.4 Sobol and FAST

In the approaches based on experimental design followed by analysis of variance or on Monte Carlo sampling followed by regression, sensitivity analysis is based on an approximation of the crop model by a simpler linear model. In the variance-based methods described in this section, the principle is to decompose the output variability $D = \text{Var}(\hat{Y})$ globally, without an intermediate simplified model (Sobol, 1993).

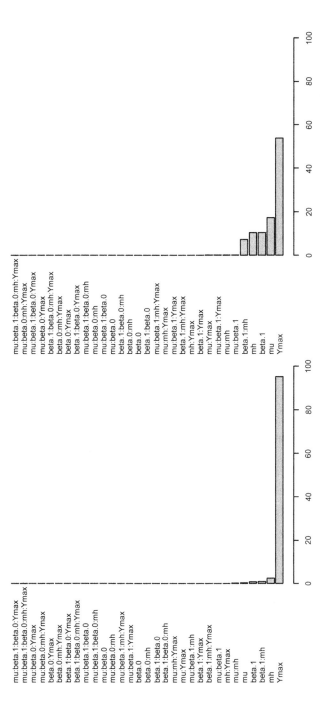

FIGURE 5.10 Main effects and interaction sensitivity indices (% of total variance) calculated with the ANOVA method for weed model parameters Ymax, mh, mu, beta.0, beta.1 with herbicide application (on the left) and without herbicide application on year 3 (on the right).

4.4.1 Sobol Decomposition of the Model

The methods are based on model and variance decompositions that are very similar to those encountered in analysis of variance. In contrast to the analysis of variance decomposition, the uncertain factors are now assumed to vary continuously within the uncertainty interval $[0, 1]$ without loss of generality. Contrary to ANOVA, this method applies to quantitative factors only.

It follows that, with s quantitative factors, the decomposition of the variance $\text{var}(Y) = \text{var}[f(Z)]$ is expressed as:

$$\text{var}(Y) = \sum_{i=1}^{s} D_i + \sum_{i<j} D_{ij} + \ldots + D_{1\ldots s}$$

In this decomposition, D_i corresponds to the main or first-order effect of Z_i denoted by

$$\text{var}[E(Y|Z_i = z_i)],$$

and s is the total number of uncertain factors. The terms

$$D_{ij}, \ldots, D_{1\ldots s}$$

correspond to the interactions between the input factors. This is very similar to the analysis of variance. However, $var(Y)$ now represents the variability of Y with respect to the overall uncertainty in the input factors, and not only over a limited number of experimental design points. This makes it more adequate for taking account of irregular and non-linear effects.

In probabilistic terms, D_i is the variance of the conditional expectation $E(Y|Z_i = z_i)$.

If Y is sensitive to Z_i, $E(Y|Z_i = z_i)$ is likely to vary a lot when z_i changes and so D_i is likely to be large. This is why D_i is also called an 'importance measure' in the vocabulary of sensitivity analysis.

4.4.2 Sensitivity Indices

Sensitivity indices are derived from the decomposition by dividing the importance measures by var(Y) :

$$S_i = \frac{D_i}{\text{var}(Y)}$$

$$S_{ij} = \frac{D_{ij}}{\text{var}(Y)}$$

. . .

Consequently, the sensitivity indices satisfy

$$S_1 + \ldots + S_s + S_{1,2} + \ldots + S_{1,2,\ldots s} = 1$$

and can be interpreted as the proportions of $var(Y)$ explained by the various factorial terms.

Two main types of sensitivity indices can be defined for each factor Z_i. The first-order sensitivity index S_i is useful for measuring the average influence of factor Z_i on the model output, but it takes no account of the interaction effects involving Z_i. The second useful index is the total sensitivity index of Z_i, equal to the sum of all factorial indices involving Z_i:

$$TS_i = S_i + \sum_{j \neq i} S_{ij} + \ldots + S_{1\ldots s}.$$

Note that TS_i is also equal to $1 - S_{-i}$, where S_{-i} denotes the sum of all indices where Z_i is not involved.

4.4.3 Estimation Based on Monte Carlo Sampling

In order to estimate the first-order sensitivity index S_i, the basic idea is to evaluate the model response at N randomly sampled pairs of scenarios $sc_{A,k}$ and $sc_{B,k}$ defined by

$$\begin{aligned} sc_{A,k} &= (z_{k,1}, \ldots, z_{k,i-1}, z_{k,i}, z_{k,i+1}, \ldots, z_{k,s}) \\ sc_{B,k} &= (z'_{k,1}, \ldots, z'_{k,i-1}, z_{k,i}, z'_{k,i+1}, \ldots, z'_{k,s}) \end{aligned}, \quad k = 1, \ldots, N$$

with the same level $z_{k,i}$ of Z_i and all other levels sampled independently. Let D denote $var(Y)$, then

$$\hat{f}_0 = \frac{1}{2N} \sum_{k=1}^{N} \left[f(sc_{A,k}) + f(sc_{B,k}) \right]$$

$$\hat{D} = \frac{1}{2N} \sum_{k=1}^{N} \left[f(sc_{A,k})^2 + f(sc_{B,k})^2 \right] - \hat{f}_0^2$$

$$\hat{D}_i = \frac{1}{N} \sum_{k=1}^{N} f(sc_{A,k}) \cdot f(sc_{B,k}) - \hat{f}_0^2$$

are unbiased estimators of, respectively, the average value of Y, its total variance, and the main effect of Z_i. An obvious estimator of S_i is then

$$\hat{S}_i = \frac{\hat{D}_i}{\hat{D}}.$$

The procedure just described requires $2N$ model simulations for the estimation of each first-order index. A more efficient sampling scheme, the winding stairs, was proposed by Jansen (1994). It is not described here for the sake of brevity.

The same principle can be generalized to the estimation of second-order or higher effects and to the estimation of total sensitivity indices. For estimating the interaction sensitivity S_{ij}, for instance, the model responses have to be calculated for pairs of scenarios $sc_{A,k}$ and $sc_{B,k}$ with the same levels of Z_i and Z_j. For estimating total sensitivity, the model responses have to be calculated for pairs of scenarios $sc_{A,k}$ and $sc_{B,k}$ with the same levels of all factors except Z_i. This allows the sensitivity index S_{-i} to be estimated, and TS_i is then estimated by

$$\hat{T}S_i = 1 - \hat{S}_{-i}.$$

4.4.4 FAST Sampling

The Fourier amplitude sensitivity test (FAST) is another method for estimating variance-based measures of sensitivity. It is inspired by the Fourier decomposition of a time series in signal theory and was developed initially for analyzing the sensitivity of chemical reaction systems to rate coefficients (Cukier et al., 1973, 1975; Schaibly and Shuler, 1978). Recently, its use has been generalized to many domains of applications and new developments have been proposed. More details can be found in Chan et al. (2000).

In the FAST method, all input factors are assumed to be quantitative and coded so that their domain of variation is [0, 1].

Then the possible scenarios belong to the multidimensional input space $[0, 1]^s$. With Monte Carlo sampling, the simulated scenarios are selected at random within $[0, 1]^s$. With the FAST method, they are selected systematically (or almost systematically) along a search trajectory which is specifically designed to explore efficiently the input space. This is illustrated in Figure 5.11 where two trajectories used by the FAST method are displayed. Each trajectory includes 100 values. The trajectory defined by the black points goes through its initial value only two times, while the trajectory defined by the white points goes through its initial value 24 times. Each trajectory is characterized by a frequency noted ω ($\omega = 1$ for the black trajectory and $\omega = 12$ for the white trajectory). In practice, the frequencies should be defined using the rules of Cukier et al. (1973).

More formally, the levels of the input factors Z_i for the simulated scenarios $\mathbf{z}_k$ ($k = 1, \ldots, N$), are given by

$$z_{k,i} = G(\sin(\omega_i u_k + \phi_i)),$$

where the scalars

$$u_k = -\pi + \frac{2k - 1}{N}\pi$$

form a regularly spaced sample of the interval $(-\pi, +\pi)$ and can be interpreted as coordinates on the search curve; $G(u)$ is a transformation

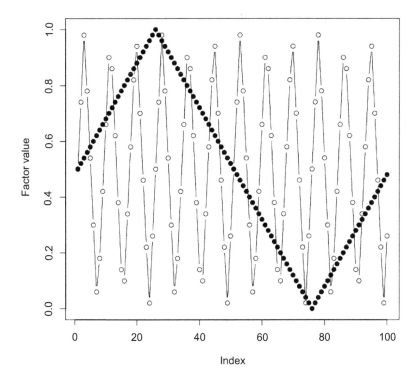

FIGURE 5.11 Two examples of trajectories including 100 factor values used by the FAST method. The uncertain factor varies between 0 and 1.

function from $[-1, 1]$ to $[0, 1]$; and the ϕ_is are optional random phase-shift parameter taking values in $[0, 2\pi)$. The transformation function

$$G(u) = \frac{1}{2} + \frac{1}{\pi}\arcsin(u),$$

proposed by Saltelli et al. (1999), ensures that the levels of each factor are uniformly, or almost uniformly, sampled.

The extended FAST method (Saltelli et al., 1999) allows the estimation of the first-order and the total sensitivity indices from the generated trajectories using a spectral decomposition of the model output variability. The idea is to have each factor vary at a different frequency. Then by looking at the response at each frequency (like in a radio tuned to different frequencies) it is possible to identify the contribution of the factor at that frequency.

In a simulation study on a crop model, it appeared more efficient than the Monte Carlo approach to estimate first and total sensitivity indices (Makowski et al., 2005). As opposed to the classical FAST, it requires separate sets of simulations for each input factor Z_i of interest. The extended FAST method can be easily implemented with the R function fast99 of the package sensitivity. See the example below.

Example

The Weed Model

The extended FAST method was implemented to calculate the first-order and total sensitivity indices of the 16 weed model parameters. The computations were performed using the fast99 function of the R package sensitivity with 1000 model runs per factor (i.e., 16,000 model runs in total). fast99 was implemented as follows:

```
fast99(model = weed.simule,    factors = paraNames,    n = 1000,
q = "qunif", q.arg = q.arg.fast, weed.deci = weed.deci.1)
```

TABLE 5.3 First-Order and Total Sensitivity Indices Estimated with the FAST Method for the 16 Parameters of the Weed Model with and Without Herbicide (16 × 1000 model runs)

Parameter	With Herbicide		Without Herbicide on Year 3	
	First order	Total	First order	Total
mu	**3.01E-03**	**6.56E-03**	**1.80E-01**	**1.85E-01**
v	2.23E-04	2.79E-03	7.26E-03	1.05E-02
phi	3.31E-04	3.03E-03	1.11E-02	1.54E-02
beta.1	**4.16E-03**	**1.09E-02**	**1.06E-01**	**1.56E-01**
beta.0	3.73E-08	2.28E-03	3.93E-07	1.26E-03
chsi.1	2.59E-05	2.33E-03	1.34E-03	2.83E-03
chsi.0	7.87E-08	2.28E-03	3.05E-07	1.27E-03
delta.new	8.16E-05	2.58E-03	1.59E-03	4.57E-03
delta.old	4.37E-05	2.37E-03	2.02E-03	3.55E-03
mh	**1.64E-02**	**2.44E-02**	**9.41E-02**	**1.45E-01**
mc	7.20E-05	2.44E-03	2.11E-03	4.11E-03
Smax.1	3.94E-05	2.34E-03	9.38E-04	3.10E-03
Smax.0	8.48E-08	2.24E-03	2.58E-08	1.38E-03
Ymax	**9.64E-01**	**9.66E-01**	**5.35E-01**	**5.38E-01**
rmax	8.88E-05	2.23E-03	6.55E-03	9.09E-03
gamma	6.90E-07	2.05E-03	6.00E-04	2.21E-03

The most influential parameters are indicated in bold.

where weed.simule is the model function, paraNames includes the names of the uncertain factors, n is the number of simulations per factor (here, 1000 simulations per factor), q defines the probability distributions of the uncertain factors (uniform here), q.arg defines the parameters of the chosen distributions (here, the lower and upper bounds of the uniform distributions), and weed.deci includes the values of the weed model input variables. See Section 6 for further details. The results are presented in Table 5.3. They confirm that the most influential parameter is Ymax and that the influence of this parameter on yield is decreased when no herbicide is applied in year 3.

5. RECOMMENDATIONS

Uncertainty analysis and sensitivity analysis do not have the same objective. Uncertainty analysis aims at propagating the uncertainties in the model inputs through the model equations in order to derive a distribution of model outputs. Such a distribution is useful for assessing the level of uncertainty associated with model predictions and for optimizing model-based decisions. The methods presented do not take into account model residual error and they assume that the model is perfect if the input variable and parameter values are correct.

The objective of sensitivity analysis is rather different; it aims to rank the uncertain factors according to their influence on the model outputs. The results of a sensitivity analysis can be used to identify the uncertain factors (input variables or parameters) that deserve a more accurate estimation.

As both approaches require intensive computations, we strongly advise modelers to define precisely their objectives before starting their analysis, i.e., to determine whether an uncertainty or a sensitivity analysis is needed, to describe the levels of uncertainty of the uncertain factors, to specify the output variables that need to be analyzed.

As mentioned in this chapter, a wide range of methods are available for both uncertainty and sensitivity analysis. It is recommended to implement several methods and to compare the results as illustrated in this chapter with the weed model. In practice, the Morris method frequently gives useful outputs even with a small number of model runs. It is also recommended to study the sensitivity of the results of any method to the number of simulations in order to avoid any wrong conclusions due to a small number of model runs.

Finally, we advise the modelers to make all the assumptions about the probability distributions of the uncertain factors as transparent as possible. It is thus important to describe all the distributions and to present their parameters. When possible, it is useful to perform the analyses with several distributions in order to assess the robustness of the results to the distribution assumptions and to assess the reliability of the conclusions.

6. R CODE USED IN THIS CHAPTER

6.1 Code Used to Draw Figure 5.1

The code shown below shows how to calculate and plot values of Poisson probability distribution density and values of triangle density (using the `triangle` R package).

```
library(triangle)
par(mfrow = c(1,2), oma = c(5,1,5,1))
x < - 0:10
plot(x,dpois(x,lambda = 3), xlab = "Number of weed plants per m2",
  ylab = "Probability", type = "b")
text(9,0.22,"A")
x < - seq(0.7,1, by = 0.01)
plot(x,dtriangle(x,a = 0.7,c = 0.95,b = 1),xlab = "Herbicide
efficacy", ylab = "Probability density", type = "l", lwd = 2)
text(0.75,6.5,"B")
```

6.2 Uncertainty Analysis with the Magarey's Model

The function of the Magarey's model is defined in the Wetness R function presented below.

```
Wetness <- function(T, Tmin, Topt, Tmax, Wmin, Wmax) {
  fT <- ((Tmax-T)/(Tmax-Topt))*(((T-Tmin)/(Topt-Tmin))^
  ((Topt-Tmin)/(Tmax-Topt)))
  W <- Wmin/fT
  W[W > Wmax | T < Tmin | T > Tmax] <- Wmax
  return(W)
  }
```

Data used to define the log normal distribution of the model parameters are read from an external file.

```
TAB <- read.table("MagareyParam.txt", header = T)
meanP <- mean(log(TAB))
SIG <- cov(log(TAB))
print(meanP)
print(cor(log(TAB)))
```

10,000 values of the model parameters are randomly generated using the function rmvnorm of the mvtnorm R package.

```
Num <- 10000
library(mvtnorm)
Z <- rmvnorm(Num,meanP,SIG)
Zexp < -exp(Z)
Tmin_vec <-Zexp[,1]
```

```
Topt_vec <- Zexp[,3]
Tmax_vec <- Zexp[,2]
Wmin_vec <- Zexp[,4]
Wmax_vec <- Zexp[,5]
```

Code Used to Draw Figure 5.2A (response curve obtained with mean parameter values)

```
W <- Wetness(seq(5,35, by = 0.1),Tmin = mean(Tmin_vec),
Topt = mean(Topt_vec), Tmax = mean(Tmax_vec), Wmin = mean(Wmin_vec),
Wmax = mean(Wmax_vec))
par(mfrow = c(2,2))
plot(seq(5,35, by = 0.1), W, xlab = "Temperature (°C)",
ylab = "Wetness duration (h)", type = "l", lwd = 3, ylim = c(0,50) )
text(7,45,"A")
```

New parameter values are generated if $Tmin > Topt$ or $Topt > Tmax$ or $Wmin > Wmax$

```
for (k in 1:Num) {
while(Tmin_vec[k] > Topt_vec[k]     |     Topt_vec[k] > Tmax_vec[k]     |
Wmin_vec[k] > Wmax_vec[k]) {
Z <- rmvnorm(1,meanP,SIG)
Zexp <- exp(Z)
Tmin_vec[k] <- Zexp[1]
Topt_vec[k] <- Zexp[3]
Tmax_vec[k] <- Zexp[2]
Wmin_vec[k] <- Zexp[4]
Wmax_vec[k] <- Zexp[5]
}
}
```

Code Used to Draw Figure 5.2B (examples of responses obtained with different sets of parameter values)

```
plot(c(0), c(0), pch = " ", xlab = "Temperature (°C)",
ylab = "Wetness duration (h)", xlim = c(5, 35), ylim = c(0, 50))
text(30,45,"B")
T_vec <- seq(from = 5, to = 35, by - 0.1)
W_mat <- matrix(nrow = Num, ncol = length(T_vec))
for (i in 1:Num) {
W_mat[i,] <- Wetness(T_vec, Tmin_vec[i], Topt_vec[i],
Tmax_vec[i], Wmin_vec[i], Wmax_vec[i])
if (i < 20) {lines(T_vec, W_mat[i,])}
}
```

Median values, 1%, 10%, 90%, and 99% percentiles are calculated using the R function `apply` and `quantile`. Results are then plotted (Figure 5.2C) and a histogram of wetness duration values is drawn for temperature $= 27°C$ (Figure 5.2D).

```
med_vec<-apply(W_mat, 2, quantile, 0.5)
Q0.01_vec<-apply(W_mat, 2, quantile, 0.01)
Q0.1_vec<-apply(W_mat, 2, quantile, 0.1)
Q0.9_vec<-apply(W_mat, 2, quantile, 0.9)
Q0.99_vec<-apply(W_mat, 2, quantile, 0.99)
plot(c(0), c(0), pch=" ", xlab="Temperature (°C)",
ylab="Wetness duration (h)", xlim=c(5, 35), ylim=c(0, 150))
text(7,140,"C")
lines(T_vec, med_vec, lwd=3)
lines(T_vec, Q0.9_vec, lty=2)
lines(T_vec, Q0.1_vec, lty=2)
lines(T_vec, Q0.99_vec, lty=9)
lines(T_vec, Q0.01_vec, lty=9)
hist(W_mat[,221],xlab=paste("Wetness duration (h) for T=",
T_vec[221],"°C"), main=" ", xlim=c(0,90))
text(80,7000,"D")
```

6.3 Sensitivity Analysis with the Magarey's Model

The code shown below calculates the correlations between the model outputs
and the model parameter values that were randomly generated for the uncertainty analysis. Correlations are allocated to a matrix and are then plotted as
functions of the temperatures (Figure 5.3).

```
Indices_mat<-matrix(nrow=5, ncol=length(T_vec))
for (i in 1:length(T_vec)) {
Indices_mat[1,i]<-cor(Tmin_vec,W_mat[,i])
Indices_mat[2,i]<-cor(Topt_vec,W_mat[,i])
Indices_mat[3,i]<-cor(Tmax_vec,W_mat[,i])
Indices_mat[4,i]<-cor(Wmin_vec,W_mat[,i])
Indices_mat[5,i]<-cor(Wmax_vec,W_mat[,i])
}
dev.new()
par(mfrow=c(1,2))
plot(T_vec,Indices_mat[1,], xlab="Temperature (°C)",
ylab=" Correlation", ylim=c(-0.6,0.6), type="l")
lines(T_vec,Indices_mat[2,], lty=4, lwd=3)
lines(T_vec, Indices_mat[3,], lwd=3)
lines(c(5,10),c(-0.3,-0.3))
text(14,-0.3, "Tmin")
lines(c(5,10),c(-0.4,-0.4), lty=4, lwd=3)
text(14,-0.4, "Topt")
lines(c(5,10),c(-0.5,-0.5), lwd=3)
text(14,-0.5, "Tmax")
plot(T_vec,Indices_mat[4,], xlab="Temperature (°C)",
ylab=" Correlation", ylim=c(0.2,1), type="l")
lines(T_vec,Indices_mat[5,], lwd=3)
lines(c(5,10),c(0.4,0.4))
text(16,0.4, "Wmin")
```

```
lines(c(5,10),c(0.3,0.3), lwd = 3)
text(16,0.3, "Wmax")
```

6.4 Sensitivity Analysis for the Weed Model Using the Morris and Fast Methods

6.4.1 A Function for Running the Weed Model That Can Be Called by Morris or Fast99

We first define the data frames that contain the input information necessary for running the weed model for 10 years. The data frame weed.deci.1 has herbicide treatment every year, the data frame weed.deci.2 has herbicide treatment every year except year 3.

Case 1. Herbicide application every year, soil tillage every two years

```
weed.deci.1 = data.frame(Soil = c(1,0,1,0,1,0,1,0,1,0),
Crop = rep(1,10), Herb = c(1,1,1,1,1,1,1,1,1,1))
```

Case 2. No herbicide application in year 3, soil tillage every two years

```
weed.deci.2 = data.frame(Soil = c(1,0,1,0,1,0,1,0,1,0),
Crop = rep(1,10), Herb = c(1,1,0,1,1,1,1,1,1,1))
```

Below is the function that runs the model. The first argument, X, is the experimental design matrix generated by the morris function or fast99. On each line it has values for each of the parameters that vary. The different lines correspond to the different parameter vectors explored by the morris function or by fast99. The user does not create this matrix, it is created automatically by the morris or fast99 functions. It must be the first argument of the function. It can have any name, but X is convenient since the matrix is called X by the morris function. It can be printed from within the model function if you want to examine the parameter values.

The morris and fast99 functions can also pass any number of additional arguments to this function. They appear in the model function after the argument X, with the same names as in the morris or fast99 functions. In the example below, there is one additional argument, the data frame with the input data for the model.

The model function must return a one-column matrix. Each line is the model output that corresponds to the corresponding line of parameter values. In the example below, the apply function runs the model for each line of the X matrix and has yield in year 3 as the output. The year 3 yield values are stored in Y, which is converted to a matrix and returned.

```
weed.simule <-function(X, weed.deci) {
    # 4th row for yield simulated in year 3
    Y = apply(X,1,function(param) weed.model(param,
weed.deci)[4,"Yield"])
    return(as.matrix(Y))
}
```

6.4.2 Implementation of the Morris Method

The sensitivity analysis is performed using the morris function which is in the sensitivity package. The example used in this chapter is given below. The first argument, weed.simule, is the name of the function that will run the model for each parameter vector. The argument paraNames is a list of names of the uncertain factors, r = 500 gives the number of replicates, levels indicates the number of divisions in the grid, and grid.jump defines the size of the jump. The argument type = "oat" indicates a standard Morris design. The arguments binf and bsup are vectors, with the minimum and maximum values of the parameters. After those arguments, one can put any number of additional arguments, which will be passed to the function that runs the model. In the example, there is one additional argument, the data frame with the input variables for the model.

The standard function plot, when applied to an object generated by morris, creates a plot of sigma versus mu*. The print function prints out those values. One can also obtain the design matrix and the vector of model responses. For the example below, these would be in output.morris1$X and output.morris1$Y.

```
paraNames < −colnames(weed.factors)
#Case 1. With herbicide
output.morris1 = morris(model = weed.simule ,
factors = paraNames,
r = 500,  design = list(type = "oat",  levels = 4  ,  grid.jump = 2),
scale = T,
binf = weed.factors["binf",], bsup = weed.factors["bsup",],
weed.deci = weed.deci.1)
plot(output.morris1, xlim = c(−0.1,1.2), main = "with herbicide")
print(output.morris1)
#Case 2. Without herbicide on year 3
output.morris2 = morris(model = weed.simule ,
factors = paraNames,
r = 500,  design = list(type = "oat",  levels = 4  ,  grid.jump = 2),
scale = T,
binf = weed.factors["binf",], bsup = weed.factors["bsup",],
weed.deci = weed.deci.2)
plot(output.morris2, xlim = c(−0.1,1.2), main = "without herbicide
on year 3")
print(output.morris2)
```

6.4.3 Implementation of the FAST Method

The fast99 function is shown below. One complication here is the way that one specifies the parameters that describe the distributions of the parameters. For example, in the case of uniform distributions, one must give the lower and upper bounds for each parameter. These values must be in the form of a list of lists. For example, if there are two parameters with uniform distributions, then the list of lists would have the form

```
q.arg.fast
= list(list(lower_param1,upper_param1),list(lower_param2,
upper_param2)).
```

With several parameters, it can be inconvenient to write this list of lists explicitly. Therefore, in the example, the list of lists is created by the function

```
q.arg.fast.runif.
q.arg.fast = q.arg.fast.runif(weed.factors)
```

The FAST method is implemented using 1000 simulations per parameter. A specific parameter design is created (see Figure 5.11), and the Weed model is then run for each generated set of parameter values using the weed.simule function.

```
#Case 1. With herbicide
output.fast99_1 = fast99(model = weed.simule, factors = paraNames,
n = 1000, q = "qunif", q.arg = q.arg.fast ,
weed.deci = weed.deci.1)
plot(output.fast99_1)
print(output.fast99_1)
#Case 2. Without herbicide on year 3
output.fast99_2 = fast99(model = weed.simule, factors = paraNames,
n = 1000, q = "qunif", q.arg = q.arg.fast ,
weed.deci = weed.deci.2)
plot(output.fast99_2, cex.lab = 0.75)
print(output.fast99_2)
```

EXERCISES

Easy

1. Consider the model $f(z_1, z_2) = z_1 + 2\,z_2$, where z_1 and z_2 are two uncertain factors. Suppose that the uncertainty about z_1 and z_2 is described by two independent normal probability distributions, $z_1 \sim N(20, 16)$ and $z_2 \sim N(60, 64)$. Determine the probability distribution describing the uncertainty about $f(z_1, z_2)$.

2. Generate 10 series of 100 values randomly sampled from a normal distribution $N(10, 16)$ with the R function rnorm. Estimate the mean of the normal distribution using each series.

3. A sensitivity analysis was performed for the dynamic crop model Azodyn simulating canola yield in order to study the sensitivity of the simulated yields to the 14 model parameters for a site in France. Two sensitivity analyses were performed, one based on the Morris method and one based on the Fast method. The results are presented in Figures 5.12 and 5.13 (the names correspond to the model parameters).

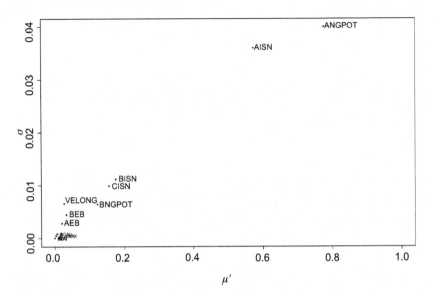

FIGURE 5.12   Results of the sensitivity analysis performed for the Azodyn model with the Morris method.

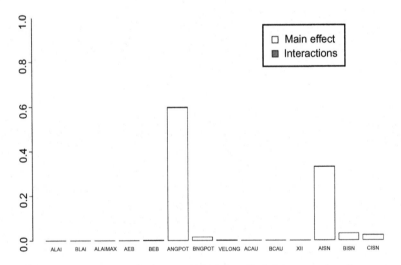

FIGURE 5.13   Results of the sensitivity analysis performed for the Azodyn model with the Fast method.

a. Describe the figures and interpret the results.
b. What are the most influential parameters according to the results?
c. What are the differences between the results obtained with the two methods?

Moderate (4.1−4.2) to High (4.3)

4. We consider a model simulating the percentage of plants with eyespot pathogen (*Pseudocercosporella herpotrichoides*) in a field as a function of cumulative degrees-days since sowing. The model is defined by

$$\hat{Y}(t) = 100 \times \frac{1 - \exp[-(c_1 + c_2)t]}{1 + \dfrac{c_2}{c_1}\exp[-(c_1 + c_2)t]}$$

where $\hat{Y}(t)$ is the predicted percentage of diseased plants when the cumulative degrees-days is equal to t, and $\theta = (c_1, c_2)$ are the model parameters.

Here, the objective is to predict the percentage for a field located in the Paris Basin at $t = 2300$ C° day. The values of the two model parameters were studied in past studies (Colbach and Meynard, 1995) but are not perfectly known. In this study, we consider that the uncertainty ranges of c_1 and c_2 are [4.5 10^{-8} − 3.5 10^{-4}] and [4 10^{-4} − 6.5 10^{-3}] respectively. The nominal values of c_1 and c_2 are equal to 1.75 10^{-4} and 3.5 10^{-3} respectively.

4.1 We assume that the uncertainty in c_1 and c_2 is modeled by uniform distributions over the parameter uncertainty ranges. A sample of ten values of $\theta = (c_1, c_2)$ is generated by Monte Carlo sampling and is reported in Table 5.4. Each value of θ defines an input scenario.

  a. Calculate $\hat{Y}(2300)$ for each one of the ten scenarios presented in Table 5.4.

TABLE 5.4 Ten Values of c_1 and c_2 Generated by Monte Carlo Sampling

c_1	c_2
1.71 10^{-4}	6.42 10^{-3}
1.25 10^{-4}	2.52 10^{-3}
9.65 10^{-5}	1.67 10^{-3}
3.38 10^{-4}	4.79 10^{-3}
2.97 10^{-4}	4.39 10^{-3}
4.88 10^{-5}	5.51 10^{-3}
1.36 10^{-4}	5.94 10^{-4}
2.99 10^{-5}	1.11 10^{-3}
1.97 10^{-4}	3.36 10^{-3}
3.17 10^{-4}	5.93 10^{-4}

b. Estimate the expected value and standard deviation of $\hat{Y}(2300)$ from the ten computed values of $\hat{Y}(2300)$.

c. Estimate the probability $P\left[\hat{Y}(2300) \geq 80\%\right]$ from the ten computed values of $\hat{Y}(2300)$.

d. The procedure described above is repeated five times leading to five samples of 10 values of $\theta = (c_1, c_2)$. Each sample is used to estimate the expected value of $\hat{Y}(2300)$. The five estimated values of $E\left[\hat{Y}(2300)\right]$ are 88.47, 95.28, 92.02, 96.48, and 79.03. How do you explain this large variability?

e. Define a procedure to choose the size of the sample of values of θ in order to estimate accurately $E\left[\hat{Y}(2300)\right]$ and $P\left[\hat{Y}(2300) \geq 80\%\right]$.

4.2 a. Perform a local sensitivity analysis of $\hat{Y}(2300)$ with respect to c_1 and c_2 at the nominal parameter values. Which parameter has the highest relative sensitivity?

b. Calculate five equispaced values of c_1 and c_2 from the minimal to the maximal parameter values.

c. Set c_2 equal to its nominal value and calculate $\hat{Y}(2300)$ for the five values of c_1 defined above. Then, set c_1 equal to its nominal value and calculate $\hat{Y}(2300)$ for the five values of c_2.

d. Calculate the sensitivity index of Bauer and Hamby (1991) (see textbook) for each parameter from the computed values obtained in 4.2c. Which parameter has the highest index?

e. Calculate the sensitivity index of Bauer and Hamby (1991) for c_2 when c_1 is set equal to its minimal value. Compare this index value with the value obtained in 4.2d.

4.3 Consider a complete factorial design with three modalities per factor.

a. How many distinct scenarios (i.e., values of $\theta = (c_1, c_2)$) are included in this design?

b. Define a complete factorial design with three modalities per factor using only the minimal, nominal, and maximal parameter values.

c. Calculate the general mean of $\hat{Y}_i(2300)$ where $\hat{Y}_i(2300)$ is the value of $\hat{Y}(2300)$ obtained with the ith scenario.

d. The total variability of $\hat{Y}(2300)$ can be measured by

$$\text{var}\left[\hat{Y}(2300)\right] = \frac{1}{N} \sum_{i=1}^{N} \left[\hat{Y}_i(2300) - \overline{Y}\right]^2$$

where $\overline{Y}$ is the mean of $\hat{Y}_i(2300)$ and N is the number of scenarios in the factorial design. Calculate $\text{var}\left[\hat{Y}(2300)\right]$.

e. Estimate $E\left[\hat{Y}(2300)|c_1\right]$ for each value of c_1 considered in the factorial design.

f. Estimate $E\left[\hat{Y}(2300)|c_2\right]$ for each value of c_2 considered in the factorial design.

g. Estimate $\text{var}\{E[\hat{Y}(2300)|c_1]\}$ and then

$$\frac{\text{var}\{E[\hat{Y}(2300)|c_1]\}}{\text{var}[\hat{Y}(2300)]}.$$

h. Estimate $\text{var}\{E[\hat{Y}(2300)|c_2]\}$ and then

$$\frac{\text{var}\{E[\hat{Y}(2300)|c_2]\}}{\text{var}[\hat{Y}(2300)]}.$$

i. To which sensitivity indices do

$$\frac{\text{var}\{E[\hat{Y}(2300)|c_1]\}}{\text{var}[\hat{Y}(2300)]} \quad \text{and} \quad \frac{\text{var}\{E[\hat{Y}(2300)|c_2]\}}{\text{var}[\hat{Y}(2300)]}$$

correspond?

j. Estimate $\text{var}[\hat{Y}(2300)|c_1]$ for each value of c_1 considered in the factorial design.

k. Estimate $\text{var}[\hat{Y}(2300)|c_2]$ for each value of c_2 considered in the factorial design.

l. Estimate $E\{\text{var}[\hat{Y}(2300)|c_1]\}$ and then

$$\frac{E\{\text{var}[\hat{Y}(2300)|c_1]\}}{\text{var}[\hat{Y}(2300)]}.$$

m. Estimate $E\{\text{var}[\hat{Y}(2300)|c_2]\}$ and then

$$\frac{E\{\text{var}[\hat{Y}(2300)|c_2]\}}{\text{var}[\hat{Y}(2300)]}.$$

Difficult

5. a. We consider the simple generic infection model for foliar fungal plant pathogens defined by Magarey et al. (2005):

$$W = \min\left\{W_{\max}, \frac{W_{\min}}{f(T)}\right\}$$

and

$$f(T) = \left(\frac{T_{\max} - T}{T_{\max} - T_{\text{opt}}}\right)\left(\frac{T - T_{\min}}{T_{\text{opt}} - T_{\min}}\right)^{(T_{\text{opt}} - T_{\min})/(T_{\max} - T_{\text{opt}})} \quad \text{if } T_{\min} \leq T \leq T_{\max} \text{ and zero otherwise}$$

where T is the mean temperature during wetness period ($°C$), W is the wetness duration required to achieve a critical disease intensity (5% disease severity or 20% disease incidence) at temperature T. $T_{\min}$, T_{opt}, $T_{\max}$ are minimum, optimal, and maximum temperatures for infection, respectively, $W_{\min}$ and $W_{\max}$ are minimum and maximum possible wetness duration requirements for critical disease intensity, respectively. This model was used to compute the wetness duration requirement as a function of the temperature for many species and was included in a disease forecast system (Magarey et al., 2005, 2007).

T_{min}, T_{opt}, T_{max}, W_{min}, and W_{max} are five species-dependent parameters whose values were estimated from experimental data and expert knowledge for different foliar pathogens (e.g., Magarey et al., 2005; EFSA 2008b). Suppose that experiments were carried out to estimate these parameters for a fungus x and a crop species y, but that the parameter values remain uncertain. Assume that the range of possible values for the parameters is:

Tmin (°C): 10–15
Tmax (°C): 32–35
Topt (°C): 25–30
Wmin (h): 12–14
Wmax (h): 35–48

Questions:
Define uniform distributions for describing the uncertainty about the parameters.

   b. Sample 1000 parameter values from these distributions using the R function runif.

   c. Create a new R function including the model equations and use this function for computing the values of W with the 1000 parameter values for a temperature of 20°C.

   d. Describe the distribution of the 1000 values of W by plotting a histogram, and by calculating the median and the 1st and 3rd quartiles.

   e. Estimate the probability that W is lower than 20h using the 1000 values of W.

6. Consider the model and the parameter values defined in Exercise 5.

Questions:

   a. Use the R function expand.grid to define a complete factorial design including all the possible combinations of the minimum and maximum parameter values.

   b. Compute W for all the combinations of parameter values included in the design.

   c. Perform an analysis of variance with the R function aov for estimating the sensitivity indices of the model parameters.

   d. Compute the sensitivity indices using the Morris method. Compare the results.

   e. Identify the most sensitive parameters from the results.

REFERENCES

Bauer, L.R., Hamby, K.J., 1991. Relative sensitivities of existing and novel model parameters in atmospheric tritium dose estimates. Rad. Prot. Dosimetry 37, 253−260.

Box, G.E.P., Draper, N.R., 1987. Empirical Model Building and Response Surfaces. Wiley, New York.

Chan, K., Tarantola, S., Saltelli, A., Sobol, I.M., 2000. Variance-based methods. In: Saltelli, A., Chan, K., Scott, E.M. (Eds.), Sensitivity Analysis. Wiley, New York, pp. 167−197.

Colbach, N., Meynard, J.-M., 1995. Soil tillage eyespot: influence of crop residue distribution on disease development and infection cycles. European Journal of Plant Pathology 101, 601−611.

Cukier, R.I., Fortuin, C., Shuler, K.E., Petshek, A.G., Schaibly, J.H., 1973. Study of the sensitivity of coupled reaction systems to uncertainties in rate coefficients. I. Theory. The Journal of Chemical Physics 59, 3873−3878.

Cukier, R.I., Schaibly, J.H., Shuler, K.E., 1975. Study of the sensitivity of coupled reaction systems to uncertainties in rate coefficients. III. Analysis of the approximations. The Journal of Chemical Physics 63, 1140−1149.

De Rocquilly, E., 2006. La maitrise des incertitudes dans un contexte industriel. 1er partie : Une approche méthodologique globale basée sur des exemples. Journal de la Société Française de Statistique 147, 33−72.

Draper, D., 1995. Assessment and propagation of model uncertainty. J.R.Statist. Soc. B 57, 45−97.

EFSA, 2008. Scientific opinion of the Panel on Plant Health on a request from the European Commission on *Guignardia citricarpa* Kiely. The EFSA Journal 925, 1−108.

Jansen, M.J.W., Rossing, W.A.H., Daamen, R.A., 1994. Monte Carlo estimation of uncertainty contributions from several independent multivariate sources. In: Gasman, J., van Straten, G. (Eds.), Predictability and Non-Linear Modelling in Natural Sciences and Economics. Kluwer, Dordrecht, pp. 334−343.

Lacroix, A., Beaudoin, N., Makowski, D., 2005. Agricultural water nonpoint pollution control under uncertainty and climate variability. Ecological Economics 53, 115−127.

Low Choy, S., O'Leary, R., Mengersen, K., 2009. Elicitation by design for ecology: using expert opinion to inform priors for Bayesian statistical models. Ecology 90, 265−277.

Magarey, R.D., Sutton, T.B., Thayer, C.L., 2005. A simple generic infection model for foliar fungal plant pathogens. Phytopathology 95, 92−100.

Makowski, D., Naud, C., Jeuffroy, M.-H., Barbottin, A., Monod, H., 2006. Global sensitivity analysis for calculating the contribution of genetic parameters to the variance of crop model predictions. Reliability Engineering and System Safety 91, 1142−1147.

Morris, M.D., 1991. Factorial sampling plans for preliminary computational experiments. Technometrics 33, 161−174.

Saltelli, A., Chan, K., Scott, E.M., 2000. Sensitivity Analysis. Wiley, New York.

Saltelli, A., Tarantola, S., Chan, K., 1999. A quantitative model-independent method for global sensitivity analysis of model output. Technometrics 41, 39−56.

Saltelli, A., Tarantola, S., Campolongo, F., 2000. Sensitivity analysis as an ingredient of modelling. Statistical Science 15, 377−395.

Saltelli, A., Tarantola, S., Campolongo, F., Ratto, M., 2004. Sensitivity Analysis in Practice. Wiley, New York.

Schaibly, J.H., Shuler, K.E., 1973. Study of the sensitivity of coupled reaction systems to uncertainties in rate coefficients. II. Applications. The Journal of Chemical Physics, 3879−3888.

Sobol, I.M., 1993. Sensitivity analysis for non-linear mathematical models. Mathematical Modelling and Computer Experiments 1, 407−414.

Stansbury, C.D., McKirdy, S.J., Diggle, A.J., Riley, I.T., 2002. Modeling the risk of entry, establishment, spread, containment, and economic impact of *Tilletia indica*, the cause of karnal bunt of wheat, using an Australian context. Phytopathology 92, 321–331.

Venables, W.N., Ripley, B.D., 1999. Modern Applied Statistics with S-PLUS. Springer, Berlin.

Vose, D., 1996. Quantitative Risk Analysis. Wiley, New York.

Yuan, Z., Yang, Y., 2005. Combining linear regression models: when and how? Journal of the American Statistical Association 100, 1202–1214.

## Chapter 6

Parameter Estimation with Classical Methods (Model Calibration)

1. INTRODUCTION

Parameter estimation for system models is often referred to as 'model calibration'. According to Wang, Kemanian, and Williams (2011), 'Model calibration is to adjust influential model parameters ... within their reasonable ranges so that the modeling results ... are comparable with observed data'. This indeed defines the major aspects of calibration of crop models or more generally system models. Calibration involves varying model parameters in order to improve the fit between measured and simulated values.

Calibration of system models is a major activity in modeling and is of major importance. It is a major activity because it is often repeated many times, perhaps every time a model is applied to a new range of conditions. Thus even model users who do not change the model equations are very often led to do model calibration. It is of major importance because the same equations with different values for the parameters can give very different results, both qualitatively and quantitatively. A model is only fully defined when both the equations and the parameter values are fully specified. Obtaining the best possible values for the parameters is fully as important as obtaining a good approximation to the correct functional form.

In this chapter we consider various aspects of model calibration. First, we consider in more detail exactly what model calibration is, and why it is necessary. Then we list a certain number of questions concerning model calibration, including both theoretical questions and practical questions. Then there will be a rather long section on the principles and practice of parameter estimation in statistics. This is essential, because model calibration is based on those principles and practices. We then return to our initial questions, and use the statistical results to offer at least partial answers to those questions. The final two sections concern algorithms for model calibration, and the use of R for model calibration.

Working with Dynamic Crop Models. DOI: http://dx.doi.org/10.1016/B978-0-12-397008-4.00006-X

2. AN OVERVIEW OF MODEL CALIBRATION

System models consist of a set of equations describing underlying processes, which are coupled together to create a model that describes a system. As a consequence, the parameters for system models may be obtained in two ways. First, one can study the underlying processes, and estimate the parameters based on those studies. Secondly, we can study the full system, and estimate parameters by fitting the full system model to the data.

Example 1 illustrates the difference between the two approaches, using the case of the simple maize model (see Chapter 1). Consider first just the biomass equation. Assume initial biomass is known. Given measurements of biomass at later times (the output), and daily measurements of solar radiation and LAI (the input variables), one can estimate *RUE* and *k*. In this case one is studying an individual process (biomass increase), and estimating the parameters in the equation for that process, without reference to the overall system model.

Next consider the full system model. Assume that initial biomass and LAI are known. Given measurements of biomass at later times (the output), daily minimum and maximum temperature, and daily solar radiation (the input variables for the system model), we can estimate the values of *RUE*, *k*, α, and *LAImax*. Here we are estimating the parameters based on the system model.

When we speak of model calibration, we will be referring to the second type of parameter estimation, where all (or at least many) of the system equations are involved. When we talk of estimating parameters based on the individual processes we are referring to studies involving only one (or a very few) equation and only a very few parameters.

In practice, one usually fixes most of the parameters in a system model based on the literature (i.e., based on studies of the individual processes), and only estimates a few parameters using the full system model (i.e., by calibration of the model).

Example 1 Two Methods for Estimating the Parameters in a System Model

The equation for biomass in the maize crop model of Chapter 1 up to crop maturity is

$$\Delta B(d) = RUE * RAD(d) * \{1 - \exp[-k * LAI(d)]\} \qquad (1)$$

We can then calculate biomass on any day D as

$$B(D) = B(0) + RUE * \sum_{d=0}^{D-1} RAD(d) * \{1 - \exp[-k * LAI(d)]\}$$

If we have a sample of measurements of *B(D)* on the one hand (the output), and of *B(0)* and *RAD* and LAI every day up to day *D-1* on the other hand (the explanatory variables), then we can estimate the parameters *RUE* and *k* by fitting the above equation to those data. This involves just the equation for a single process (biomass increase).

Alternatively, consider the system model, which involves the two equations for LAI and biomass:

$$\Delta LAI(d) = \alpha * \Delta TT(d) * [LAImax - LAI(d)] \tag{2}$$

$$\Delta B(d) = RUE * RAD(d) * \{1 - \exp[-k * LAI(d)]\} \tag{3}$$

Here we suppose that we don't have measurements of LAI each day, but we do have daily measurements of maximum and minimum temperature. Then we can use Equation (2) to simulate LAI each day and insert that into Equation (3). The resulting model has two outputs (biomass and LAI). The explanatory variables are $LAI(0)$, $B(0)$, daily RAD, and daily maximum and minimum temperature, and four parameters (α, $LAImax$, RUE, and k). Given a sample of data of outputs and these explanatory variables, we could estimate the four parameters by fitting the two coupled equations to the data.

We thus see that there are two different models and two different types of data that could be used to estimate RUE and k.

2.1 Why Is Calibration Necessary?

Why can't one estimate all the parameters based on studies of the individual processes? In that case, there would be no need for calibration. This is after all in keeping with the basic philosophy behind system modeling, which is that we can obtain an understanding of a complex system by studying the underlying components and then coupling them together appropriately.

There has in fact been vigorous criticism of the practice of calibration. Sinclair and Seligman (2000) make the argument that it is important to base a crop model on process equations that can be studied independently, and that the crop model parameters should come from such studies. They say

'Calibration is sometimes necessary to accommodate characteristic soil or cultivar traits but when conducted on the model as a whole, it raises serious problems. This is especially so in crop models that have a large number of site- or crop-specific empirical coefficients dispersed in various sublevel models. Some models, for example, require that more than 20 coefficients need to be calibrated to define the characteristics of the genotype to be simulated [...] Further, it is virtually impossible to obtain all of these coefficients in independent experimental tests. Therefore, software has been developed to use the complete model in an iterative procedure to obtain coefficients that optimize simulated results so that they match the final observed results [...] In effect, much of the power of the mechanistic model structure is reduced to an empirical exercise'.

They cite one of the forefathers of crop modeling, Case de Wit, who famously remarked that crop model calibration is 'the most cumbersome method of curve fitting devised'.

There are indeed models which are based exclusively on processes which can be studied independently. An example is a disease model to study worldwide risk of rice diseases. It was felt that the parameters determined from study of the processes were sufficiently reliable for the purpose of the model, which was to compare risks under different climates (Savary et al., 2012).

The other side of the argument is that, in most cases, one simply cannot obtain satisfactory values for all the system model parameters from studies of the processes. Some calibration is then necessary. Fath and Jorgensen (2011) list reasons for this. They refer specifically to system models in ecology, but the same reasons apply in agro-ecology:

1. Most parameters in environmental science and ecology are not known exactly. Parameter estimation methods must be used when no literature values can be found. Also parameters are not constant but change in time or situation.
2. Models are simplifications. Models compensate for approximations elsewhere. As a result, parameters may differ somewhat from real, but unknown, values in nature.
3. Most models are lumped models. Each parameter is an average over species.

In short, they say that calibration is necessary because some parameter values are not available from the literature, and because some of the values that are available are not exactly those required for the system model.

Below we will argue that this should not be understood to mean that calibration gives more precise values of the parameters. After all, calibration only concerns a few of the system model parameters. Rather, calibration is necessary because we need to compensate for the errors in all the parameters that are not estimated in calibration. We will discuss this in some detail below.

2.2 Current Methods of Calibration

The book Methods of Introducing System Models into Agricultural Research (Ahuja and Ma, 2011) discusses calibration methods for nine agro-ecological models. The approaches that are presented are quite representative of system model calibration.

The models are HERMES, a crop and soil model designed to operate under the data-limited conditions of agricultural practice (Kersebaum, 2011); LEACHM, a model that predicts nitrate leaching through the root zone and into underlying unsaturated soil (Jabro et al., 2011); DAYCENT, a biogeochemical model used to simulate flows of carbon and nutrients for crop, grassland, forest, and savanna ecosystems (Del Grosso et al., 2011); EPIC, a model developed to assess the effect of soil erosion on soil productivity and the version APEX which is aimed at use on multiple fields (Wang et al., 2011); GPFARM, a rangeland model for simulating rangeland productivity (Adiku et al., 2011); WOFOST, a crop model used in the MARS Crop Yield Forecasting System that provides yield forecasts for the main crops over Europe (Wolf et al., 2011); RZWQM2, a model that simulates N cycling processes, the fate and transport of agricultural chemicals and crop growth

(Nolan et al., 2011); models in DSSAT, a suite of cropping system models (Jones et al., 2011); and STICS, a crop model that simulates the behavior of the crop-soil system over one or several crop cycles (Buis et al., 2011).

All of these models have a large number of parameters. For example, DAYCENT has more than 1000 parameters. In all cases it is expected that calibration of the model will be necessary. For the HERMES model it is noted that 'Putting the model into a new environment or applying it to a new crop requires the adaptation and calibration of model parameters'. The GPFARM model explains that the model needs to be reparameterized for locations different from those where the model is currently used. The LEACHM model says that 'A universal model is a misnomer; no model can work under all conditions without calibration'. The DSSAT model discussion focuses on the cultivar specific parameters (CSPs) of the model. 'Because the CSPs are not known for most cultivars, model users need to estimate them using field data'. In all cases, only a relatively small subset of model parameters is to be adjusted.

The majority of the studies suggest that calibration should be done in several stages. At each stage specific data are used or specific processes are targeted, and the most influential parameters are adjusted. For example, for the WOFOST model, it is stated that calibration should first use phenology data, then data on total biomass and maximum LAI under optimal conditions, then data on LAI and weights of specific organs over time under optimal conditions, and finally biomass, LAI, and organ weights under water-limited conditions. At each stage several parameters are estimated. For example, in the first step seven parameters are estimated; temperature sum from emergence to anthesis, temperature sum from anthesis to maturity, temperature sum from sowing to emergence, lower temperature threshold for emergence, optimal day length for development, critical day length for development, and the maximum effective temperature for emergence.

Six of the studies talk of fitting the data, without indicating precisely what criterion of fit should be used, nor how the fitting should be done in practice. The expectation seems to be that parameter adjustment will be done manually, based either on a visual comparison of the model to the data or on various quantitative measures of agreement (root mean squared error, a measure of modeling efficiency, model bias, or other).

The other three studies propose a specific criterion of fit between observed and simulated values that is to be minimized, an algorithm for finding the parameter values that minimize the criterion, and software for implementing the algorithm. The calibration proposed for the RZWQM2 model uses a weighted least squares (WLS) criterion and a Gauss-Newton algorithm to search for the optimal parameter values. The weights are chosen so that all output variables make a similar contribution to the sum of squared errors. The calibration proposed for WOFOST allows the user to choose between three criteria: ordinary least squares (OLS); WLS where the weights are the inverse squares of the simulated values and; a third criterion called the shape criterion. The STICS model proposes software that can be used to estimate parameters in several different ways, including WLS, maximum likelihood (ML) using the concentrated likelihood method, and finally, Bayes estimation using an importance sampling algorithm. The DSSAT model proposes software that can do maximum likelihood parameter estimation or a type of Bayesian parameter estimation, using the GLUE method.

Several of the studies suggest that parameters should be limited to a biologically or physically reasonable range of values. Most of the studies also suggest that the model after calibration should be tested on independent data.

To summarize, in all the models it is accepted that calibration is a necessary practice. Also, in all cases calibration only involves a relatively small fraction of the parameters in the model. The practice of calibration, on the other hand, is variable. In most cases it is suggested to calibrate the model in several stages. At each stage different outputs are used and different parameters are estimated. However, the division of the problem into stages, or even whether or not the problem should be divided into stages, and the goodness-of-fit criteria on which to base calibration differ for each model. In most cases calibration is done in an ad hoc manner, but in a few cases calibration is closely based on statistical principles of nonlinear regression.

2.3 Questions About Model Calibration

As the above survey of current methods shows, there is no general consensus about the best approach to calibration for system models. Here we list a number of the specific questions that arise. First, there are two fundamental theoretical questions concerning model calibration:

- What is the effect of model calibration on model predictive accuracy? The operational objective of model calibration is to improve the goodness-of-fit of the model. However, the real objective is to improve model predictions for new situations. Can we make any general statements about whether calibration will improve predictions?
- What can we say about the parameter values that result from model calibration? Are these values better estimates of some underlying true parameter values, or are they simply adjustment factors with no physical or biological meaning?

In addition, there are practical questions concerning model calibration:

- What data should be used for calibration? In particular, if we are mainly interested in some specific responses (for example, yield and final grain protein content), but we have data for other responses (for example, in-season measurements of LAI, biomass, etc.), should we fit the model to all the data or just to the responses of primary interest?
- What goodness-of-fit criterion should be used to measure how well the model fits the data? Current practice centers on the sum of squared errors, but different weightings are used.

- If the model is fit to several different responses, should one fit to all the responses simultaneously, or fit to one response at a time (sequential calibration)?
- Which parameters should be adjusted? System models often have a very large number of parameters, but only a small number are changed in model calibration. On what basis should one choose the subset of parameters that will be used for calibration?
- How many parameters should be adjusted? This is a question that is seldom addressed in calibration of crop models, but we will argue that it is very important to avoid estimating too many parameters.
- How exactly should one combine the two sources of information about the parameters, namely the information based on studies of the individual processes and the information based on studies of the overall system?

3. THE STATISTICS OF PARAMETER ESTIMATION

We review here the principles and practice of parameter estimation in regression that will be useful for answering our questions about calibration of system models. To illustrate many of the notions, we will use the simple model given in Example 2.

Example 2 Wheat Grain Weight Versus Degree Days after Anthesis

Darroch and Baker (1990) studied grain filling in three spring wheat genotypes. The data for cultivar Neepawa in 1988 are shown in Figure 6.1.

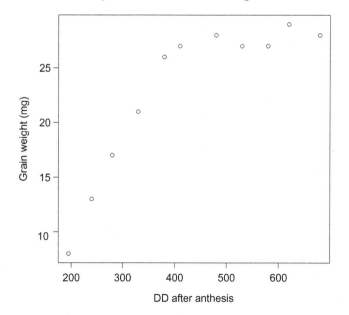

FIGURE 6.1   Data of grain weight versus degree days after anthesis.

They proposed a logistic function to model their data:

$$seed.weight = \frac{W}{1 + e^{B - C*DD}} \qquad (4)$$

This model has a single input variable, degree days after anthesis noted *DD*, and three parameters, noted *W*, *B*, and *C*. Parameter estimation here concerns estimation of the values of those three parameters.

3.1 Ordinary Least Squares Parameter Estimation

A common approach to parameter estimation in regression is ordinary least squares (OLS). We first review the underlying assumptions about the relationship between observed and simulated values. Then we explain how to estimate parameters by OLS.

3.1.1 The Assumptions Behind OLS

We assume that our data sample is chosen at random from the range of conditions of interest (the 'target population'). We will refer to each element of the sample as an 'individual'; for example, the individual could be a specific field in a specific year in the case of a crop model. The ith individual has explanatory variables x_i.

We can always write the following relation between the measured value and the corresponding value calculated by the model:

$$y_i = f(x_i; \theta) + \varepsilon_i \qquad (5)$$

where y_i is the measured value for the ith individual and $f(x_i; \theta)$ is the corresponding model prediction, which depends on explanatory variables x_i and a parameter vector θ. We assume that the value of the input variable, x_i, is fixed for this individual, but other than that the individual is chosen at random from the target population. If the experiment were redone, a different individual, but with the same value of x_i, would be chosen. Thus the model prediction $f(x_i; \theta)$ is fixed for given i, since x_i is fixed. However, y_i and therefore ε_i are random variables, since the choice of the ith individual is random. Equation (5) simply assigns a symbol, ε_i, to the difference between the measured value and the value calculated using the model. We refer to ε_i as model residual error.

We now make three assumptions about residual error (Table 6.1). The important information about the relation of simulated to observed values is in these assumptions.

TABLE 6.1 Three Assumptions About Model Residual Error

For some $\theta = \theta^{true}$
Assumption 1. Expected residual error equals 0.
$$E(\varepsilon_i) = 0 \qquad (6)$$
Assumption 2. Constant residual variance.
$$var(\varepsilon_i) = \sigma^2 \qquad (7)$$
Assumption 3. Independent residual errors
$$\text{All } \varepsilon_i \text{ independent} \qquad (8)$$

Equation (6) says that for some parameter vector $\theta = \theta^{true}$, the residual error has expectation 0, for each x_i in the sample. Taking the expectation on both sides of Equation (5), we see that this implies that

$$E(y_i|x_i) = f(x_i; \theta^{true}) \qquad (9)$$

where the notation $y_i|x_i$ means y_i given that $x = x_i$. Equation (6) or equivalently Equation (9) define what we mean when we say that a model is 'correct'. We mean that there exists some parameter vector $\theta = \theta^{true}$ such that, if we average the measured values over all individuals in the target population for any specific value of x_i, the result is equal to $f(x_i; \theta^{true})$.

This also defines what is meant by 'true' parameter values. They are the parameter values such that Equation (9) holds, i.e., such that, for every x_i, the expectation of the measurements is equal to the value calculated using the model.

Note that a 'correct' model is not necessarily a model that is a good predictor. The residual error has expectation zero, but for each individual it will not, in general, be zero. The variance of residual error is a measure of how much variability there is in the difference between measured and simulated values. If that variance is large then the difference between any particular measurement and the corresponding model prediction will in general be large.

The second assumption, in Equation (7), says that the residual variance is the same for all the x_i.

Finally, Equation (8) says that all the residual errors are independent. That is, knowing the residual error for one individual in the sample gives us no information about the residual error for another individual. If each individual in the sample is chosen at random in the population, then clearly the residual error for individual $i' \neq i$ will have no relation to the residual error for individual i. Thus this assumption is automatically fulfilled if we have a fully random sample.

Example 3 A Large Sample Simulated on the Basis of the Three Assumptions Given Above

Figure 6.2 shows what a large sample from the target population of Example 2 would look like, if the assumptions of Equations (6)–(8) are satisfied. The true parameter values are $(W^{true}, B^{true}, C^{true}) = (28.2, 4.16, 0.0165)$. As can be seen, the model with the true parameter values goes through the middle of the points for each value of DD (Assumption 1). The spread around the model is the same for all values of DD (Assumption 2). Finally all the residual errors are independent, though this cannot be seen from the figure.

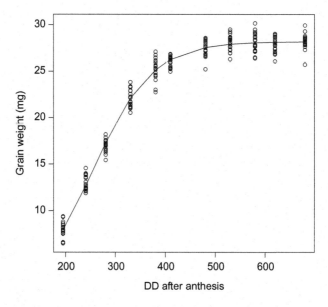

FIGURE 6.2 Large sample of measured values in the case where the three assumptions about residual error in Table 6.1 are satisfied.

3.1.2 The OLS Equations

According to this approach, the parameters are chosen to minimize the sum of squared errors (SSE) between simulated and measured values:

$$SSE = \sum_{i=1}^{n} [y_i - f(x_i; \theta)]^2 \tag{10}$$

where n is the number of data points. We can say the same thing with an explicit equation for the OLS parameters:

$$\hat{\theta}_{OLS} = \underset{\theta}{\mathrm{argmin}} \left\{ \sum_{i=1}^{n} [y_i - f(x_i; \theta)]^2 \right\} \tag{11}$$

which says that the OLS parameters are the arguments of the model that minimize the sum of squared errors.

There is no general analytical solution to Equation (11) when f is a nonlinear function of the model parameters. In general, the solution must be calculated using an algorithm, which approaches the solution in successive iterations. The R statistical language has special functions that do these calculations. The algorithms that are used, and the details of the functions, are presented below.

OLS also furnishes an estimate of residual variance, which is

$$\hat{\sigma}^2_{OLS} = \left\{ \sum_{i=1}^{n} \left[y_i - f(x_i; \hat{\theta}_{OLS}) \right]^2 \right\} / (n - p) = SSE / (n - p) \tag{12}$$

where p is the number of estimated parameters.

The result of OLS parameter estimation for Example 2 is shown in Example 4.

Example 4 OLS Parameter Estimation for the Model and Data of Example 2

The R function nls was used to estimate the parameters of the logistic model of Example 2. The OLS parameter values are $\hat{W}_{OLS} = 28.17$, $\hat{B}_{OLS} = 4.160$, $\hat{C}_{OLS} = 0.01649$. The estimated residual variance is $\hat{\sigma}^2_{OLS} = 0.8364$. The fit of the model to the data and the model residuals (observed minus simulated values) are shown in Figure 6.3.

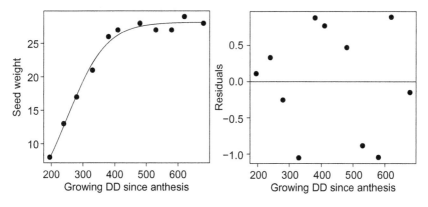

FIGURE 6.3 Left: Data (dots) and model (line) for logistic model with parameters estimated using OLS. Right: residuals.

Given the estimated parameters, we can plug them into the model in order to make predictions for any new value of the input variables. To predict for say x^*, we would use $f(x^*; \hat{\theta}_{OLS})$.

3.1.3 The Properties of OLS

We have explained that since the sample is considered to be chosen randomly, y_i is a random variable. According to Equation (11), $\hat{\theta}_{OLS}$ is then also a random variable since it is calculated from the y_i. When we want to insist on the fact that $\hat{\theta}_{OLS}$ is a random variable, we call it the OLS 'estimator'. When we just want to talk about the specific numeric values obtained for a particular sample, we will talk of the OLS 'estimate'.

We consider first the case of a linear model. Linear here means that the parameters to be estimated enter linearly in the model. For a linear model, the OLS estimator has several optimal properties (Graybill, 1976). First, the OLS estimator for a linear model is unbiased. That is, the expectation of $\hat{\theta}_{OLS}$ over samples is exactly equal to θ^{true}. It has minimum variance among estimators that are unbiased and that can be written, like the OLS estimator, as linear functions of the measured values. These are the optimal properties concerning how well OLS estimates the true parameter values. Also, $\hat{\sigma}^2_{OLS}$ is an unbiased estimator of σ^2. Finally, the model $f(x; \hat{\theta}_{OLS})$ tends toward the model with minimum mean squared error of prediction (MSEP). This follows because the model $f(x; \theta^{true})$ has smallest MSEP for models with the explanatory variables x, and $\hat{\theta}_{OLS}$ tends toward θ^{true}.

Similar properties hold for a nonlinear model (Seber and Wild, 1989 p. 565), but only asymptotically, that is, as the size of the sample goes toward infinity. There are also additional constraints on the model, essentially that it must be well behaved. In this case $\hat{\theta}_{OLS}$ is asymptotically unbiased and consistent (tends toward θ^{true} with less and less variability). The estimator of residual variance $\hat{\sigma}^2_{OLS}$ is also consistent. The difference $\hat{\theta} - \theta^{true}$ has a normal distribution asymptotically, with

$$\hat{\theta} - \theta^{true} \sim N(0, \sigma^2(F'F)^{-1}) \qquad (13)$$

where

$$F = \begin{pmatrix} \left.\frac{\partial f(x_1;\theta)}{\partial \theta_1}\right|_{\theta^{true}} & \cdots & \left.\frac{\partial f(x_1;\theta)}{\partial \theta_p}\right|_{\theta^{true}} \\ \vdots & \ddots & \vdots \\ \left.\frac{\partial f(x_n;\theta)}{\partial \theta_1}\right|_{\theta^{true}} & \cdots & \left.\frac{\partial f(x_n;\theta)}{\partial \theta_p}\right|_{\theta^{true}} \end{pmatrix}$$

Equation (13) says that the asymptotic variance-covariance matrix of $\hat{\theta} - \theta^{true}$ is $\sigma^2(F'F)^{-1}$, which can be approximated by replacing σ^2 and θ^{true} by $\hat{\sigma}^2_{OLS}$ and $\hat{\theta}_{OLS}$, respectively. We can use this approximation to calculate the variances and covariances of the estimator $\hat{\theta}_{OLS}$.

One should neither be too impressed nor too unimpressed by the fact that these properties only hold asymptotically for nonlinear models. One should not be too impressed, because samples are always finite in size, often quite small, and we have promised no good qualities for OLS for nonlinear models for finite sample size. On the other hand, asymptotic properties are far from useless. Imagine an estimator that did not promise to converge to the true parameter values asymptotically. We would have serious doubts about using such an estimator.

Example 5 illustrates some of the above properties using simulated data, where the assumptions in Table 6.1 are satisfied. We drew repeated samples from the population, and for each sample calculated the OLS estimates for the parameter values and for the residual variance. In this artificial example we know the true values θ^{true} and σ^2, and so can compare the OLS estimates with the true values. Since we know exactly the distribution of y in the population, we can also calculate MSEP for any value of $\hat{\theta}_{OLS}$.

Example 5 OLS Estimates for Samples Drawn from the Population of Example 3

The artificial data for this example were generated using

$$y_i = \frac{W^{true}}{1 + \exp^{B^{true} - C^{true} * DD_i}} + \varepsilon_i \tag{14}$$

with $(W^{true}, B^{true}, C^{true}) = (28.2, 4.16, 0.0165)$. The sample has 11 different values of DD; the same values as in Figure 6.1. The residual errors were generated according to

$$\varepsilon_i \underset{iid}{\sim} N(0, \sigma^2) \tag{15}$$

with $\sigma^2 = (0.84)^2 = 0.706$. Equation (15) says that the residual errors are independent and identically distributed (iid) with a normal distribution of expectation 0 and variance σ^2. Thus the assumptions of Table 6.1 are satisfied (residual error has expectation 0, constant variance, and all residual errors are independent).

We drew 500 different samples of size 11 (one individual at each value of DD), of size 33 (3 individuals at each value of DD) or of size 110 (10 individuals at each value of DD) from the population described by Equation (14). The estimators $\hat{\theta}_{OLS}$ and $\hat{\sigma}^2_{OLS}$ were calculated for each sample using the R function nls. Samples where nls failed to converge were eliminated.

The first three lines of Table 6.2 show the mean (averaged over 500 samples) of the parameter estimates. As sample size increases from 11 to 33 to 110, the means closely approach θ^{true}, as expected since the OLS estimator is asymptotically unbiased. In fact, they are already quite close to θ^{true} even for the smallest sample size. The next three lines show that the standard deviations of the estimators diminish substantially with larger sample size. This is an expression of the fact that the estimator is consistent; as sample size increases, the estimated values cluster more and more closely around the true value. The line 'mean $(\hat{\sigma}^2_{OLS})$' shows the mean over samples of estimated residual variance, which is quite close to the true value for all sample sizes.

TABLE 6.2 Effect of Increasing Sample Size on OLS Estimators

Sample size	11 (1 measurement at each value of DD)	33 (3 measurements at each value of DD)	110 (10 measurements at each value of DD)
Mean $(\hat{W}_{OLS})$ $W^{true} = 28.2$	28.28	28.20	28.20
Mean $(\hat{B}_{OLS})$ $B^{true} = 4.16$	4.158	4.148	4.157
Mean $(\hat{C}_{OLS})$ $C^{true} = 0.0165$	0.01650	0.01646	0.01649
$\sqrt{var(\hat{W}_{OLS})}$	0.394	0.232	0.131
$\sqrt{var(\hat{B}_{OLS})}$	0.369	0.210	0.111
$\sqrt{var(\hat{C}_{OLS})}$	0.00144	0.00082	0.00044
Mean $(\hat{\sigma}^2_{OLS})$ $\sigma^2 = 0.706$	0.711	0.706	0.706
mean (MSEP) minimum MSEP = 0.706	0.873	0.766	0.724

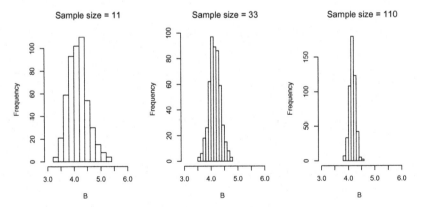

FIGURE 6.4 Histograms of the estimated values of parameter B in different samples, for increasing sample size.

The last line shows the average of MSEP. The minimum possible value, attained for $\theta = \theta^{true}$, is $MSEP = \sigma^2 = 0.706$. For small sample size, average MSEP is substantially larger, because the estimated parameter values are often quite far from the true values. As the sample size increases, the estimated parameter values are always closer to the true value, and average MSEP diminishes.

Figure 6.4 shows histograms of $\hat{B}_{OLS}$ for the different sample sizes. This again shows that as sample size increases, there is less and less variability in the estimated value around B^{true}.

3.2 Maximum Likelihood

There is a second major way of obtaining estimates of parameter values, other than least squares, called maximum likelihood (ML). This is an extremely general procedure, which can be used whenever one can make reasonable assumptions about the distribution of model error. (Note that for least squares, we did not need to make assumptions about the distribution of residual error, other than constant variance.)

The likelihood that $\theta^{true} = \theta$ for some parameter vector θ, written $L(\theta|Y)$, is defined as the probability that θ would give rise to the Y vector that was actually measured. The ML estimator of θ^{true} is then the value of θ that maximizes $L(\theta|Y)$. That is, the estimator is the parameter vector that maximizes the probability of obtaining the values that were in fact measured.

Consider the simple case where the residual errors are all independent and normally distributed with the same variance:

$$y_i = f(x_i; \theta^{true}) + \varepsilon_i$$
$$\varepsilon_i \underset{iid}{\sim} N(0, \sigma^2) \tag{16}$$

The assumptions about residual error here are the same as in Table 6.1, but with the additional assumption that the residual errors have a normal distribution.

According to Equation (16), if $\theta^{true} = \theta$ then the residual errors are $\varepsilon_i = y_i - f(x_i; \theta)$. The probability of obtaining those residual error values is

$$Prob(\varepsilon_1, \ldots, \varepsilon_n) = L(\theta|Y) = \prod_{i-1}^{n} \frac{1}{(2\pi\sigma^2)^{1/2}} e^{(y_i - f(X_i, \theta))^2 / (2\sigma^2)}$$
$$= \frac{1}{(2\pi\sigma^2)^{n/2}} e^{-\sum (y_i - f(X_i, \theta))^2 / (2\sigma^2)} \tag{17}$$

where we have used the probability for a normal distribution with expectation 0 and variance σ^2, and the fact that for independent variables, the joint probability is the product of the individual probabilities.

The ML estimator is the value of θ that maximizes the above expression. The only part of the expression that involves θ is the exponent, and the expression is maximized when the exponent is maximized or equivalently

when the negative of the exponent is minimized. Then finally the ML estimator in this case is the parameter vector that minimizes

$$SSE = \sum_{i=1}^{n} (y_i - f(X_i, \theta))^2$$

This is exactly the same expression as for the OLS estimator. In this simple
case, the OLS estimator and the ML estimator are identical.

It is also possible to show that the value of σ^2 that minimizes Equation (17) is

$$\hat{\sigma}_{ML}^2 = \sum_{i=1}^{n} (y_i - f(X_i, \theta))^2 / n$$

It can be shown that this is a biased estimator, so it is usually replaced by
the unbiased estimator

$$\hat{\sigma}_{ML,unbiased}^2 = \sum_{i=1}^{n} (y_i - f(X_i, \theta))^2 / (n - p) \tag{18}$$

In the asymptotic limit where n tends toward infinity, the difference between
the biased and unbiased expressions becomes negligible. Equation (18) is
identical to the OLS estimator of the residual variance.

If the assumptions in Equation (16) are valid, the ML estimator is asymptotically consistent and asymptotically has minimum variance for an unbiased estimator (Berger and Casella, 1990).

3.3 Suppose the Assumptions Behind OLS and ML Aren't Satisfied

We have said that OLS and ML have desirable properties when the assumptions in Table 6.1 are satisfied (plus, for ML, the assumption that the residual
errors have a normal distribution). Often, however, this will not be the case,
in particular for system models. It is important then to consider these
assumptions in more detail. When are they likely to be violated? How can
one know if they are violated? What are the consequences of ignoring violations of the assumptions and using OLS or ML anyway? What remedies are
there if the assumptions are violated?

3.3.1 Assumption 1: The Expectation of Residual Error Is 0 for All x

Assumption 1 is that there are some parameter values (the 'true' parameter
values noted θ^{true}) such that $E(y|x) = f(x; \theta^{true})$. That is, for each value of the
explanatory variables x, the model is equal to the expectation of the response.
In essence, this says that the model has the correct dependence on x. If this
assumption is satisfied, we say that the model is 'correctly specified' or

simply 'correct'. If the assumption is not satisfied, we say that the model is 'misspecified'.

For fairly simple models, with few explanatory variables, one can plot residual error versus each explanatory variable and see if the residuals seem to be centered at 0 for all x.

We illustrate this in Example 6 using the data of Example 2. In Example 4 we saw that the logistic model gives a good fit to these data. Suppose that instead of the logistic model we try a two parameter exponential model

$$f(x; \theta) = a(1 - e^{-b*DD})$$

The fit of this model to the data using OLS and the plot of the residuals versus DD are shown in Figure 6.5. The residuals are systematically negative for small and large values of DD, and positive in between. It looks like we could propose a model to relate residual error to DD. This implies that the model $f(x; \theta)$ does not include all the dependence of y on DD, since there is still some structured dependence left in the residual error. It is this structure of the residuals that indicates that the model does not correctly take into account the effect of the input variable.

Example 6 An Exponential Model Fit to the Data of Example 2

The two parameters a and b of the exponential model $f(x; \theta) = a(1 - e^{-b*DD})$ were fit to the data of Example 2 using OLS. The calculations were done using the R function nls. The left-hand graph in Figure 6.5 shows the fit of the model to the data. The right-hand graph shows the residuals.

There is some clear dependence on DD of the residuals. That is, the model does not correctly describe the effect of DD on the response. The model is misspecified.

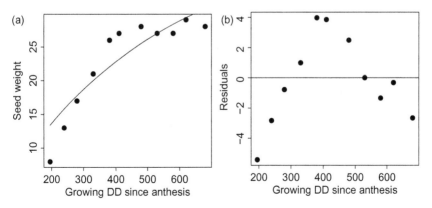

FIGURE 6.5   a: Misspecified model. Data (dots) and model (line); b: Misspecified model. Plot of residuals.

What should be done when Assumption 1 is violated? One should change to a model which gives a better fit to the data. In the above example the logistic model gives a better fit, with residuals that seem random, and so that model would be preferred. For fairly simple models, with few input variables, the model can be chosen to fit the data. That is, the model is specifically chosen to satisfy Assumption 1. For such data-driven models, we can in general avoid major violations of Assumption 1.

However, this is no longer the case with complex system models. For such models, the form of the model is not directly chosen to fit the data. Furthermore, these models often have a large number of explanatory variables, so that it is difficult to see if there is structure in the residuals versus each explanatory variable. The result is that for system models, there is a real possibility that Assumption 1 is severely violated.

3.3.1.1 Complex System Models Are Probably Misspecified

Wallach (2011) argued that, in fact, fairly complex system models probably are misspecified. He started from the assumption that the individual equations in the system model are each correctly specified. Each individual equation usually involves only a small number of explanatory variables, and so can often be fit directly to the data. However, even if the individual equations are correctly specified, the system model may well be severely misspecified. There are two reasons for this. First, when the individual equations are coupled, as in a system model, the residual variances from one equation get mixed with the explanatory variables in other equations. The result is an overall residual error that is no longer independent of the explanatory variables.

There is in addition a second cause of misspecification in system models. In calibration, most of the parameters are set to fixed values obtained from the literature. However, these values are not perfectly well known; at best they are only approximations to the true parameter values for each individual equation. If they are fixed at incorrect values, then the model is misspecified. This concerns each individual equation, and also the system model.

3.3.1.2 Consequences of Violating Assumption 1

White (1981) studied the consequences of using OLS to estimate model parameters for misspecified models. The major assumptions he made were that the input variables are random variables and residual errors are all independent and identically distributed. The results concern the asymptotic properties of OLS estimation.

A first conclusion concerns the values of the estimated parameters. Since violation of Assumption 1 means that the model does not have the correct functional form, it may make no sense to talk of true values for the parameters, since the true response might not involve the same parameters as the model. Even if the true response does involve the same parameters as the model, the OLS estimators will not converge toward the true parameter values.

We consider what this implies in the case where the individual process models do satisfy Assumption 1, but the system model is misspecified. There are now true values of the parameter values; they are the parameter values such that Assumption 1 is verified for the individual process equations. However, model calibration for the system model will not give those values, even asymptotically.

Assuming that complex system models are misspecified, we conclude that the parameter values estimated by calibrating the system model using OLS should be considered as just empirical constants, which compensate for the errors of the model. Calibration of a system model is not a way to obtain improved estimates of the true parameter values.

A second conclusion of White (1981) concerns MSEP for a misspecified model. We define the 'optimal' parameter values of a model as the values that minimize MSEP of the model for the specified target population. For a correctly specified model, the true and optimal parameter values are identical. For a misspecified model, there may or may not be any true parameter values, but if the model is reasonably well behaved there are in any case optimal parameter values.

White (1981) showed that for a misspecified model, the OLS estimators tend toward the optimal parameters, as the amount of data tends toward infinity. On the one hand, this is perhaps not so surprising. OLS minimizes the sum of squared errors. If we do this for larger and larger samples, it is not surprising that we approach the parameters that minimize mean squared error for the entire population. On the other hand, this is a very powerful result. It says that even if the model is misspecified, the OLS parameters will make the model into the best it can be. There is therefore a rationale for estimating parameters in a system model. It is an effective way of improving prediction accuracy (but don't forget that this has only been shown to hold 'asymptotically').

Example 7 illustrates the above two effects of misspecification with an artificial example. Two different misspecified models are shown. In the first, the functional dependence on x is wrong. The asymptotic parameter values in this case are not comparable to the true parameter values; in the true response, the parameters describe the exponential increase of y, while in the model the parameters are the coefficients of a linear and quadratic term. In the second misspecified model, the dependence on x has the correct form, but one of the model parameters is fixed at an incorrect value. Even in the second case, the OLS estimator of the remaining parameter is not equal, asymptotically, to the true parameter value. It is different, in order to compensate for the error in the fixed parameter.

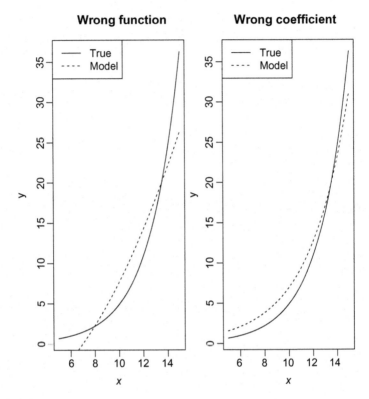

FIGURE 6.6 Two misspecified models, showing the true expectation of the response (solid line) and simulated values using parameter values estimated by OLS on a sample of size 30,000 (dashed line).

Figure 6.6 shows the fit of the two misspecified models, using the parameters estimated by OLS. The fit is reasonable. OLS parameter estimation has made the misspecified models into as close approximations to the true response as possible.

Example 7 Simple Examples of Misspecified Models

Suppose that the true response is

$$y_i = 0.09\exp(0.4x) + e_i$$
$$e_i \sim N(0, 0.1^2)$$

(19)

The distribution of x values in the target population is assumed to be $x \sim U(5, 15)$ (the x values are drawn at random from a uniform distribution with lower and upper limits of 5 and 15, respectively). All the errors and explanatory variables are mutually independent. The correctly specified model is then

$$h(x; \theta) = \theta_1 \exp(\theta_2 x)$$

with true parameter values

$$(\theta_1^{true}, \theta_2^{true}) = (0.09, 0.4)$$

We consider two different misspecified models:

$$h_1(x; \theta) = \theta_1 + \theta_2 x^2$$
$$h_2(x; \theta) = \theta_1 \exp(0.3x)$$

The first model has the wrong functional form (quadratic rather than exponential). The second has the correct functional form, but one of the parameters is set to an incorrect value. The parameters for these two models were estimated using a sample of size 30,000, which should be a good approximation to the asymptotic situation. The results, shown in Table 6.3, show that the estimated parameter values are not equal to the true parameter values.

The fitted models are shown in Figure 6.6. OLS parameter estimation has made the models as close as possible to the true response.

TABLE 6.3 OLS Parameters Estimated for Misspecified Models

Model	$\hat{\theta}_{OLS,1}$ ($\theta^{true} = 0.09$)	$\hat{\theta}_{OLS,2}$ ($\theta^{true} = 0.4$)
$h_1(x; \theta)$ (wrong function)	-7.298	0.1498
$h_2(x; \theta)$ (wrong coefficient)	0.3464	fixed at 0.3

Another important consequence of violation of Assumption 1 concerns calibration for different target populations. Suppose that we have two different target populations, where the same underlying process models, with the same parameters, apply. This might be, for example, corn fields in two different regions. For the correct model, which satisfies Assumption 1, the OLS estimator will tend asymptotically toward the same true parameter vector in both populations. However, for a misspecified model, the OLS estimator will not tend toward the same value for the two populations because the estimator compensates for errors in the model and the errors depend on the input variables.

This implies that it is dangerous to extrapolate a calibrated system model to a new target population, since the optimal parameters for the initial target population will not, in general, be optimal for the new target population. In fact, this explains why calibration is necessary for complex system models. Such models are often misspecified, and the optimal parameter values of a misspecified model are different for each target population. Calibration is necessary because for a misspecified model the optimal parameter values are different for each target population.

Example 8 illustrates this. The example shows that the OLS parameter estimate is different for two different target populations, even though the true response is the same in both cases.

Example 8 The Effect of the Target Population on the Estimated Parameters in a Misspecified Model

The true response is assumed to be

$$y_i = 0.09\exp(0.4x) + e_i$$
$$e_i \sim N(0, 0.1^2) \tag{20}$$

We consider the misspecified model

$$h(X_i; q) = q_1 \exp(0.3X_i)$$

We consider two different target populations, namely $x \sim U(5, 15)$ and $x \sim U(10, 20)$, where the notation $x \sim U(a, b)$ means that x has a uniform distribution with lower limit a and upper limit b. In each case we draw a sample of size 30,000 from the target population, and estimate θ_1 using OLS. The estimated parameter value in each case is shown in Table 6.4.

TABLE 6.4 OLS Parameters for Two Different Target Populations

Target Population	$\hat{\theta}_{OLS,1}$
$x \sim U(5, 15)$	0.3464
$x \sim U(10, 20)$	0.5706

The results show that even though the true response is the same for all values of x, the estimated parameters are different, depending on which range of x is sampled. This is a result of misspecification.

In a misspecified model, the estimated parameters compensate for errors in the model. As we have just shown, the errors are different for different target populations, and so the optimal parameter values are different for different target populations. Analogously, the optimal parameter values are different for different responses. The parameter values that minimize MSEP for one response (say yield) will not in general be the same as the parameter values that minimize MSEP for a different response (say, in-season leaf area index (LAI)).

As a result, for a misspecified model, it is not necessarily true that estimating the parameters for one response will improve predictions for a different response.

3.3.2 Assumption 2: Constant Residual Variance

Assumption 2 is that residual variance is the same for all values of the input variables. If this is not the case, then Assumption 2 is violated.

One common type of violation is where the residual variance increases as the size of the response increases. For example, consider biomass over the growing season. Often model error will be small early in the season, when biomass is small, and increase with larger biomass values.

Violation of Assumption 2 can often be detected by first doing OLS parameter estimation and then plotting residual error versus predicted response. The spread of the residual errors should be comparable for all values of the predicted response.

Example 9 illustrates a case where this assumption seems to be violated. The data are wheat grain weights, from an experiment where grain weights were measured at numerous dates in a range of environments. The model is a logistic model. It is quite clear that variability of the residuals increases as the predicted value increases. That is, the variance of residual error does not seem to be constant, so Assumption 2 is not satisfied.

Example 9 Grain Weights from Numerous Fields

Grain weight was measured at numerous dates in various fields at Clermont-Ferrand in France (Robert et al., 1999). The results concern sixteen different wheat cultivars, a range of levels of nitrogen fertilization, and two different years.

A logistic function was fit to the data, using OLS. Figure 6.7 shows standardized residuals for grain weight as a function of degree days after anthesis (DD). (A standardized residual is $\hat{\varepsilon}_i^s = (\hat{\varepsilon}_i - \bar{\hat{\varepsilon}})/\sigma_{\hat{\varepsilon}}$ where $\hat{\varepsilon}_i = y_i - f(x_i; \hat{\theta}_{OLS})$, $\bar{\hat{\varepsilon}}$ is the

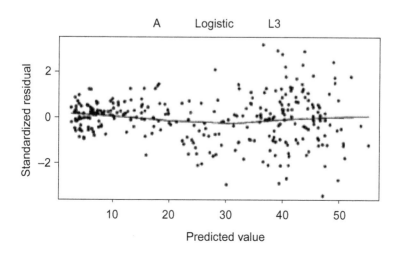

FIGURE 6.7   Plot of standardized residuals for data and model of Example 9.

mean of the $\hat{e}_i$ and σ_e is the standard deviation of the $\hat{e}_i$.) The solid line is a smoothed approximation to average residual error. The average of the standardized residuals is close to 0 throughout, so Assumption 1 can be assumed to hold. However, the spread of the residuals increases with increasing predicted value, so Assumption 2 is violated.

Another situation that leads to unequal variances is the case where there is more than one type of output from a model (for example, the responses might be LAI and biomass). This is often the case with system models. In this case the different outputs may not even have the same units. In that case, it is essential to take into account differences in the residual variances; the assumption of constant variance makes no sense.

What if one ignores unequal variances and estimates the parameters using OLS anyway? If Assumption 1 is satisfied, the OLS estimator is then still asymptotically unbiased and consistent. That is, the expectation of the parameter estimators will still be equal to the true parameter values, and the estimators will tend to be closer and closer to the true values as the sample size increases. However, there will be more variability in the estimators around the true values than if we had equal residual variances.

There are several possible remedies available for the problem of unequal residual variances. These are described below.

3.3.2.1 Weighted Least Squares

Suppose that we know or can estimate residual variance at each x_i, noted σ_i^2. For example, this might be based on measurements on multiple individuals at the same x_i. If Assumption 1 is satisfied, then the variability between individuals with the same x is a measure of residual variance. If we divide each y_i and each $f(x_i; \theta)$ by σ_i, then the residual error after this transformation is

$$\varepsilon_i^* = y_i/\sigma_i - f(x_i; \theta)/\sigma_i = \varepsilon_i/\sigma_i$$

The variance of this new residual error is

$$\text{var}(\varepsilon_i^*) = (1/\sigma_i^2)\text{var}(\varepsilon_i) = 1$$

Thus after the transformation the residual variance is identical (and equal to 1) for all x_i. Now Assumption 2 is satisfied and we can use OLS.

The sum of squared errors (SSE) to be minimized is now

$$SSE = \sum_{i=1}^{n} \left[y_i/\sigma_i - f(X_i; \theta)/\sigma_i\right]^2 = \sum_{i=1}^{n}(1/\sigma_i^2)[y_i - f(X_i; \theta)]^2 \qquad (21)$$

This procedure of minimizing the above SSE is called weighted least squares (WLS), because it is equivalent to weighting each term in OLS by

$1/\sigma_i^2$. The R function nls allows one to specify a weight for each measurement, and thus can be used for WLS.

According to Equation (21), we should give less weight to x_i values with large residual variance. This is logical. We should attach less importance to x_i values where there is more variability around $f(x_i; \theta)$, because then the measured value contains less information about the true parameter values. If we don't use different weights, then we give as much importance to points with little information as to points with high information.

3.3.2.2 Power Transformation

Suppose that the residuals after parameter estimation by OLS indicate that residual variance is approximately proportional to some power of the expected value. In this case, we can use a power transformation of both measured and simulated values, as proposed by Box and Cox (1964). The measured and simulated values are transformed as

$$y_i^* = \begin{cases} (y_i^\lambda - 1)/\lambda & \lambda \neq 0 \\ \log(y_i) & \lambda = 0 \end{cases}$$

$$f^*(x_i; \theta) = \begin{cases} (f(x_i; \theta)^\lambda - 1)/\lambda & \lambda \neq 0 \\ \log(f(x_i; \theta)) & \lambda = 0 \end{cases}$$

where λ is an additional parameter whose value needs to be chosen by the user or estimated from the data.

The quantity to be minimized by OLS then becomes

$$SSE = \sum_{i=1}^{N} \left[y_i^* - f^*(x_i; \theta) \right]^2 \tag{22}$$

The variance of the transformed variable (Seber and Wild, 1989, p. 69) is approximately

$$\text{var}(y_i^*) \approx \text{var}(y_i)[E(y_i|x_i)]^{2\lambda-2} \tag{23}$$

Two cases are of particular interest. If the variance of y_i is proportional to $E(y_i|x_i)^2$, then according to Equation (23) choosing the transformation with $\lambda = 0$ (the log transformation) will make $\text{var}(y_i^*)$ approximately constant. If the variance of y_i is proportional to $E(y_i|x_i)$, then choosing the transformation with $\lambda = 0.5$ will make $\text{var}(y_i^*)$ approximately constant.

In the general case, one needs to estimate the additional parameter λ. One possibility is to try various values of λ and choose the transformation that gives residual variances that seem more or less constant for all x_i.

Alternatively, in the maximum likelihood framework, one can estimate simultaneously the parameter vector θ and the variance parameter λ using maximum likelihood. This involves solving for the vector $(\theta, \sigma^2, \lambda)$ that maximizes the likelihood. This is illustrated in Example 10.

Example 10 Estimation of a Variance Parameter Using ML

Using the data presented in Example 9, Robert et al. (1999) tested the logistic equation with two alternative assumptions concerning the residual variance. The residual variance was either considered to be constant or to be of the form $\text{var}(\varepsilon_i) = \sigma^2 f(x_i; \theta)^\lambda$. The parameters of the logistic equation, and also σ^2 and λ, were estimated using ML. The estimated value of λ was 1.29. Figure 6.8 shows the standardized residuals for the raw data (left) and after the power transformation (right). The residuals on the right have a more uniform distribution, so it seems that the power transformation has allowed us to satisfy the assumption of constant residual variance.

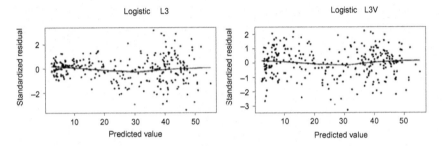

FIGURE 6.8 Residual errors without power transformation (left) and with power transformation with $\lambda = 1.29$ (right).

3.3.2.3 Concentrated Likelihood

Another method of handling unequal variances is the method of concentrated likelihood, a maximum likelihood method (Seber and Wild, 1989). This method is adapted to the case where the model calculates multiple responses, each with a different residual variance.

Suppose that all the residuals are independent and normally distributed, and that the residual variances are all equal for a single response but different for different responses (for example, for yield and LAI). The likelihood is then

$$L(Y|\theta, \Sigma) = \prod_{ij} \frac{1}{\left(2\pi\sigma_j^2\right)^{1/2}} e^{\left[y_{ij}-f_j(x_i;\theta)\right]^2/(2\sigma_j^2)} \tag{24}$$

where the product is over the different responses (j) and over measurements of that response (i).

The concentrated likelihood method proceeds in two stages. First, treating θ as fixed, one calculates the residual variances that maximize the likelihood. This gives

$$\hat{\sigma}_j^2(\theta) = (1/n_j) \sum_{i=1}^{n_j} \left[y_{ij}-f_j(x_i; \theta)\right]^2 \tag{25}$$

where n_j is the number of measurements of response $_j$. Then Equation (25) is substituted into Equation (24) to give a 'concentrated' likelihood that involves only θ. Maximizing that likelihood is equivalent to minimizing the expression below:

$$\hat{\theta} = \arg \min_{\theta} \prod_{j} \left\{ (1/n_j) \sum_{i=1}^{n_j} \left[y_{ij} - f_j(x_i; \theta) \right]^2 \right\}^{n_j/2} \qquad (26)$$

Equation (26) gives the ML estimator of θ. The minimization could be done, for example, using the R function `optim`. The estimated residual variances can be calculated from Equation (25) , by putting $\hat{\theta}$ in place of θ.

The criterion to be minimized in Equation (26) is a product of squared differences and not a sum, like OLS. Note that this criterion is invariant to changes in units. Changing the units of any of the outputs would simply multiply the criterion by a constant, which would not change the estimated parameter values. This is what we want. We do not want the estimated parameters to change because we express biomass in g/m^2 instead of tons/ha, for example.

Example 11 illustrates the use of concentrated likelihood for a crop model with two different output variables, LAI and aboveground biomass.

Example 11 Concentrated Likelihood Estimation of Model Parameters

Wallach et al. (2011) used data consisting of LAI and aboveground biomass measurements of maize, in six different fields, to estimate two parameters of the STICS crop model (Brisson et al., 1998). There were overall $n_1 = 57$ LAI measurements and $n_2 = 17$ biomass measurements.

The estimated parameters were the maximum rate of LAI increase (dlaimax) and the maximum radiation use efficiency during the vegetative stage (efcroiveg). Equation (26) was used. The estimated parameter values are shown below.

Parameter	Estimated Value
dlaimax	0.00132
efcroiveg	3.38

3.3.3 Assumption 3: All Residual Errors Are Independent

Assumption 3 is that all residual errors are independent. That is, knowing residual error for one measurement gives no information about residual errors for other measurements.

Violations of this assumption occur if there is some dependence between residual errors. This occurs when sampling is not random, in particular, when the same individual is sampled several times. For example, the data might be biomass measured several times during the course of the season in each of several fields. The fields might be chosen at random, but once the field is chosen, then multiple measurements are done in the same field. The particulars of that field could then influence all of the measurements, which would lead to some relation between residual errors within the same field.

Non-independent residual errors show up on residual plots in that repeated measurements on the same individual have related residuals. Example 12 illustrates this. Seed weights measured in three different years are presented. All the residuals except the first for the year 1986 are positive, and all those for 1988 are negative. This type of similarity between different residuals for the same field is a typical symptom of non-independence of residual error.

Example 12 Grain Weight in Three Different Years

The data here are grain weights of the spring wheat cultivar Neepawa in three different years (Darroch and Baker, 1990). The parameters of the logistic model

$$seed.weight = \frac{W}{1 + e^{B - C \cdot DD}}$$

were estimated by OLS using all the data. The left-hand panel of Figure 6.9 shows the fit of the model to the data, the right-hand panel shows the residuals.

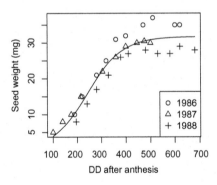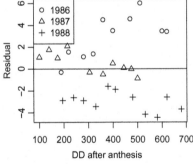

FIGURE 6.9 Grain weight data for three years and logistic model fitted using OLS (left), residuals (right).

It can be seen that residuals for the same year tend to have similarities; for example, all are negative for 1988. This is an indication that Assumption 3 is violated.

What if one ignores correlations and estimates the parameters using OLS? If Assumption 1 is satisfied, the OLS estimator is then still asymptotically unbiased and consistent, as for violations of Assumption 2. That is, asymptotically the expectation of the parameter estimators will still be equal to the true parameter values, and the estimators will tend to be closer and closer to the true values as the sample size increases.

There are, however, some important disadvantages associated with ignoring non-independence in the residual errors. Firstly, there will be more variability (between possible samples) in the parameter estimators than if we had independent residual errors. In addition, we are then ignoring the fact that the actual amount of information in the data is less than with independent residual errors. That is, the effective sample size is smaller than the number of data points (see Chapter 2). As a result, estimates of the standard error of the parameter estimators which ignore non-independence will be unrealistically small.

Below we present ways of treating the situation of non-independent residuals.

3.3.3.1 The Variance-Covariance Matrix

The variance-covariance matrix of residuals is central to consideration of non-independence of residuals. The variance-covariance matrix of the residuals, noted Σ, is a square matrix of size n by n, where n is the number of measurements. This matrix has the variances of the residual errors on the diagonal and the covariances in the off-diagonal positions.

When Assumptions 2 and 3 of Table 6.1 are satisfied, Σ has the form

$$\Sigma = \begin{pmatrix} \sigma^2 & 0 & \cdots & 0 & 0 \\ 0 & \sigma^2 & \cdots & 0 & 0 \\ \vdots & \vdots & \ddots & \vdots & \vdots \\ 0 & 0 & \cdots & \sigma^2 & 0 \\ 0 & 0 & \cdots & 0 & \sigma^2 \end{pmatrix} \tag{27}$$

There are no off-diagonal terms because all residuals are independent and therefore all the covariances are zero. All the elements on the diagonal are identical because we are assuming equal variance for every residual.

Now consider a situation like that in Example 12, where there are multiple measurements in each of several fields. If the fields are chosen at random, then residual errors for measurements in different fields are independent. However, residual errors for different measurements in the same field will not be independent. Then the variance-covariance matrix Σ has the form

$$\Sigma = \begin{pmatrix} \Sigma_1 & 0 & \cdots & 0 & 0 \\ 0 & \Sigma_2 & \cdots & 0 & 0 \\ \vdots & \vdots & \ddots & \vdots & \vdots \\ 0 & 0 & \cdots & \Sigma_{m-1} & 0 \\ 0 & 0 & \cdots & 0 & \Sigma_m \end{pmatrix} \qquad \Sigma_i = \begin{pmatrix} \sigma^2_{(i)1,1} & \cdots & \sigma_{(i)1,n_i} \\ \vdots & \ddots & \vdots \\ \sigma_{(i)n_i,1} & \cdots & \sigma^2_{(i),n_i,n_i} \end{pmatrix}$$

$$\tag{28}$$

The variance-covariance matrix is now made up of blocks (left-hand matrix), where each block corresponds to one of the m fields sampled. The covariances between blocks are 0, because residual errors in different fields are assumed to be independent. Each block, however, (right hand matrix) has non-zero covariances.

In Example 12, the variance-covariance matrix has three blocks, corresponding to the three years with measurements. The blocks for 1986 and 1987 (10 measurements in each year) would have size 10 × 10, and the block for 1988 (11 measurements) would have size 11 × 11. The covariance σ_{ij} in a block is the covariance between the residual errors for the ith and j^{th} measurement times for that block. Since $\sigma_{ij} = \sigma_{ji}$, the number of different covariances in a block with n measurements is n*(n − 1)/2.

A major problem is the fact that the full variance-covariance matrix often has a large number of different covariances. In Example 12, the blocks for 1986, 1987, and 1988 have respectively 45, 45, and 55 covariances. Since the measurement times are different in the different years, all of these covariances can be different. For example, σ_{12} for 1986 is the covariance between residual errors for measurements at times 189 and 220 DD after anthesis, σ_{12} for 1987 is the covariance between residual errors for measurements at times 100 and 140 DD after anthesis, and σ_{12} for 1988 is the covariance between residual errors for measurements at times 195 and 240 DD after anthesis. In general these will all be different values. Thus for all three years there is a total of 145 covariances.

3.3.3.2 Generalized Least Squares

Suppose that one knows or can estimate the variance-covariance matrix Σ. If we have multiple measurements at each value of x, those values could be used to estimate the variances and covariances. For example, if in all three fields in Example 12 the measurements were made at the same DD values, then we could estimate the variance-covariance matrix using those values.

Given Σ, one can transform the measured and simulated values so that the transformed variables are independent. The first step is to express the variance-covariance matrix as $\Sigma = R^T R$, where R is an upper triangular matrix. (This is known as the Cholesky decomposition and is done by the R function chol.) Let Y be the vector of all measurements, so $Y - (y_1, \ldots, y_n)$, let F be the corresponding vector of simulated values, so $F = (f(x_1; \theta), \ldots, f(x_n; \theta))$, and E the vector of residuals, so $E = (\varepsilon_1, \ldots, \varepsilon_n)$. At the second step, the measured and simulated values are transformed as follows:

$$Y^* = (R^T)^{-1} Y$$
$$F^*(\theta) = (R^T)^{-1} F(\theta)$$

(29)

The transformed residuals are $E^* = Y^* - F^*(\theta) = (R^T)^{-1}E$. The variance-covariance matrix of the transformed residuals is

$$\mathrm{var}(E^*) = \mathrm{var}((R^T)^{-1}E) = (R^T)^{-1}\Sigma R^{-1} = (R^T)^{-1}R^T R R^{-1} = I$$

where I is the identity matrix (the diagonal elements are all 1, the off-diagonal elements are all 0). Thus, after transformation, the variance-covariance matrix of the residuals has the form of Equation (27) (with $\sigma^2 = 1$). This means that we can now use OLS on the transformed data.

The SSE criterion to be minimized is

$$\begin{aligned}
SSE &= [Y^* - f^*(x;\theta)]^T[Y^* - f^*(x;\theta)] \\
&= [Y - f(x;\theta)]^T R^{-1}(R^T)^{-1}[Y - f(x;\theta)] \qquad (30) \\
&= [Y - f(x;\theta)]^T \Sigma^{-1}[Y - f(x;\theta)]
\end{aligned}$$

The first line of Equation (30) says that we can use OLS on the transformed measurements and simulated values. The last line of Equation (30) says that, equivalently, we can work with the original measurements and simulated values, but then the SSE to be minimized includes the inverse of the variance-covariance matrix. Minimizing this last expression is called generalized least squares.

3.3.3.3 Determinant Criterion

Suppose that the blocks in the variance-covariance matrix are all identical. For example, suppose that we have measurements in multiple fields, and all measurements are done at the same times. This could, for example, be measurements of LAI at certain times, biomass at other times, and yield at harvest. Then all fields would have the same variance-covariance matrix. We also assume that residual errors have a normal distribution. (This is a maximum likelihood technique.) In this case we can estimate parameters using the determinant criterion (Bates and Watts, 1988).

We write the measured values and the model values as a matrix. Each column corresponds to a single individual (for example, a single field). Thus

$$Y = \begin{pmatrix} y_{1,1} & \cdots & y_{N,1} \\ \vdots & \ddots & \vdots \\ y_{1,M} & \cdots & y_{N,M} \end{pmatrix} \quad F = \begin{pmatrix} f(x_{1,1};\theta) & \cdots & f(x_{N,1};\theta) \\ \vdots & \ddots & \vdots \\ f(x_{1,M};\theta) & \cdots & f(x_{N,M};\theta) \end{pmatrix} \quad Z = Y - F$$

where M is the number of measurements per individual and N is the number of individuals. Z is the matrix of residuals. In this case the likelihood can be shown to be

$$L(Y|\theta, \Sigma) \propto \frac{1}{|\Sigma|^{N/2}} e^{tr[(Z^T \Sigma^{-1} Z]/2} \qquad (31)$$

where Σ is the variance-covariance matrix for each individual (the same for all individuals), the vertical bars indicate a determinant and the trace (tr) is the sum of the diagonal elements of a matrix.

This likelihood can be maximized in two steps. First, one solves for the value of Σ that maximizes the likelihood for given θ. The result is

$$\hat{\Sigma}(\theta) = \frac{Z^T Z}{N}$$

Then one substitutes that into Equation (31). The resulting expression is

$$L(Y; \theta, \hat{\Sigma}(\theta)) = k' - (N/2)\ln|Z^T Z| \tag{32}$$

where k' is a constant. Thus, the maximum likelihood estimator of θ is the value of θ that minimizes the determinant $|Z^T Z|$.

This result is remarkably general. It applies whatever the structure of the variance-covariance matrix, as long as it is the same for all individuals. The other requirement is that the number of individuals must be greater than M (the number of measurements per individual) and greater than p (the number of parameters).

3.3.3.4 Random Parameter Models

Up to now, we have concentrated on describing and then estimating the elements of the variance-covariance matrix. An alternative approach is to assume that the model parameters are random variables. They have fixed values for a given individual, but can vary between individuals. Given the specific parameters for an individual, the residual errors for that individual are assumed to be independent. In this case, covariances between residual errors for different measurements on the same individual arise because measurements on the same individual share the same parameter values. One can use the R function `nlme` to estimate the parameters for random parameter models.

As a simple example, we return to the data in Example 12. We assume that the residual errors are all independent but that in the logistic model the parameter W is a random variable. The values of W for different individuals are assumed to be independent. However, all measurements on the same individual have the same value of W. The covariance between two measurements on the same individual is then

$$\text{cov}(y_{ij}, y_{ij'}) = \text{cov}\left(\frac{W_i}{1 + e^{B - C^* DD_{ij}}} + \varepsilon_{ij}, \frac{W_i}{[1 + e^{B - C^* DD_{ij'}}]} + \varepsilon_{ij'}\right)$$

$$= \frac{\text{var}(W)}{[1 + e^{B - C^* DD_{ij}}][1 + e^{B - C^* DD_{ij'}}]}$$

where the subscript ij refers to the jth measurement on individual i. The random variables here are W, ε_{ij}, and $\varepsilon_{ij'}$. The residual errors have been assumed to be mutually independent and independent of W, so all the covariances involving the residual errors are zero. However, there remains the term cov(W,W) = var(W), so cov($y_{ij}, y_{ij'}$) is not zero. Thus, the fact of having a random parameter introduces covariances between different measurements for the same individual.

Random parameter models are often an appealing way to treat non-independence of residuals. First, the assumption that differences between individuals arise from differences in their parameter values can often seem quite reasonable. In addition, random parameter models can be a parsimonious way of describing non-independence, if the model has relatively few parameters. In Example 12, even if we treat all three parameters as random, there are only six variances and covariances to estimate, namely var(W), var(B), var(C), cov(W,B), cov(W,C), and cov(B,C). This can be compared to the 145 covariances that are needed if we make no assumptions about the structure of the variance-covariance matrix.

An example of a random parameter model is presented in Example 13. The model here is a static model that calculates N uptake, yield, grain protein content, and residual soil N at harvest. It was found that the random parameter approach led to substantially larger gross margins than using OLS. Here, a more realistic treatment of the structure of error has practical consequences, in terms of farmer profit.

Example 13 Nitrogen Response Model with Random Parameters

In Makowski and Wallach (2002) the following equations were used to model N uptake (N_{up}) as a function of applied N (N_{app}) and end of winter soil mineral N (N_{winter}), yield (*Yld*) as a function of N uptake and applied N, grain protein content (*Prot*) as a function of N uptake and yield, and finally, residual soil nitrogen at harvest (*Res*) as a function of applied N.

$$N_{up} = \begin{cases} CN_{app} + a_1 + a_2 N_{winter} & N_{app} < N_{max} \\ BY_{max} + \dfrac{C(N_{app} - N_{max})}{1 + T(N_{app} - N_{max})} & N_{app} \geq N_{max} \end{cases}$$

$$Yld = \begin{cases} Y_{max} + A(N_{up} - BY_{max}) & N_{app} < N_{max} \\ Y_{max} & N_{app} \geq N_{max} \end{cases}$$

$$Prot = P_1 + P_2 \frac{N_{up}}{Yld}$$

$$Res = \begin{cases} R_{min} & N_{app} < N_{max} \\ R_{min} + R(N_{app} - N_{max}) & N_{app} \geq N_{max} \end{cases}$$

The model has 11 parameters. All residual errors were assumed to be independent. Parameter estimation was done in two ways. In the first approach, all

parameters were treated as fixed. Estimation then concerned the 11 fixed parameters and four residual variances (for N uptake, yield, grain protein content,
and residual soil N). In the second approach, the nine parameters a_1, C, Y_{max},
A, B, P_1, P_2, R_{min}, and R were treated as random parameters. Estimation then
concerned the expectations and variances of these nine random parameters,
seven covariances between them (the remaining covariances were assumed to
be 0), the values of the two fixed parameters and the four residual variances.
Parameter estimation for the random parameter model used the S-PLUS function
nlme, which is analogous to the R function of the same name.

The parameter vector obtained in each approach was used to calculate optimal fertilizer rates for each field in the data set. Then observed yields were used
to estimate the gross profit (yield*price − amount of N*cost) that would have
been earned for each field, if the calculated optimal N dose had been applied. It
was found that the random parameter model led to substantially higher gross
margins than treating the parameters as fixed.

3.4 How Many Parameters to Estimate?

A major decision in model calibration is the number of parameters to estimate.
In Chapter 9, we show that estimating too many parameters can result in a
model that fits the data well but is a very poor predictor. It is thus essential to
avoid overparameterization. The methods we propose are generally applicable
to the problem of model choice. Here we apply them to the specific problem of
choosing between models with different numbers of estimated parameters.

The first two criteria assume that the parameters have been ordered, so
that we have already decided which will be the first parameter to estimate,
the second, etc. The only question is where to stop; how many parameters
should be estimated? These methods assume that the parameters are estimated using maximum likelihood. The third method uses cross validation to
determine the number of parameters to estimate. This approach can be used
with any method of parameter estimation.

3.4.1 Likelihood Ratio Test

Suppose that we have decided on an order for the parameters to be esti-
mated, noted $\theta_1, \ldots, \theta_k, \ldots$. That is, if only one parameter is estimated it will
be the parameter noted θ_1, if two are estimated they will be θ_1 and θ_2, etc.
At each stage of the process of deciding how many parameters to estimate,
we must choose between estimating k parameters or k + 1 parameters. If we
choose k, then the process stops. We estimate the parameters $\theta_1, \ldots, \theta_k$. If
we choose k + 1, then we continue and next compare k + 1 and k + 2
parameters.

At each stage, we are comparing two different models, namely the model where $\theta_1, \ldots, \theta_k$ are fit to the data and all other parameter values are fixed, and the model where $\theta_1, \ldots, \theta_k, \theta_{k+1}$ are fit to the data and the other parameter values are fixed. The first model is a special case of the second, where the parameter θ_{k+1} has been fixed to some value. In this case we say that the models are 'nested'.

To compare the two models we can use the likelihood ratio test (Seber and Wild, 1989). The likelihood ratio is

$$LR = -2\left[\log(L(Y|\hat{\theta}_{k,ML})) - \log(L(Y|\hat{\theta}_{k+1,ML}))\right] \qquad (33)$$

where $\log(L(Y|\hat{\theta}_{k,ML})$ and $\log(L(Y|\hat{\theta}_{k+1,ML}))$ are, respectively, the natural logarithms of the likelihood for the model where k or k + 1 parameters are estimated. The hypothesis we test is whether θ_{k+1} has the fixed value assumed by the simpler model. (We assume that all the other assumptions required by ML are correct.) If the hypothesis is verified, then LR has approximately a chi-squared distribution with $\Delta p = 1$ degrees of freedom (since the more complex model has one more estimated parameter than the simpler model). We reject the hypothesis with a type 1 error probability of 5% if $LR > \chi^2_{0.95,\Delta p}$, where $\chi^2_{0.95,\Delta p}$ is the 95th percentile of a chi-squared distribution with Δp degrees of freedom.

The likelihood ratio test can also be used to test if it is worthwhile to use a more complex model for residual variance, since this also involves adding more estimated parameters. Example 14 illustrates the use of the likelihood ratio test in this case. In this example, the models differ only in that the more complex model has an extra parameter to describe how residual variance varies with the expected value. The conclusion is that the more complex model should be used.

Example 14 Likelihood Ratio Test for Extra Residual Variance Parameter

Example 9 and Example 10 both involve fitting a logistic model, and both use the same data. In Example 9 it is assumed that residual variance is constant $(\text{var}(\varepsilon_i) = \sigma^2)$, while in Example 10 it is assumed that residual variance is proportional to some power of the expected response $(\text{var}(\varepsilon_i) = \sigma^2 f(x;\theta)^\lambda)$. In both cases the estimators are the ML estimators (using the fact that OLS is equivalent to ML). The constant variance model is nested within the other, since it can be obtained by setting $\lambda = 0$. The difference in the number of estimated parameters between the two models is 1.

Table 6.5 shows the likelihood ratio for results for each of nine different environments (Robert et al., 1999). The likelihood ratio values were compared to $\chi^2_{0.95,1} = 3.84$. For all environments the likelihood ratio is greater than this

TABLE 6.5 Likelihood Ratio for Testing for an Additional Parameter to Describe Residual Variance

Environment	Likelihood Ratio
1	90.42
2	170.36
3	66.69
4	133.27
5	89.42
6	166.41
7	125.36
8	72.85
9	92.51

value, and so in each case the values are significant at the 5% level. We therefore reject the simpler model with $\lambda = 0$ and prefer the model where λ is estimated.

3.4.2 Akaike Information Criterion and Bayesian Information Criterion

A popular criterion of model choice, which can be used for nested or non-nested models, is the Akaike Information Criterion (AIC) (Burnham and Anderson, 1998). AIC for a model is

$$AIC = -2\log L(Y|\hat{\theta}_{ML}) + 2p$$

where $\log L(Y|\hat{\theta}_{ML}, \hat{\sigma}^2_{ML})$ is the natural logarithm of the likelihood, evaluated using the maximum likelihood parameter values, and p is the number of estimated parameters (including the parameters to describe residual variance). The first term in the equation measures how well the model fits the data (the better the fit, the smaller this term). The second term is a penalization that increases as the number of parameters increases. To choose between models one calculates AIC for each; the best model has the smallest value of AIC.

For a random sample where residual error has a normal distribution, AIC can be written in terms of squared errors. The likelihood in this case is

$$L(Y|\hat{\theta}_{ML}, \hat{\sigma}^2_{ML}) = \frac{1}{\sqrt{2\pi\hat{\sigma}^2_{ML}}^n} e^{\frac{-\sum\limits_{i=1}^{n}(y_i-\hat{y}_i)^2}{2\hat{\sigma}^2_{ML}}} \tag{34}$$

where n is the number of observations, y_i are the observed values, $\hat{y}_i$ are the simulated values and $\hat{\sigma}^2_{ML}$ is the maximum likelihood estimator of the residual variance. Since

$$\hat{\sigma}^2_{ML} = (1/n)\sum_{i=1}^{n}(y_i-\hat{y}_i)^2$$

we have

$$AIC = n\log(2\pi) + n\log\left[\sum_{i=1}^{n}(y_i-\hat{y}_i)^2/n\right] + n + 2p \tag{35}$$

There is a version of AIC which is corrected for small samples. It is

$$AIC_C = -2\log L(Y|\hat{\theta}_{ML}) + 2p + \frac{2p(p+1)}{n-p-1} \tag{36}$$

The correction factor becomes important for n/p < 40. In fact, one can always use the corrected form; for larger sample sizes it will be very close to the uncorrected form.

Another criterion for comparing models that is quite widely used is the Bayesian Information Criterion (BIC) defined as

$$BIC = -2\log(L(Y|\hat{\theta}_{ML})) + p\log n \tag{37}$$

The difference between this criterion and AIC is in the penalization term. In general log(n) > 2, and so the penalization in BIC is greater than in AIC. BIC then will favor smaller models than AIC.

3.4.3 Cross Validation

The goal in model choice is to choose the model that gives the best predictions. It therefore seems logical to estimate MSEP for each model, and to choose the model with the smallest estimated value of MSEP. A general approach for estimating MSEP is cross validation (see Chapter 9). The approach here then is to use cross validation to estimate MSEP for different numbers of estimated parameters, and to fix the number of estimated parameters at the number that minimizes estimated MSEP (Linhart and Zucchini, 1986).

This approach is shown in Example 15. The choice is between estimating different numbers of parameters for a crop model, the remaining parameters being fixed at their default values. The parameters to estimate are chosen using a stepwise procedure; at each step, the next parameter is that which leads to the greatest reduction in squared error. This gives the order of parameters to estimate. Cross validation is used to decide how many parameters to estimate. In this example, minimum estimated MSEP is attained for three estimated parameters.

Example 15 Using Cross Validation to Choose Between Models

Wallach et al. (2001) used cross validation to determine the number of parameters to estimate for calibration of a maize crop model. Parameters to estimate were chosen by forward regression. The results are shown in Table 6.6. MSE_{comb} and $MSEP_{comb}$ refer, respectively, to mean squared error and mean

TABLE 6.6 Prediction Error of a Maize Crop Model for Different Numbers of Estimated Parameters

Number of Adjusted Parameters	0	1	2	3	4
Adjusted parameters	–	p2logi	p2logi, r2hi	p2logi, r2hi, himax	p2logi, r2hi, himax, p2evap
Adjusted values	–	0.0085	0.0085, 0.0048	0.0087, 0.0030, 0.5044	0.0086, 0.0355, 0.4992, 0.5193
$\sqrt{MSE_{comb}}$	0.72	9.15	8.01	7.68	7.49
$\sqrt{MSEP_{comb}}$	0.72	9.55	8.39	8.20	8.34

squared error of prediction (estimated by cross validation) for a weighted combination of yield, biomass, and LAI.

The best model (smallest estimated value of MSEP) is that with three estimated parameters. Note that MSE_{comb} on the other hand continues to decrease even for four parameters. This illustrates the fact that MSE should not be used for model choice.

3.4.4 You Cannot Estimate Unidentifiable Parameters

Regardless of the method for choosing the parameters to estimate, it is necessary to be sure that all of the parameters are identifiable (i.e., that they can be estimated from the available data). To be identifiable a parameter must have an effect on the response variables in the data, and must not appear in a group of parameters which could be replaced by a single parameter. Example 16 presents an illustration of unidentifiable parameters.

Example 16 An Illustration of Unidentifiable Parameters

For this example, we will use a sub-model of the AZODYN crop model (Jeuffroy and Recous, 1999). This sub-model includes three output variables, namely above-ground winter wheat dry matter (dm, kg.ha^{-1}), leaf area index (lai, m^2.m^{-2}), and nitrogen uptake (nu, kg.ha^{-1}). These variables are simulated daily starting at the end of winter until flowering. It is assumed that there is no nitrogen stress, water stress, pests, or diseases.

Dry matter is calculated as follows:

$$B(d) = B(d - 1) + EBMAX \times f(T(d - 1))$$
$$\times EIMAX[1 - \exp(-K \times LAI(d - 1))] \times C \times RAD(d - 1)$$

where $B(d)$ and $B(d - 1)$ represent biomass on days d and $d-1$, respectively, $f(T(d - 1))$ is a function of temperature on day $d - 1$, $RAD(d - 1)$ is the global radiation on day $d - 1$ (MJ.ha^{-1}) and C, $EBMAX$ (kg.MJ^{-1}), $EIMAX$, and K are four parameters. Before flowering, $LAI(d - 1)$ is calculated as a function of the critical nitrogen uptake level on day $d - 1$ ($CNU(d - 1)$, kg.ha^{-1}) as follows:

$$LAI(d - 1) = D \times CNU(d - 1)$$

where D is an additional parameter. Consequently, $B(d)$ is related to $CNU(d - 1)$ by

$$B(d) = B(d - 1) +$$
$$EBMAX \times f(T(d - 1)) \times EIMAX[1 - \exp(-K \times D \times CNU(d - 1))]$$
$$\times C \times RAD(d - 1)$$

This last equation shows that, when only dry matter measurements are available, it is not possible to simultaneously estimate K and D and it is also impossible to simultaneously estimate $EBMAX$, C, and $EIMAX$. Only the products $EBMAX \times C \times EIMAX$ and $K \times D$ can be estimated because simulated dry matter depends only on these two products. An infinite number of sets of parameter values gives identical values of $EBMAX \times C \times EIMAX$ and $K \times D$: for example, we obtain $K \times D = 0.02016$ with $K = 0.72$ and $D = 0.028$ but also with $K = 0.6$ and $D = 0.0336$. Only two parameters can be estimated from biomass measurements, one parameter among ($EBMAX$, C, $EIMAX$) and one parameter among (K, D). When both dry matter and LAI measurements are available, it is possible to estimate one additional parameter. Since D has an influence on LAI, it is possible to simultaneously estimate K and D if biomass and LAI measurements are available. However, it is still impossible to estimate $EBMAX$, C, and $EIMAX$. It is necessary to fix two of the three; then one can estimate the remaining parameter.

4. APPLICATION OF STATISTICAL PRINCIPLES TO SYSTEM MODELS

We return now to the theoretical and practical questions of model calibration. Based on the statistical theory of parameter estimation presented above, we can now propose at least partial answers to many of those questions.

4.1 The Fundamental Questions

Does calibration improve predictive accuracy? Do the parameter values estimated by calibration tend toward the true values of the parameters as sample
size increases?

We have argued that complex system models, such as crop models, are
probably misspecified, even if the individual process models are correctly specified. That is, these models do not have the correct dependence on the explanatory variables. A major reason is that most of the system model parameters are
fixed at values from the literature, and those values are only approximate.
Metselaar (1999) did a detailed study of the information in the literature concerning parameter values. He found quite a large variability between different
sources; the average coefficient of variation for a given parameter was found to
be 38%. That is, one does not really know the exact values of the parameters,
and the uncertainty in their values based on the literature can be quite large.
Other studies which have considered uncertainty in crop model parameters have
reported similarly large uncertainties (for example, Confalonieri et al., 2009).

The estimated parameter values have no meaning for a misspecified
model, even in the limit of very large amounts of data. In particular, the
parameters do not tend toward the true parameter values for the individual
process models. The estimated parameter values are simply correction factors,
that compensate for the multiple errors throughout the model (parameters
fixed at incorrect values, possibly errors in the forms of the equations, mixing
of residual error with explanatory variables).

However, even for a misspecified model, estimating the parameters by
OLS still makes the model tend asymptotically toward the model that minimizes MSEP (assuming some other assumptions are satisfied). That is, even
if the model does not have exactly the correct dependence on the explanatory
variables, calibration using least squares should, in general, tend to improve
prediction accuracy.

We have also seen that for a misspecified model, the parameters that
minimize prediction error depend on the range of conditions of the sample,
even if the true model is the same for the different conditions. This suggests
that model calibration is not only useful for improving predictions, but is
necessary. When using a model under new conditions, one needs parameter
values adapted to those conditions.

4.2 Practical Questions About Model Calibration

4.2.1 What Data Should Be Used for Calibration?

Suppose that we are primarily interested in some specific response, say final
yield. Suppose further that we have data on other responses such as LAI, biomass, soil water, plant and soil N, etc. Should we fit the model to all the
data, or only to the data for the response of primary interest?

A common assumption is that a system model can be improved by improving prediction of all responses. The implicit assumption is that the closer the model is to the true behavior of the system, the better.

In fact, the situation is not so clear. It is true that for a correctly specified model, more data is always better. If several different outputs all depend on the same parameters, then it is best to use data from all the responses to estimate those parameters. However, this is not necessarily the case for a misspecified model. In that case, the parameter values that minimize prediction error for one response are not the same as the parameter values that minimize prediction error for another response. As a result, fitting the model to one response can degrade the predictive quality for a different response. Thus if major interest is in one particular response (e.g., yield), it might be best to fit the model only to that response and not to other responses (e.g., in-season LAI). Thus, accepting that complex system models are misspecified, we must accept at least the possibility that fitting the model to other variables may worsen prediction of the response of primary interest.

There have only been a few studies which have compared model calibration using different responses. Example 17 presents one such study. In this case, it was found best to fit the model only to the responses of major interest, and not to other responses. However, this is only one study, and the results no doubt depend on the details of the way calibration was performed. This question of which data to fit should probably be considered an open question.

Example 17 Sequential Parameter Estimation for the STICS Model. A Comparison of Three Different Sequences

The STICS model (Brisson et al., 1998) was calibrated, using data on multiple outputs (Guillaume et al., 2011). The data came from 373 different fields. They were split, with approximately half being used for parameter estimation and the rest to estimate MSEP. The outputs of major interest were yield and grain N concentration. Three different calibration procedures were compared. The first had 12 steps (method I12). At each step only a few (usually one) response variables were used, and a relatively small number of parameters were adjusted to that response, using a least squares criterion. In the first ten steps, data for responses other than yield and grain N concentration were fit. The last two steps minimized the SSE for grain yield and grain N concentration, respectively. The second procedure had three steps (method I3). In the first step, the goodness-of-fit criterion involved soil N and soil water, aboveground plant biomass, LAI, and aboveground plant N. The last two steps involved yield and grain N concentration, as in I12. The last procedure had only two steps (method I2), namely minimization of the goodness-of-fit criterion for yield and grain N concentration. When two or more response variables were fit simultaneously, the concentrated likelihood method was used.

The results for yield and grain N concentration are shown in Table 6.7. RMSE is root mean squared error, calculated using the same data used for calibration. RMSEP is root mean squared error of prediction, estimated using the data that were not used for calibration. The smallest errors of prediction were obtained using scheme I2. The conclusion in this case is that it is best to use only the responses of major interest for calibration.

TABLE 6.7 Results of Parameter Estimation for the STICS Model Using Three Different Calibration Schemes

Calibration Scheme	Yield		Grain N Concentration	
	RMSE	RMSEP	RMSE	RMSEP
Before calibration	2.01	2.33	4.20	3.6
I12 (use all responses, many stages)	0.85	1.13	3.4	5.1
I3 (use all responses, few stages)	0.85	1.26	2.6	3.8
I2 (use only yield and grain N concentration)	0.74	1.06	2.6	3.6

4.3.1 What Goodness-of-Fit Criterion Should Be Used for Model Fitting

Consider first the case where the model is fit to a single type of response. Suppose also that the assumptions of Table 6.1 are satisfied. Then we would use the OLS criterion for model fitting (minimize the sum of squared errors).

If examination of the residuals after OLS indicates that residual variance does not seem to be constant, then we could use weighted least squares (Equation (21)) or a power transformation (Equation (22)) to achieve constant variance. If residual variance seems to be proportional to the square of the predicted value, then we could do a log transformation of the measured and simulated values (a special case of the power transformation).

Consider next the case with multiple responses. Suppose that for each response, we have constant residual variance (perhaps after some transformation). Suppose again that the assumptions of Table 6.1 are satisfied, except that different responses have different residual variances. Suppose, in addition, that the residual errors have a normal distribution. In this case we can use the concentrated likelihood approach to estimate the parameters (Equation (26)).

The approaches proposed above are expected to give good results when the underlying assumptions of Table 6.1 are satisfied. However, we have argued that these assumptions are probably not satisfied for complex system models. In particular, the assumption that the system model is correctly specified is probably not satisfied. In addition, there will often be non-independence of residuals. Also, the assumption that residual errors have a normal distribution, which we have used in deriving the ML methods, is often violated.

What then is the justification for basing calibration on the statistical results? It is true that we no longer have any of the optimality properties that depend on the assumptions. Nevertheless, the statistical approaches still seem appealing because they have properties that seem useful. For a single response, variable, it seems logical to give less weight to errors associated with a larger residual variance as weighted least squares and the power transformation do, because such residuals provide less information about the response of the system. For multiple responses, we certainly want a criterion that is invariant to changes in units, as is the case for the concentrated likelihood.

We have also proposed ways of dealing with the non-independence of residuals, including generalized least squares, the determinant criterion, and random parameter models. Once again, these solutions should probably be used if possible. However, these solutions are not always applicable, and often one simply ignores the non-independence of the residuals. If so, one should be aware of the problems this entails. In particular, the actual amount of information in the data is less for non-independent residuals than if the residuals are independent. Also, if some individuals have many more measurements than others, then one is giving greater weight to those individuals than is warranted for non-independent residuals.

4.3.2 Simultaneous or Sequential Parameter Estimation?

If we decide to use several different responses for model calibration, there is the question of whether to combine all the responses in a single criterion (as in the concentrated likelihood criterion) or to fit the responses sequentially one at a time. For example, one could first fit a few parameters to LAI, then other parameters to biomass, etc.

An advantage of the sequential procedure is its simplicity. One can use a simple sum of squared errors as the goodness-of-fit criterion. Also, one fits few parameters at each stage, which helps avoid numerical problems. However, the sequential procedure has an important disadvantage. The problem is that the parameters that are fit to one response often also affect other responses. Thus, in the sequential procedure, fitting some response can degrade the fit to responses fit previously. This was in fact the case in the study reported in Example 17. In the procedure with multiple steps, it was found that in many cases one step degraded the fit to responses treated in previous steps.

It seems preferable to estimate all parameters simultaneously, using all the responses that one wants to take into account. We have already seen that one can use the concentrated likelihood criterion to do so.

There is a specific case which merits mention, namely estimation of the parameters related to phenology in a crop model (for example, degree days to flowering and to maturity). It is often argued that the first stage of calibration should estimate just those parameters.

Often phenology only depends on daily temperature and perhaps day length, but not on other variables calculated in the system model (for example, if stresses do not change the timing of phenological events). In this case the estimated phenology parameters will not depend on the values of the other parameters in the model. We can estimate the phenology parameters using just the process equation or equations for phenology. The parameters will be estimated using data on the timing of phenological events as outputs and data for daily temperature and day length as inputs. In this case, estimation of the phenology parameters is not really part of the calibration process as we have defined it. It is rather parameter estimation for one of the process models in the overall model. It should indeed be done as a separate preliminary step.

4.3.3 Which Parameters Should Be Estimated?

The choice of which parameters to estimate can have an important influence on the results of model calibration. There are actually two questions here; in what order should parameters be considered for estimation, and then how many parameters should one estimate?

In practice, one often simply decides on a list of parameters to estimate, without considering separately the question of how many parameters to estimate. We will argue that the decision as to the number of parameters is extremely important, and should be based on statistical considerations. Therefore we treat separately the choice of parameters that are candidates for estimation and the choice of the number of parameters to estimate.

We consider first the choice of which parameters to select for estimation. Various approaches have been proposed.

One approach, perhaps the most common, is to choose the parameters that seem most important in determining the measured responses. This can be based on expert knowledge of the equations of the model. A justification for this approach is that it is more efficient to adjust a few important parameters rather than many minor parameters.

Another method for selecting parameters is to perform a sensitivity analysis. The principle is to calculate a sensitivity index for each parameter (see Chapter 5) and to select parameters with the highest sensitivity index values. This method allows modelers to identify the parameters that have a strong influence on the output variables of a model. An application is presented by Harmon and Challenor (1997), where they defined a sensitivity index which was used to identify the important parameters in an oceanic ecosystem model.

By choosing parameters with high sensitivity, one is insuring that the responses will vary quite a bit as the parameter values vary. However, there is no assurance that varying those parameters will result in a good fit of the model to the data. Also, one should be careful not to choose parameters whose estimators are highly correlated.

A third approach specific to crop models is to define a list of varietal parameters, supposedly specific to each variety, and to estimate those parameters. The idea here is that the parameters common to all varieties are relatively well known, but for each new variety one needs to estimate the parameters that are specific to that variety. However, this separation of parameters into two groups, one with well-known values and the other with unknown values, is probably an over-simplification. Many parameters that are treated as common to all varieties are in fact only poorly known (and are perhaps variety-specific) while some parameters that are treated as variety-specific could be estimated from similar varieties.

A final approach is to proceed as in forward regression. At the first stage, the entire list of potential candidate parameters is tested, estimating one parameter at a time. The parameter that leads to the smallest SSE is chosen as $\theta_{(1)}$, the highest priority parameter. Then each remaining candidate is tested in conjunction with $\theta_{(1)}$. The parameter that leads to the smallest SSE is $\theta_{(2)}$, etc. This approach must then be coupled with a method for deciding when to stop entering parameters (see Example 15).

The advantage of this method is that it chooses the parameters that are most useful for reducing MSE. It also avoids estimating parameters with highly correlated estimators. One disadvantage of this method is computer time. In a model with p parameters, choosing $\theta_{(1)}$ requires p estimation runs, then choosing $\theta_{(2)}$ requires an additional p − 1 runs, etc. Another drawback is the difficulty of estimating MSEP for a model calibrated in this way. For example, cross validation (see Chapter 9) becomes very computationally intensive.

4.3.4 How Many Parameters to Estimate?

Once one has chosen the parameters which are candidates for estimation, one must decide on the number of parameters to estimate. It is well known that estimating too many parameters relative to the amount of data leads to overparameterization, where predictive quality is worse than if fewer parameters had been estimated (see Chapter 9).

Common practice is to simply decide on the number of parameters to estimate, without any procedure for determining whether that number is justified or not. The danger of overparameterization is probably very important.

To decide on the number of parameters to estimate one can use a likeli-hood ratio test (Equation (33)), AIC (Equation (36)), BIC (Equation (37)), or a cross validation procedure.

Even though the underlying assumptions for these procedures (except cross validation) are probably not satisfied, they should at least provide some protection against overparameterization. In fact, this may be the area of model calibration where it is most important to base the approach on results in statistics.

4.3.5 How Should Prior Information Be Used?

In system models, there is information about most or perhaps all parameter values based on studies of the individual processes. We have referred to this as information from the literature. It can also be termed 'prior' information (prior to using data on the overall system for calibration).

Prior information about parameter values is typically used in two ways in model calibration. First, most of the parameters in the system model are fixed at the values given by the prior information. Secondly, for the remain-ing parameters, those that are adjusted during calibration, one often puts upper and lower limits on the parameter values, based on 'reasonable' ranges as indicated by the prior information. Those parameters are not allowed to take values outside that range. This seems like a reasonable procedure. Each of the estimated parameters in fact has two roles. On the one hand, it is a correction factor that compensates for errors throughout the model. On the other hand, it enters in some specific way in the model, and determines how some input variable affects a model response. For example, the radiation use efficiency parameter determines how intercepted solar radiation affects bio-mass increase. If a parameter is allowed to take highly unrealistic values, the model will give highly unrealistic predictions as to the effect of the associ-ated input variable.

Prior information is thus either totally accepted (the parameter is set to the value provided by the prior information), or is totally ignored, except for the upper and lower bounds (the parameter value is re-estimated using only system data). Bayesian parameter estimation (see Chapter 7) is an alternative approach which makes different use of the prior information. In this approach, parameter estimation involves finding a compromise between the parameter values indicated by the prior information and the values that give the best fit to the system data.

4.3 Adding New Parameters for Calibration

Up to now we have considered the situation where calibration involves esti-mating some of the parameters of the system model. However, we also argued that the estimated values are just empirical correction factors, and should not be considered as approximations to the true parameter values in the process equations.

That being the case, it might be of interest to take a radically different approach to parameter estimation. Here we do not alter the parameter values based on prior information; we add simple new parameters that can be used to improve the fit of the model to the data.

A very simple example, for yield prediction, would be

$$f^*(x; \theta, a, b) = a + bf(x; \theta) \tag{38}$$

where $f(x; \theta)$ is yield prediction by the uncalibrated system model, a and b are the new parameters to be estimated, and $f^*(x; \theta, a, b)$ is yield prediction by the modified model. Thus, the new parameters a and b simply enter as additive and multiplicative constants, respectively.

If OLS is used to estimate the parameters a and b, then the mean of the predictions of $f^*(x; \theta, a, b)$ will be equal to the mean of the measured values (bias will be 0). Thus this simple approach allows us to eliminate bias. In some cases, this may be as good as estimating multiple parameters in the model. For example, Guillaume et al. (2011), in their sequential parameter estimation study (Example 17), found that most of the reduction in SSE that resulted from calibration was due to reduction of bias.

The linear model in Equation (38) uses yield simulated by the uncalibrated model as the unique explanatory variable. A different possibility would be to include other simulated outputs as explanatory variables. Example 18 illustrates this procedure in the case of a model for predicting sunflower yield and oil content.

Example 18 Using System Model Outputs as Inputs to a Linear Equation

The SUNFLO model (Casadebaig et al., 2011) uses a linear model approach to predict harvest index and oil content at harvest in sunflower. There is a system model that predicts crop dynamics (noted here $\text{SUNFLO}^{\text{system}}$), and then the outputs of that model are used as inputs to a linear model. The linear model for oil content for instance is

Oil content $= \theta_0 + \theta_1 *$ (sum of intercepted radiation after flowering)
$+ \theta_2 *$ (temperature sum after flowering) $+ \theta_3 *$ (stress during maturity)
$+ \theta_4 *$ (leaf area duration) $+ \theta_5 *$ (nitrogen nutrition at flowering)
$+ \theta_6 *$ (thermal date of flowering) $+ \theta_7 *$ (light extinction coefficient)
$+ \theta_8 *$ (potential oil content)

The first six explanatory variables are values simulated by $\text{SUNFLO}^{\text{system}}$. The last two input variables (light extinction coefficient and potential oil content) are taken from field experiments for the genotype in question. The parameters $\theta_0 - \theta_8$ are estimated by OLS.

Example 19 describes a study where a similar approach (use of uncali-brated simulated values in a linear regression model) was used to estimate regional yield.

Example 19 Use of the CROPGRO-soybean Model for Assessing Climate Impacts on Regional Soybean Yields

In this study (Irmak et al., 2005) 17 years of data were used for parameter esti-mation in each of eight regions, and 7 years of data were used to evaluate model performance. Overall, 10 different approaches to regional yield forecast-ing were compared. The worst approach was to use the uncalibrated model. If the uncalibrated model was multiplied by a constant that was fit to the data for each region, estimated MSEP was greatly reduced. If, in addition, two of the model parameters were estimated for each region using the data, MSEP was fur-ther reduced, but very slightly compared to calibration with a single parameter per region. The best approach did not involve the crop model at all.

The study shows the usefulness of a simple linear correction to predicted yield (Table 6.8), although other approaches may be even better.

TABLE 6.8 Prediction Error of Different Models for Estimating Regional Yield

Predictor	$RMSEP = \sqrt{MSEP}$
Mean of historical yields	318
Uncalibrated model	1487
$\hat{Y}_{ij} = b_j f(X_{ij}; \theta)^1$	353
$\hat{Y}_{ij} = a_j f(X_{ij}; \theta, C\hat{N}_j, D\hat{U}L_j)^2$	327

[1] $\hat{Y}_{ij}$ is predicted yield for region j, year i. $f(X_{ij}; \theta)$ is yield simulated by the uncalibrated CROPGRO-soybean model. The parameter b_j was estimated from the data for region j.
[2] $\hat{Y}_{ij}$ is predicted yield for region j, year i. $f(X_{ij}; \theta, C\hat{N}_j, D\hat{U}L_j)$ is yield simulated by the CROPGRO-soybean model with default values for all parameters except curve number (CN) and drained upper limit (DUL). The parameters CN_j, DUL_j and a_j were estimated from the data for region j.

---

5. ALGORITHMS FOR OLS

The advantage of having software with built-in functions for parameter esti-mation, like R, is that one doesn't need to program an algorithm that will actually do the parameter estimation. As long as the algorithms work well, there is no need to know how the algorithms operate. But when something goes wrong, it is useful to have some basic understanding of how the algo-rithms operate.

We consider here two commonly used algorithms: the Gauss-Newton algorithm (which is the default algorithm for the `nls` function in R) and the Nelder-Mead or simplex algorithm (which is the default algorithm for the `optim` function in R). These two algorithms operate very differently. The Gauss-Newton algorithm is specifically built to find the parameter values that minimize a sum of squared errors. The Nelder-Mead algorithm on the other hand can be used to find the parameters that minimize any function.

5.1 Gauss-Newton Algorithm

We first present the OLS solution for a linear model (i.e., a model where the parameters enter linearly). A linear model can be written $Y = X\theta$ in matrix notation, where Y is the vector of n measured values, θ is the vector of p parameters, and X is the matrix of input variables. In terms of individual elements

$$\begin{pmatrix} y_1 \\ \vdots \\ y_n \end{pmatrix} = \begin{pmatrix} x_{1,1} & \cdots & x_{1,p} \\ \vdots & \ddots & \vdots \\ x_{n,1} & \cdots & x_{n,p} \end{pmatrix} \begin{pmatrix} \theta_1 \\ \vdots \\ \theta_p \end{pmatrix} \tag{39}$$

where $x_{i,j}$ is the value of input variable j for individual i. For the linear model there is an analytical matrix expression for the least squares parameters:

$$\hat{\theta}_{OLS} = (X^T X)^{-1} X^T Y \tag{40}$$

The basic Gauss-Newton algorithm approximates a model that is non-linear in the parameters by a linear model and then uses Equation (40) to find new parameter values. Since the model is not really linear, the resulting parameters are not the exact solution. Therefore, one repeats the procedure starting at the new parameter values. The procedure is iterated until the changes in the parameter values become very small. In more detail, the algorithm works as follows:

a. Define an initial estimation for θ, noted $\hat{\theta}^{(i)}$ with $i = 0$.

b. Use a Taylor series expansion to derive a linear approximation to the nonlinear model. The approximation is:

$$f(x; \theta) \approx f\left(x; \hat{\theta}^{(i)}\right) + \sum_{j=1}^{p} \frac{\partial f(x; \theta)}{\partial \theta_j}\bigg|_{\hat{\theta}^{(i)}} \left(\theta_j - \hat{\theta}_j^{(i)}\right)$$

where $\frac{\partial f(x;\theta)}{\partial \theta_j}\big|_{\hat{\theta}^{(i)}}$ is the derivative of $f(x; \theta)$ with respect to the jth parameter, evaluated at $\hat{\theta}^{(i)}$.

We now have an equation like Equation (39), with the replacements

$$\begin{pmatrix} y_1 - f(x_1; \hat{\theta}^{(i)}) \\ \vdots \\ y_n - f(x_n; \hat{\theta}^{(i)}) \end{pmatrix} \underset{replaces}{\longrightarrow} \begin{pmatrix} y_1 \\ \vdots \\ y_n \end{pmatrix}$$

$$\begin{pmatrix} \dfrac{\partial f(x_1; \theta)}{\partial \theta_1}\Big|_{\hat{\theta}^{(i)}} & \cdots & \dfrac{\partial f(x_1; 0)}{\partial \theta_p}\Big|_{\hat{\theta}^{(i)}} \\ \vdots & \ddots & \vdots \\ \dfrac{\partial f(x_1; \theta)}{\partial \theta_1}\Big|_{\hat{\theta}^{(i)}} & \cdots & \dfrac{\partial f(x_1; 0)}{\partial \theta_p}\Big|_{\hat{\theta}^{(i)}} \end{pmatrix} \underset{replaces}{\longrightarrow} \begin{pmatrix} x_{1,1} & \cdots & x_{1,p} \\ \vdots & \ddots & \vdots \\ x_{n,1} & \cdots & x_{n,p} \end{pmatrix}$$

$$\begin{pmatrix} \theta_1 - \hat{\theta}_1^{(i)} \\ \vdots \\ \theta_p - \hat{\theta}_p^{(i)} \end{pmatrix} \underset{replaces}{\longrightarrow} \begin{pmatrix} \theta_1 \\ \vdots \\ \theta_p \end{pmatrix}$$

c. Use Equation 40 with the above replacements to calculate the ordinary least squares estimate of $\theta - \hat{\theta}^{(i)}$. This gives the next approximation, $\hat{\theta}^{(i+1)} = \hat{\theta}^{(i)} + (\theta - \hat{\theta}^{(i)})$.

d. Repeat steps b–c until convergence is achieved (i.e., until there is very little change in the parameters or in the SSE).

The behavior of this algorithm is illustrated in Example 20, using the logistic model and data of Example 2. Note that the first iteration leads to values which are much worse (higher MSE) than the initial values. Both parameters after the first iteration are negative. In this particular case the model can still be evaluated with negative parameter values, and the algorithm goes on to find the OLS solution. With other models, negative parameter values might cause an error. To avoid this problem, it is important to start with good initial estimates of the parameter values. Often, one can roughly estimate some or all of the parameters directly from the data. It is worthwhile doing so. If the algorithm doesn't converge, one can try different starting values.

Example 20 Path of Algorithm Used by *nls*

Consider the data and model of Example 2. In order to work in only two dimensions (much easier to visualize), here we fix the parameter W = 28.171259, which is its OLS value, and estimate the two remaining parameters of the logistic curve, B and C, using the Gauss-Newton algorithm of the R function nls.

Figure 6.10 shows a contour plot of MSE, and the trajectory of the `nls` iterations. The starting point is indicated. The first iteration leads to a value
(B = −1.698, C = −0.014) which is in fact much worse (higher SSE) than the
starting value. The algorithm than reduces the step size, then reduces it again to
find a smaller SSE value than the initial value. The algorithm then changes direction and goes quite directly to the OLS solution. Overall, eight iterations were
required.

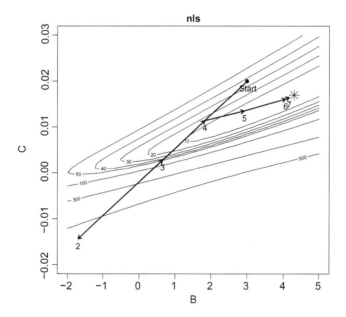

FIGURE 6.10   Path of Gauss-Newton algorithm for estimation of two parameters of logistic model.
The numbers indicate the number of the iteration. The final value is indicated by a star.

5.2 Nelder-Mead algorithm

The Nelder-Mead algorithm (also called the simplex algorithm) does not use
derivatives and is considered more robust than the Gauss-Newton algorithm.
It may converge when the Gauss-Newton algorithm doesn't. This is the
default algorithm of the R function `optim`.

As with the Gauss-Newton algorithm, the user must give initial estimates
of the parameter values. This algorithm then begins by creating a simplex,
which is a geometric figure with p + 1 vertices in p-dimensional space,
where p is the number of parameters. For two parameters this is just a triangle. For three parameters it is a tetrahedron, etc.

The algorithm evaluates the SSE at each vertex of the simplex. Then the vertex with the highest SSE (the worst point) is reflected through the centroid of the other points (reflection). If this new point is better than the current best point, the algorithm elongates the distance of reflection to go further (expansion). If the new point is no better than the second worst point, then the distance of reflection is reduced (contraction). If no new point better than the second worst point is found, the original simplex is shrunk toward the best point (shrinkage).

Example 21 Steps in the Execution of the Nelder-Mead Algorithm

The first steps of the Nelder-Mead algorithm applied to the same model and data as Example 20 are shown in Figure 6.11. Once again we estimate two parameters, B and C, of the logistic function Equation (4). The initial point given by the user is indicated by 'start'. First (upper left panel), the simplex (here a triangle) is created by the algorithm. Then (upper right panel) the worst point (highest SSE) is reflected through the center of the line connecting the other two points (reflection). This new point is worse (higher SSE) than the other two points, so it is not accepted. The distance of reflection is reduced until an SSE value lower than the second worst point is found (lower left panel). The new point is used to create a new triangle, where the previously second worst point is now the worst (lower right panel).

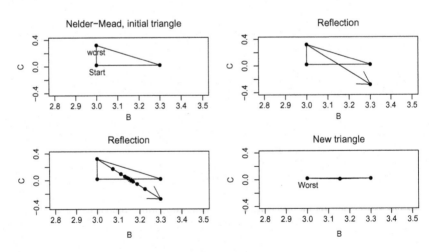

FIGURE 6.11   Steps in the Nelder-Mead algorithm.

5.3 Comparison of Algorithms

Example 22 shows a comparison of the Gauss-Newton and Nelder-Mead algorithms as implemented in R. Here all three parameters of the logistic model were estimated, using the data of Example 2. The final parameter values calculated by the two algorithms are very similar. Nelder-Mead requires many more model evaluations than Gauss-Newton, which is fairly typical behavior.

Example 22 Comparison of the Gauss-Newton and Nelder-Mead Algorithms as Implemented in R

The three parameters of the logistic model were estimated using the data of Example 2. The estimation was done using the R function `nls` with its default algorithm (Gauss-Newton), and with the R function `optim` using its default algorithm (Nelder-Mead). The `nls` function announced 'Number of iterations to convergence: 8', but this doesn't count the function evaluations needed to estimate the derivatives. Each iteration requires one evaluation to evaluate SSE plus three evaluations to estimate the derivatives with respect to the three parameters W, B, and C. (In addition there are a few extra function evaluations.) Both algorithms converged to essentially the same point, but Nelder-Mead required almost eight times as many evaluations of the model.

Method	W	B	C	Number of Evaluations of the Model
Gauss-Newton (nls)	28.1713	4.16022	0.0164920	37
Nelder-Mean (optim)	28.1712	4.15961	0.0164895	285

There are many variants of the Gauss-Newton and simplex algorithms, and also many other types of algorithms, some of which are available with `nls` or `optim` as options. For example, an option with `optim` is to use a simulated annealing algorithm, which is a global optimization algorithm (Goffe et al., 1994). Global algorithms are less sensitive to initial values than Gauss-Newton type algorithms and are more likely to converge to the global optimal solution. An important drawback of these algorithms is that they often require a long calculation time. An application of simulated annealing to a crop model is described by Mavromatis et al. (2001).

5.4 Potential Problems

There are two different types of problems that arise with these algorithms. The first is that the algorithm may fail to converge. A commonly encountered error message for nls is of the form

'Error in nls(y ~ W/(1 + exp(B − C * x)), obs, list(W = 15, B = 1, C = 0.001)) : singular gradient'.

The 'singular gradient' message means that one of the derivatives is zero, and so the algorithm can't continue. This may be because the parameters are unidentifiable, or because the starting values are poor and lead the algorithm to evaluate the model at parameter values that create errors. Nelder-Mead may also fail to converge if the model is evaluated at a point that causes an error.

The second problem is that the algorithm may seem to converge, but the result is not the true minimum value of the function, but rather a local minimum. This is more pernicious, since one might not even suspect that there is a problem. A protection here is to start the algorithm at several different initial parameter values.

It is important to keep the ratio of the number of estimated parameters to the number of observations relatively low. Many numeric problems result from trying to estimate numerous parameters simultaneously from a small number of data. This is in addition to the problem of overparameterization which we have already emphasized.

6. R FUNCTIONS FOR PARAMETER ESTIMATION

We discuss here two R functions that can be used for parameter estimation. The first is nls (for nonlinear least squares), which is specifically designed to find parameter values that minimize a sum of squared errors between observed values and values simulated using a nonlinear model. The second is optim, which is an R function that can be used to find the parameter values that minimize any function.

6.1 The nls Function When the Model Consists of a Single Equation

The simplest general form of the nls function is the following:

$$nls(formula, data, start)$$

where formula is a formula expression that relates the measurements to the model, data gives the data (measured values and explanatory variables of the model) and start gives the initial guesses for the parameter values. The

following example calculates the OLS parameters for the grain weight data
and logistic model of Example 2.

```
obs <- c(8,13,17,21,26,27,28,27,27,29,28)
DD <- c(195,240,280,330,380,410,480,530,580,620,680)
data <- data.frame(y = obs,x = DD)
start <- list(W = 29,B = 4.2,C = 0.017)
logistic <- nls(y~W/(1 + exp(B−C*x)),data,start)
```

The first three lines store the measured grain weight values in the vector
obs, the values of the explanatory variable DD in the vector DD, and then create a data frame with two columns; the values of obs are in a column named
y and the values DD in a column named DD. The names of the columns are
important. They must be the same names that appear in formula. The fourth
line creates a list called start, with initial guesses of the value of each
parameter. Here again the names must be the same as the names that appear
in formula.

The last line is the call to the nls function. The first argument, the formula, says that 'y is modeled as W/(1 + exp(B − C * x))'. The tilde indicates that this is a formula. The R objects that appear in this expression are
y, W, B, C, and x. Every one of these objects must appear by name, and be
assigned a value, in either the second or third argument. If the name appears
in the third argument (the object start in our example) it is treated as a
parameter to be estimated. In our case the parameters are W, B, and C. The
remaining names in formula are y and DD; they are defined in the second
argument (the object data in our example).

In fact, you do not really have to define every variable that isn't a parameter in data. If objects used in the formula are not found in data, then R will
search for these objects among the objects that have been defined previously.
However, it is good programming practice to include all the variables in
data, to ensure that R is using the values you want.

In the above program, the result of executing nls is stored in the object
logistic. The basic results of nls can be printed by simply putting logistic on a line by itself. The result is as follows, where the values given for
W, B, and C are their estimated values:

```
logistic #print the results of nls
Nonlinear regression model
model: y~W/(1 + exp(B−C * x))
data: data
W B C
28.17126 4.16022 0.01649
residual sum−of−squares: 5.596
Number of iterations to convergence: 5
Achieved convergence tolerance: 3.181e−06
```

One can obtain more detailed information by using the instruction
`summary(logistic)`:

```
summary(logistic,correlation=T)
Formula: y~W/(1+exp(B-C*x))
Parameters:
Estimate Std. Error t value Pr(>|t|)
W 28.171259 0.408285 69.00 2.17e-12 ***
B 4.160219 0.360084 11.55 2.86e-06 ***
C 0.016492 0.001415 11.65 2.68e-06 ***
---
Signif. codes: 0 '***' 0.001 '**' 0.01 '*' 0.05 '.' 0.1 ' ' 1
Residual standard error: 0.8364 on 8 degrees of freedom
Correlation of Parameter Estimates:
W B
B -0.43
C -0.53 0.98
Number of iterations to convergence: 5
Achieved convergence tolerance: 3.181e-06
```

Here we have as additional information the standard deviations of the parameter estimators and the statistical significance of the parameters (are they significantly different than 0 or not, usually not a very useful question for non-linear models). We have asked for the correlations between parameters ('correlation = T' in the summary function) and so we also have the correlations. The correlation between the estimators of B and C is quite high (0.98).

6.2 Recovering and Analyzing the nls Results

The result of executing the nls function is an object (an nls object, logically enough) that contains the results of the estimation. There are specialized instructions for recovering or analyzing the results of nls.

The function coef operates on nls objects and returns the estimated values of the parameters.

```
# get coefficients
coef(logistic)
W B C
28.17125873 4.16021931 0.01649193
```

The function fitted operates on nls objects and returns the values calculated using the calibrated model. Below we use this function to obtain a graph of observed versus predicted values, to which we add the 1:1 line (Figure 6.12).

```
fitted(logistic)
[1] 7.888914 12.666716 17.252816 22.052749 25.116203 26.226212
27.527684 27.885441 28.045235 28.105961 28.146948
# graph observed versus predicted values and add 1:1 line
```

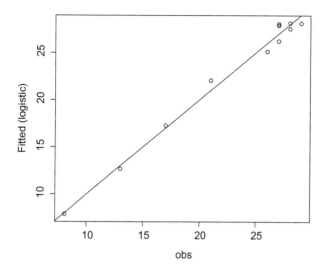

FIGURE 6.12   Observed grain weight values versus simulated values, using the `fitted` function.

```
plot(DD,fitted(logistic))
abline(0,1)
```

The function `residuals` returns the differences between the observed and calculated values. Below we plot residuals versus predicted values, and add the line at 0. We also calculate MSE (Figure 6.13).

```
# graph residuals
dev.new()
plot(DD,residuals(logistic), xlab = "growing DD since anthesis",
ylab = "residulas")
abline(0,0)
```

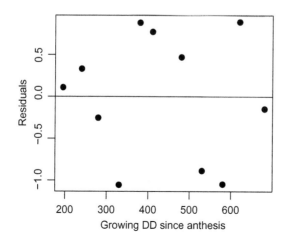

FIGURE 6.13   Residuals of logistic model fitted to grain weight data.

```
MSE <-mean(residuals(grain.fill)^2)
MSE
[1] 0.5087239
```

The function `predict` uses the regression formula, with the estimated parameters, to evaluate the model for new values of the explanatory variables. The first argument is the name of the `nls` object. The second argument is a data frame or list with the values of the new explanatory variables. In the following code, we use the calibrated model to predict for each integral value of DD from 195 to 680. Then we plot the observed values and the predicted values (Figure 6.14).

```
# plot observed values and predicted response for each integer value
of DD
new.DD <-195:680
new.pred <- predict(logistic,list(x = new.DD))
plot(DD,obs, xlab = "growing DD since anthesis",ylab = "seed weight")
lines(new.DD,new.pred)
```

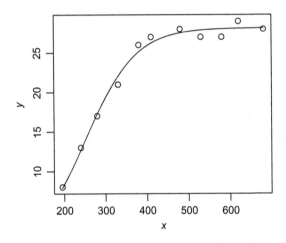

FIGURE 6.14 Measured grain weight values and curve of values simulated by logistic model.

The function AIC operates on an `nls` object and calculates the value of AIC for the model with the estimated parameters.

```
# AIC for logistic model with p = 4 parameters (W, B, C and residual
variance)
AIC(logistic)
[1] 31.7823
```

6.3 `nls` **for More Complex Models**

Often one wants to estimate parameters for more complex models that do not have the form of a single equation. This is obviously the case for complex system models. The structure of the `nls` instruction is then the same as for a one equation model, but in place of the explicit formula one has a call to the function that evaluates the model. That function must calculate and return a vector of simulated values that corresponds to the vector of measured values.

To illustrate this, we use the same example as above, which only has a single equation, the logistic equation, but we put it into a function. The program below first has the function that evaluates the model and then the main program with the `nls` instruction.

```
# function to calculate grain weight using a logistic equation
# x is a vector, so pred is a vector of simulated values
logistic.nls <- function(x,W,B,C) {
pred <- W/(1 + exp(B−C*x)) # the vector of predicted values
return(pred)
}
# main program
obs <- c(8,13,17,21,26,27,28,27,27,29,28)
DD <- c(195,240,280,330,380,410,480,530,580,620,680)
data <- data.frame(y = obs, x = DD)
start <- list(W = 29, B = 4.2, C = 0.017)
logistic <- nls(y ~ logistic.nls(x,W,B,C), data, start)
```

In the first argument of the `nls` function, the formula says that the model is evaluated in the function `logistic.nls`, which has four arguments. Therefore the definition of `logistic.nls` must also show four arguments. The arguments are passed by position (the first argument in the call from `nls` is used as the first argument in `logistic.nls`, etc). The names do not need to be identical in the call from `nls` and in the function definition, though in this case they are. As in the simple case, all of the object names that appear in the formula (in our case y, x, W, B, and C) must be assigned values either in the second or third arguments to `nls`.

6.4 `nls` with a Wrapper Function

Suppose that the model has already been programmed as a function in R. Suppose further that the model function simulates for just a single context (soil, year, management), and for that context it outputs daily values of all of the state variables. This is typical of crop models. The measured data on the other hand typically come from several different contexts, and for each context there are only data for selected days and certain output variables. Thus the model function is not adapted to `nls`, which requires that the model function calculates a vector that is directly comparable to the vector of measurements.

One possibility would be to modify the model function. But in general, we want to keep the model function unchanged, since it will also be used for other purposes than parameter estimation. The solution here is to write a second function that is called by the `nls` function and, in turn, calls the model function. This new function has three main responsibilities. First, at each iteration (each new parameter vector sent by `nls`) it must fix the values of the model parameters, those that are kept fixed and those that are being varied. Secondly, it must call the model function several times, once for each context for which there are data. Thirdly, it must separate out from the results of the model function just the simulated values that correspond to the measured values, and return those values. This intermediary function is a 'wrapper' function, in the sense that it is covering the model function so that the `nls` instruction doesn't directly see it.

In the following example, we estimate the parameters for the simple maize model. Often for system models we only want to estimate a subset of all the parameters. Here we will estimate just two of the parameters, namely LAImax and alpha.

The data are LAI values at 60 days after emergence for 10 different fields in 2005. They are stored in the vector `LAI60.obs`. The input variables are daily weather (there are 15 different weather variables). Weather data are stored in a 3 dimensional array (days by climate variables by site) called `weather`. The model parameter values are stored in the vector `param`. These are the fixed values for those parameters that are not being estimated, and the starting guesses for the parameters that are being estimated.

```
LAI60.obs
[1] 6.087119 5.191289 3.618078 3.536665 4.709734 7.004336 3.648694
6.958858
[9] 6.219985 8.472625
weather[1:5,,1]
[,1] [,2] [,3] [,4] [,5] [,6] [,7] [,8] [,9] [,10] [,11] [,12] [,13]
[,14] [,15]
[1,] 10 43.4 −0.55 2005 1 6.5 12.6 4.3 0 2.2 5.1 7.3 85.6 145 0.67
[2,] 10 43.4 −0.55 2005 2 3.2 12.3 5.1 0 3.3 5.1 8.0 81.6 145 0.98
[3,] 10 43.4 −0.55 2005 3 7.5 9.2 1.4 0 1.8 1.0 4.0 80.8 145 0.56
[4,] 10 43.4 −0.55 2005 4 8.2 7.4 0.1 0 2.4 -0.5 2.7 79.3 145 0.61[5,]
10 43.4 −0.55 2005 5 7.6 6.6 0.6 0 2.1 -0.4 2.5 80.6 145 0.5
param
Tbase RUE K alpha LAImax TTM TTL
7.00e+00 1.85e+00 7.00e−01 2.43e-03 7.00e+00 1.20e+03 7.00e+02
```

The wrapper function that calls the model to predict values of LAI60 is shown below. It first sets the values of the parameters to the new values sent from `nls`, for those parameters that are being estimated. Then there is a loop over sites. For each value of *i*, the wrapper prepares the weather data for the ith site, then calls the maize model and extracts from the results the calculated LAI on day 60 after sowing. It returns a vector of predicted values, in the same order as the observed values.

```
# function called by nls to calculate LAI on day 60
maize.wrapper <-function(nsites,param,weather,sdate,ldate,LAImax,alpha)
{
# replace the default values of LAImax and alpha with the current values from nls
param["LAImax"] <-LAImax
param["alpha"] <-alpha
for (i in 1:nsites)
{
# extract weather for current site, and give names to climate variables
weather.site <-data.frame(weather[,c(1,4:8),i])
names(weather.site) <-c("site","year","day","I","Tmax","Tmin")
# Call the maize model for this site-year, and extract LAI on day 60
all.results <-maize.model(param,weather.site,sdate,ldate)
LAI60[i] <-all.results$LAI[all.results$day = =(sdate+60)]
}
# return the vector of predicted values
return(LAI60)
}
```

The following shows the `nls` instruction that calls the wrapper function, and the results.

```
# nls calls the maize.wrapper wrapper function
# lines that start with + are continuations of previous instruction
nls.LAI60 <-nls(LAI60.obs ~maize.wrapper(nsites,param,weather,sdate,ldate,LAImax,alpha),
+ list(nsites = nsites,param = param,weather = weather,sdate = sdate,ldate = ldate),
+ list(LAImax = 7,alpha = 0.0024))
summary(nls.LAI60)
Formula: LAI60.obs ~maize.nls(nsites, param, weather, sdate, ldate, LAImax, alpha)
Parameters:
Estimate Std. Error t value Pr(>|t|)
LAImax 7.9410018 0.7813556 10.163 7.52e-06 ***
alpha 0.0021693 0.0002872 7.554 6.58e-05 ***
- - -
Signif. codes: 0 '***' 0.001 '**' 0.01 '*' 0.05 '.' 0.1 ' ' 1
Residual standard error: 0.8102 on 8 degrees of freedom
Number of iterations to convergence: 7
Achieved convergence tolerance: 2.034e-06
```

6.5 The `optim` Function for Function Minimization

There are cases where the `nls` function is unable to converge to the OLS parameter values. This may be a problem of the starting values, or because the function is ill-behaved. In this case it is worthwhile to attempt to use the `optim` function with a simplex algorithm, which is generally more robust than the Gauss-Newton algorithm of nls. The optim function is a general purpose function minimization algorithm. It is not specifically designed for OLS problems.

The general form of the *optim* function is

optim(initial.values, model.function, , other.arguments.to.model.function,
   method = "Nelder − Mead")

The first argument is a vector of initial values of the parameters to be estimated, the second argument is the name of the function that evaluates the
model and returns the sum of squared errors for the current parameter values,
the third argument is not needed here (hence the two commas), and the next
argument or arguments are passed to model.function. The method argument
specifies the minimization algorithm to be used.

We illustrate using the grain weight data and the logistic model of
Example 2. Below is the function (logistic.optim) that calculates the SSEs
in this case. The first argument of the function must be the vector of values
of the parameters that are to be estimated. At each iteration, optim calls the
function with the current values of those parameters. The function can have
any other arguments. They must be in the same order as in the optim instruction. In the example, the additional arguments are x (growing degree days,
needed to evaluate the logistic model) and y (observed grain weights, needed
to calculate the SSEs).

```
logistic.optim<-function(params,x,y) {
W< params[1]
B<-params[2]
C<-params[3]
pred<-W/(1+exp(B-C*x)) # the vector of predicted values
SS<-sum((y-pred)^2) # the sum of squared errors
return(SS)
}
```

Then the use of optim is as follows:

```
# estimate parameters using optim
params<-c(29,4.2,0.017) # initial guesses of parameter values
logistic<-optim(params,logistic.optim,,DD,obs,method = "Nelder-Mead")
logistic
logistic$par
[1] 28.17115372 4.16033032 0.01649234
$value
[1] 6.696963
$counts
function gradient
204 NA
$convergence
[1] 0
$message
NULL
```

The value of $value is the final best value of the quantity that is minimized, in our case the SSE. $counts gives the number of times the model function was called (204), and $convergence = 0 indicates successful convergence. Note that the parameter values are in the object logistic$par.

The function that evaluates the model and that is called by optim cannot be the same as the function that is called by nls, because the quantities returned are different in the two cases. nls requires that the model function return a vector of predicted values, while optim requires the sum of squared differences. That means, unfortunately, that one cannot simply try nls first and then just replace nls by optim if nls doesn't converge. One must replace the function or the wrapper used with nls by one adapted to optim.

The optim function is not as convenient for OLS parameter estimation as nls. It does not produce information about the accuracy of the parameter estimates as nls does. Furthermore, the summary function, the predict function, and other useful functions that take nls objects as arguments do not accept optim objects as arguments.

There is thus a real advantage to using nls if possible in preference to optim. If, however, nls doesn't converge, and one uses optim, one can still obtain the additional information that nls provides, though with additional effort. One possibility is to use the parameter values estimated by optim as initial guesses for nls. Perhaps with these presumably very good initial guesses nls will converge, and one can take advantage of the information it provides. If not, one can write R code to obtain the additional information. We show below how to calculate several of the quantities of interest.

6.6 Additional Information When optim Is Used

Since the optim function doesn't provide a simple way to obtain all the results of interest after OLS parameter estimation, these must be programmed by the user. Here we show how to calculate the estimator of residual variance, the vector of residuals, and the standard errors and correlations of the parameter estimators.

The example here uses the grain weight data and logistic model of Example 2. We suppose that we have run optim, and that the estimated values of the three parameters of the logistic model are in the object logistic$par.

For these calculations we need to have a function that evaluates the model for all the contexts for which we have data, and returns the simulated values. For our example that means that the function must evaluate the model for each value of DD for which we have data.

```
# function that evaluates the model
logistic.jacobian <-function(params,x)
{
```

```
W <- params[1]
B <- params[2]
C <- params[3]
pred <- W/(1 + exp(B-C*x))
return(pred)
}
```

We can calculate the vector of residuals and the estimated residual variance and standard deviation as follows:

```
# residuals. logistic is the optim object with results of OLS
# obs is vector of observed grain weights. DD is vector of degree days after anthesis
residualsOptim <- obs - logistic.jacobian(logistic$par,DD)
# Calculate estimated residual variance and residual standard deviation
sig2 <- sum(residualsOptim^2)/(length(DD)-3)
sig2[1]
0.6994954
sig <- sqrt(sig2)
sig
[1] 0.8363584
```

The estimated standard deviations of the parameter estimators, and their correlations, are given by Equation (13). This involves numerical derivatives, for which we use the R package numDeriv. This package has a function called jacobian, which calculates the matrix of derivatives we need. This function has three arguments of interest to us. The first is the name of the function that calculates the model predictions, in our case logistic.jacobian. The second is the vector of parameter values. The derivatives will be calculated with respect to each parameter, at the point represented by the values of the vector. In our case, this second argument is logistic$par, which contains the OLS estimates provided by optim. The third argument (three commas later, since we're skipping two arguments) has the other arguments required by the function logistic.jacobian. In our case, there is only one, namely DD. If there were other additional arguments, they would appear in order, separated by commas. The R program is then

```
library(numDeriv)
# X is a matrix of derivatives. Size n (data points) x p (parameters)
X <- jacobian(logistic.jacobian,logistic$par,,,DD)
# Calculate var-covar matrix Sigma
inverseXtX <- solve(t(X)%*%X)
Sigma <- inverseXtX*(sig2)
# Standard deviations of estimators are square roots of diagonal elements
stdevs <- sqrt(diag(Sigma))
stdevs
[1] 0.408279270 0.360092422 0.001415028
# Correlations
correlations <- ((1/stdevs)%*%t(1/stdevs))*Sigma
```

```
correlations
[,1] [,2] [,3]
[1,] 1.0000000 -0.4311114 -0.5336516
[2,] -0.4311114 1.0000000 0.9771435
[3,] -0.5336516 0.9771435 1.0000000
```

The results here are identical to those furnished by `nls`, as they should be.

EXERCISES

1. Define 'model calibration'.
2. What are the two ways of estimating parameters for a system model?
3. Consider the cotton model (Model 1 below). Describe two ways of estimating the parameter P2. In each case, what are the explanatory variables and what are the observed variables?
4. Suppose that in the cotton model (Model 1 below) each individual equation is correctly specified. Consider in particular the equation for ΔNM for $TSQ + AM < t \leq TF$. What exactly does the fact that this equation is correctly specified mean?
5. Suppose that in the sugar cane model (Model 2 below) the observed yields are noted y_i, $i = 1, \ldots, 6$, and the corresponding model predictions are noted $f(x_i; \alpha, \kappa, x0)$. What is the quantity to be minimized for OLS parameter estimation? Assuming that the assumptions of Table 6.1 are satisfied and that the residuals have a normal distribution, what quantity does the maximum likelihood estimator minimize?
6. In the cotton model (Model 1 below) suppose that it is decided to estimate only two parameters for calibration, namely P2 and P5; all the other parameters are fixed at the values in Table 6.9. It is found that calibration substantially improves the fit of the model to the data. Can you assume that the estimated values of P2 and P5 are better estimates of the true values of those parameters than the values in Table 6.9? Explain.
7. In the cotton model (Model 1 below), suppose that there are two sets of data available from two different environments. Suppose that the model is calibrated using data from the first environment. The data set from the first environment is very extensive, so that the standard errors of the parameter estimators are very small (the parameters are very well estimated). Can you then assume that the calibrated model should also perform well in the second environment? Explain.
8. Consider the graph of Figure 6.15 for residuals of some model, as a function of the values simulated by the model. Which of the assumptions underlying OLS (Table 6.1) seem to be satisfied? Which assumptions don't seem to be satisfied? What action would you suggest? Explain.

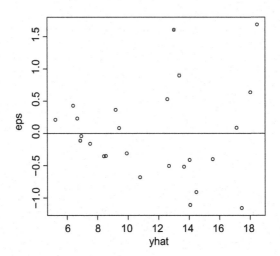

FIGURE 6.15 Graph of residuals of a model versus simulated values.

9. Consider the graph of Figure 6.16 for residuals of a model for four different fields, with 10 measurements per field.
   a. Which of the assumptions underlying OLS (Table 6.1) seem to be
   satisfied? Which assumptions don't seem to be satisfied?
   b. What would be the form of the variance-covariance matrix here?
   c. What action would you suggest to correct for non-respect of the
   assumptions? Explain.
   d. Can you use the determinant criterion here? If not, why not?

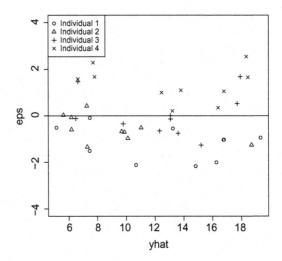

FIGURE 6.16 Residuals for multiple measurements for each of four individuals.

10. In the cotton model (Model 1 below) suppose that there are measurements as shown in Table 6.10. What criterion would you suggest minimizing for estimating model parameters? Give the explicit formula.

11. Suppose that we compare two models, the first with two estimated parameters and the second with three estimated parameters. The SSE for the first model is 27 and for the second it is 21.

    a. If the number of data points is 10, what are the values of AIC for these two models? Which model would you choose?

    b. If the number of data points is 5, what are the values of AIC for these two models? Which model would you choose?

12. In the cotton model (Model 1 below) we decide to only estimate a limited number of parameters; the others will be kept at the values given in the table. How would you decide on the most important parameters to estimate? Justify your answer.

13. Consider the sugar cane model and data (Model 2 below).

    a. Use the nls function to estimate the parameters α, κ, and $x0$. (Reasonable starting values are around $\alpha = 200$, $\kappa = 0.01$, $x0 = -20$.) Use at least two different starting vectors. What are the OLS parameter estimates?

    b. What is the estimated value of residual variance?

    c. Create a graph with simulated values every day and observed values as points.

    d. Create a graph of observed versus simulated values. Add the diagonal line with intercept 0 and slope of 1.

    e. Create a graph of residuals. Add the horizontal line for residuals $= 0$.

14. In the cotton model (Model 1 below), we will estimate the two parameters P2 and P5, keeping all other parameters at their nominal values. The observed data are given in Table 6.3. The R code for the model is obtainable from the ZeBook R package.

    a. Estimate the parameters using the OLS criterion. What are the parameter values? What are the assumptions that would make this a reasonable approach?

    b. Estimate the parameters using the concentrated likelihood criterion. What are the parameter values? What are the assumptions that would make this a reasonable approach?

MODELS FOR EXERCISES

Model 1: Dynamic Model for Numbers of Cotton Fruiting Points

As an example of a system model, we will use a model of fruiting form production in cotton. The fruit buds in cotton are called squares, and the fruits are called bolls. Cotton is an indeterminate plant. Square production and boll growth occur simultaneously. Young squares can be shed naturally, and small bolls can also be shed naturally. The shedding percentages increase with the number of large bolls on the plant, which have priority for photosynthates.

A simple model of square and boll production was proposed by Wallach (1980). The state variables in this model are the number of squares produced (N), the number of squares marked for shedding (NM), and the number of bolls not marked for shedding (NSET). (It is assumed that the transformation that leads to shedding occurs at a certain age for squares and bolls, and that actual shedding occurs a fixed time later.) From these state variables one can calculate the number of squares (NSQUARE), the number of small bolls (bolls that can be shed, NSMALL), the number of large bolls (bolls beyond the age where shedding is determined, NLARGE) and the number of open bolls (NOPEN) on the plant. Those values can be compared to field measurements. The only explanatory variable in the model is physiological days after emergence, noted t, where one physiological day corresponds to 14°D above a threshold of 12°C. The equations below hold before the first bolls open.

The production of squares starts at time t = TSQ and is proportional to t at first. Once set bolls appear, the rate of square production decreases. Square production is 0 if $t \leq TSQ$ or $NSET \geq PMAX$ or $t > TESQ$.

$$\Delta N(t) = P1^*(t - TSQ) \quad TSQ < t \leq TF$$
$$\Delta N(t) = P2^*(1 - NSET(t)/PF) \quad TF < t$$

The first squares are marked at time t = TSQ + AM. Initially a fixed percentage of squares, produced a time AM earlier, is marked for shedding. After the first set bolls appear, the fraction marked for shedding increases as the number of set bolls increases. No squares are marked if $t \leq TSQ + AM$. All squares are marked if $NSET \geq PMAX$.

$$\Delta NM(t) = \Delta N(t-AM)^*P3 \quad TSQ + AM < t \leq TF$$
$$\Delta NM(t) = \Delta N(t-AM)^*P4 \quad TF < t \leq TF + AM$$
$$\Delta NM(t) = \Delta N(t-AM)^* \left[P4 + (1 - P4)(1 - NSET(t)/PF)\right] \quad TF + AM < t$$

The potential number of set bolls is the number of squares produced a time AF earlier, minus those marked for shedding a time AM after being produced. The percentage of those fruiting points that become set bolls depends on the number of set bolls on the plant. No bolls are set for $t \leq TSQ + AF$ or if $NSET \geq PMAX$.

$$\Delta NSET(t) = [\Delta N(t-AF) - \Delta NM(t-AF+AM)]^*P5^*(1 - NSET(t)/PF)$$
$$t > TSQ + AF$$

The observable quantities can be calculated as

$$NSQUARE(t) = N(t) - NFLOWER(t) - NM(t - AW)$$
$$NLARGE(t) = NSET(t - AL)$$
$$NSMALL(t) = NFLOWER(t) - NLARGE(t) -$$
$$[NFLOWER(t - (AF - AL) - AW) - NSET(t - AW)]$$

where

$$NFLOWER(t) = N(t - AFL) - NM(t - AFL + AM)$$

is the cumulative number of flowers up to t.

The model parameters and estimated values are shown in Table 6.9. The model is coded in R in the function `cotton.model` of the ZeBook R package (see Appendix 2).

TABLE 6.9 Values of Cotton Model Parameters, Based on Process Equations

Parameter Name	Value Estimated from Process Equation	Explanation of Parameter
PMAX	85 bolls/m-row	Maximum number of large bolls
P1	0.87 squares/m-row/p.day^2	Rate of square production before first large bolls
P2	16 squares/m-row/p. day	Parameter in equation for rate of square production after first large bolls
P3	0.47	Fraction of squares shed before first large bolls
P4	0.71	Parameter in equation for fraction of squares shed after first large bolls
P5	2.0	Parameter in equation for fraction of small bolls not shed
PF	111 bolls/m-row	Parameter that determines effect of number of large bolls on square production and square shedding
PSF	200 bolls/m-row	Parameter that determines effect of number of large bolls on boll shedding
TSQ,TF,AM,AF, AFL,AW,AL	15.1, 40.9, 8, 25.8, 20.8, 5, 4.3 p. days	Parameters that determine times of events

Model 2: Mitscherlich Equation for Sugar Cane Yield

Table 6.10 has data for sugar cane yield as a function of fertilizer amount from Schabenberger and Pierce (2002, p. 238). They propose to model this data using the Mitscherlich equation:

$$y = \alpha\{1 - \exp[-\kappa(x - x0)]\}$$

TABLE 6.10 Observed Numbers of Fruiting Bodies (Each Measurement in a Different Field)

Type of Fruiting Body	Measurement Time t	Number of Fruiting Bodies
Squares	25	72.68
Squares	50	229.2
Squares	60	95.85
Small bolls	50	36.67
Small bolls	60	54.43
Small bolls	70	15.90
Large bolls	50	3.386
Large bolls	60	44.83
Large bolls	70	89.78

where α is the asymptotic limit for y, and $x0$ is the value of x where $y = 0$. Table 6.11

TABLE 6.11 Sugar Yield from Sugar Cane for Various Amounts of Applied Nitrogen

Nitrogen (kg/ha)	Sugar Cane Yield (Mg/ha)
0	72
25	106
50	133
100	157
150	184
200	191

REFERENCES

Adiku, S.G.K., Ahuja, L.R., Dunn, G.H., Derner, J.D., Andales, A.A., Garcia, L., et al., 2011. Parameterization of the GPFARM-Range model for simulating rangeland productivity. In: Ahuja, L.R., Ma, L. (Eds.), Methods of Introducing System Models Into Agricultural Research. American Society of Agronomy, Madison, pp. 209–228.

Ahuja, L.R., Ma, L. (Eds.) (2011). *Methods of Introducing System Models Into Agricultural Research.*

Bates, D.M., Watts, D.G., 1988. Non linear regression analysis and its applications. Wiley, New York.

Berger, R.L., Casella, G., 1990. Statistical Inference. Wadsworth and Brooks, Pacific Grove, CA.

Box, G.E.P., Cox, D.R., 1964. The analysis of transformations. Journal of the Royal Statistical Society: Series B (Statistical Methodology) 26, 211−252.

Brisson, N., Mary, B., Ripoche, D., Jeuffroy, M.H., Ruget, F., Nicoullaud, B., et al., 1998. STICS: a generic model for simulating crops and their water and nitrogen balances. I. Theory and parameterization applied to wheat and corn. Agronomie 18, 311−346.

Buis, S., Wallach, D., Guillaume, S., Varella, H., Lecharpentier, P., Launay, L., et al., 2011. The STICS crop model and associated software for analysis, parameterization, and evaluation. In: Ahuja, L.R., Ma, L. (Eds.), Methods of Introducing System Models into Agricultural Research. American Society of Agronomy, Madison, pp. 395−426.

Burnham, K.P., Anderson, D.A., 1998. Model Selection and Inference. Springer-Verlag, New York.

Casadebaig, P., Guilioni, L., Lecoeur, J., Christophe, A., Champolivier, L., Debaeke, P., 2011. SUNFLO, a model to simulate genotype-specific performance of the sunflower crop in contrasting environments. Agr. Forest. Meteorol. 151 (2), 163−178. 10.1016/j.agrformet.2010.09.012.

Confalonieri, R., Acutis, M., Bellocchi, G., Donatelli, M., 2009. Multi-metric evaluation of the models WARM, CropSyst, and WOFOST for rice. Ecol. Modell. 220 (11), 1395−1410.

Darroch, B.A., Baker, R.J., 1990. Grain filling in three spring wheat genotypes: statistical analysis. Crop. Sci. 30 (3), 525−529.

Del Grosso, S.J., Parton, W.J., Keough, C.A., Reyes-Fox, M., 2011. Special features of the DayCent modeling package and additional procedures for parameterization, calibration, validation, and applications. In: Ahuja, L.R., Ma, L. (Eds.), Methods of Introducing System Models into Agricultural Research. American Society of Agronomy, Madison, pp. 155−176.

Fath, B., Jorgensen, S.E., 2011. Fundamentals of Ecological Modelling: Applications in Environmental Management and Research, *(4th ed)* Elsevier, Amsterdam.

Goffe, W.L., Ferrier, G.D., Rogers, J., 1994. Global optimization of statistical functions with simulated annealing. J. Econom. 60, 65−99.

Graybill, F.A., 1976. Theory and Application of the Linear Model. Belmont: Wadsworth and Brooks.

Guillaume, S., Berez, J.-E., Wallach, D., Justes, E., 2011. Methodological comparison of calibration procedures for durum wheat parameters in the STICS model. Euro. J. Agron 35, 115−126.

Harrel, F.E.J., 2001. Regression Modeling Strategies. Springer, New York.

Irmak, A., Jones, J.W., Jagtap, S.S., 2005. Evaluation of the CROPGRO-soybean model for assessing climate impacts on regional soybean yields. Transactions of the ASAE 48, 2343−2353.

Jabro, J.D., Hutson, J.L., Jabro, A.D., 2011. Parameterizing LEACHM model for simulating water drainage fluxes and nitrate leaching losses. In: Ahuja, L.A., Ma, L. (Eds.), Methods of Introducing System Models into Agricultural Research. American Society of Agronomy, Madison, pp. 95−116.

Jeuffroy, M.-H., Recous, S., 1999. Azodyn: a simple model for simulating the date of nitrogen deficiency for decision support in wheat fertilization. Eur. J Agron. 10, 129−144.

Jones, J.W., He, J., Boote, K.J., Wilkens, P., Porter, C.H., Hu, Z., 2011. Estimating DSSAT cropping system cultivar-specific parameters using Bayesian techniques. In: Ahuja, L.R.,

Ma, L. (Eds.), Methods of Introducing System Models into Agricultural Research. American Society of Agronomy, Madison, pp. 365–394.

Kersebaum, K.C., 2011. Special features of the HERMES model and additional procedures for parameterization, calibration, validation, and applications. In: Ahuja, L.R., Ma, L. (Eds.), Methods of Introducing System Models into Agricultural Research. American Society of Agronomy, Madison, pp. 65–94.

Linhart, H., Zucchini, W., 1986. Model Selection. Wiley, New York.

Makowski, D., Wallach, D., 2002. It pays to base parameter estimation on a realistic description of model errors. Agronomie 22, 179–189.

Mavromatis, T., Boote, K.J., Jones, J.W., Irmak, A., Shinde, D., Hoogenboom, G., 2001. Developing genetic coefficients for crop simulation models with data from crop performance trials. Crop. Sci. 41, 40–51.

Metselaar, K. (1999, February 2). Auditing Predictive Models: a Case Study in Crop Growth. WAU Dissertation no. 2570.

Nolan, B.T., Malone, R.W., Ma, L., Green, C.T., Fienen, M.N., Jaynes, D.B., 2011. Inverse modeling with RZWQM2 to predict water quality. In: Ahuja, L.R., Ma, L. (Eds.), Methods of Introducing System Models into Agricultural Research. American Society of Agronomy, Madison, pp. 327–364.

Robert, N., Huet, S., Hennequet, C., Bouvier, A., 1999. Methodology for choosing a model for wheat kernel growth. Agronomie 19, 405–417.

Savary, S., Nelson, A., Willoquet, L., Pangga, I., Aunario, J., 2012. Modeling and mapping potential epidemics of rice diseases globally. Crop Prot. 34, 6–17.

Schabenberger, O., Pierce, F.J., 2002. Contemporary Statistical Models for the Plant and Soil Sciences. CRC Press, Boca Raton.

Seber, G.A.F., Wild, C.J., 1989. Nonlinear Regression. Wiley, New York.

Sinclair, T.R., Seligman, N., 2000. Criteria for publishing papers on crop modeling. Field Crops Res. 68, 165–172.

Wallach, D., 1980. An empirical mathematical model of a cotton crop subjected to damage. Field Crops Res. 3, 7–25.

Wallach, D., Buis, S., Lecharpentier, P., Bourges, J., Clastre, P., Launay, M., et al., 2011. A package of parameter estimation methods and implementation for the STICS crop-soil model. Environ. Model. and Softw. 26, 386–394.

Wallach, D., Goffinet, B., Bergez, J.E., Debaeke, P., Leenhardt, D., Aubertot, J.-N., 2001. Parameter estimation for crop models: a new approach and application to a corn model. Agron. J. 93, 757–766.

Wang, X., Kemanian, A., Williams, J., 2011. Special features of the EPIC and APEX modeling package and procedures for parameterization, calibration, validation, and applications. In: Ahuja, L.R., Ma, L. (Eds.), Methods of Introducing System Models into Agricultural Research. American Society of Agronomy, Madison, pp. 177–208.

White, H., 1981. Consequences and detection of misspecified nonlinear regression models. J. Am. Stat. Assoc. 374, 419–433.

Wolf, J., Hessel, R., Boogaard, H., De Wit, A., Akkermans, W., Van Diepen, K., 2011. Modeling winter wheat production across Europe with WOFOST—The effect of two new zonations and two newly calibrated model parameter sets Joost Wolf, Rudi Hessel, Hendrik Boogaard, Allard de Wit, Wies Akkermans, and Kees van Diepen. In: Ahuja, L.R., Ma, L. (Eds.), Methods of Introducing System Models into Agricultural Research. American Society of Agronomy, Madison, pp. 297–326.

## Chapter 7

Parameter Estimation with Bayesian Methods

1. INTRODUCTION

Bayesian methods are attractive when parameters have a biological or physical meaning because they allow modelers to estimate model parameters from two different types of information, namely a sample of data (like the frequentist estimation methods) and prior information about parameter values. In crop models, prior information on parameter values can be obtained from past studies carried out to analyze crop growth and development, from published papers, and from expert knowledge. Dynamic crop models are based on equations that describe the processes involved in crop growth and development, and there is in general information available about these processes. For example, there might be information about the thermal time to flowering or information about maximum rate of root elongation. Bayesian methods combine this type of prior information with experimental data to estimate model parameters.

The result of the application of a Bayesian method is a probability distribution of parameter values. All Bayesian methods proceed in two steps. The first step is to define a parameter probability distribution based on literature or expert knowledge. This distribution is called the *prior parameter distribution* and reflects the initial state of knowledge about parameter values. The prior distribution can be, for example, a uniform distribution with lower and upper bounds derived from expert knowledge or a normal distribution with some mean and variance. The second step consists of calculating a new parameter probability distribution from both the prior distribution and the available data. This new distribution, called the *posterior parameter distribution*, is computed by using Bayes' theorem. The posterior distribution can be used in different ways. Point estimates of parameters can be taken as their expected values or, alternatively, the modes of the posterior distribution. The posterior parameter distribution can also be used for generating the probability distribution of the model outputs, for instance, the distribution of yield or leaf area index (LAI).

Working with Dynamic Crop Models. DOI: http://dx.doi.org/10.1016/B978-0-12-397008-4.00007-1

Bayesian methods are becoming increasingly popular for estimating parameters of complex mathematical models (e.g., Campbell et al., 1999). This is because the Bayesian approach provides a coherent framework for dealing with uncertainty. This is also due to the increase in the speed of computer calculations and the recent development of new algorithms (Malakoff, 1999). The principle is initially a prior probability distribution of the model parameters whose density is noted $P(\theta)$. This prior distribution describes our belief about the parameter values before we observe the set of measurements Y. In practice, $P(\theta)$ is based on past studies, expert knowledge, and literature. The Bayesian methods then tell us how to update this belief about θ using the measurements Y to give the posterior parameter density $P(\theta|Y)$ (density of θ given the data Y). What we now believe about θ is captured in $P(\theta|Y)$.

All the estimation methods described in Chapter 6 are called frequentist methods. With those methods, the parameters θ are fixed, but the parameter estimators are random because they depend on observations. The variances of these estimators can also be computed (for example, the variance of an ordinary least squares estimator) and reflect the variability of the data we might have observed in other samples. In the Bayesian approach, the parameters are defined as random variables and the prior and posterior parameter distributions represent our belief about parameter values before and after data observation. This approach has several advantages:

• parameters can be estimated from different types of information (data, literature, expert knowledge)
• the posterior probability distribution can be used to implement uncertainty analysis methods (see Chapter 5)
• the posterior probability distribution can be used for optimizing decisions in the face of uncertainty.

The purpose of the Bayesian methods presented in this section is to approximate and describe $P(\theta|Y)$. We will see that, in simple special cases, it is possible to derive an analytical expression of the posterior density but that, in most cases, $P(\theta|Y)$ can only be approximated using iterative algorithms that require a large number of model simulations.

This chapter is organized as follows. The ingredients for implementing a Bayesian estimation method are presented in Section 2. It is then shown, in Section 3, how the mode of the posterior distribution can be calculated using optimization procedures similar to those used by frequentist parameter estimation methods. Section 4 describes two important algorithms for estimating posterior parameter distributions, namely importance sampling and Metropolis-Hastings.

2. INGREDIENTS FOR IMPLEMENTING A BAYESIAN ESTIMATION METHOD

2.1 Prior Probability Distribution

2.1.1 Definition

The purpose of a prior parameter probability distribution is to summarize the initial state of knowledge about parameter values before using the data. Different approaches have been proposed to define prior parameter distributions. An important preliminary step is to identify all existing sources of information on parameter values. In many cases, information can be found in published papers. In some situations, possible ranges of parameter values can be defined from basic knowledge of the system process (e.g., a growth rate should be constrained between zero and one). Expert knowledge is also an important source of information and various procedures have been defined for expert elicitation such as the SHELF method developed by Tony O'Hagan and Jeremy Oakley (http://www.tonyohagan.co.uk/shelf/). This method is based on a questionnaire where experts are asked to define plausible ranges and quantiles about parameter values. It is then possible to fit some probability distributions to the ranges of values and quantiles provided by the experts.

In some applications, there is almost nothing known about the parameter values. In such cases, the benefit resulting from the use of a Bayesian method compared to frequentist techniques is not obvious. However, even in such cases, it is possible to estimate parameter values with a Bayesian method if a non-informative prior distribution is used. A non-informative prior distribution aims at giving (almost) equal weight to all possible values. In practice, all prior distributions bring information about parameter values, and it is more realistic to consider these distributions as 'weakly informative' rather than non-informative. A uniform distribution with a large range of values or a Gaussian distribution with large variance is frequently used when almost nothing is known about parameter values. Jeffreys distributions can also be used in this case; they are based on the Fisher information and are invariant when the scale of the parameter is changed (Carlin and Louis, 2008; Gelman et al., 2008).

Before implementing a Bayesian estimation method, it is necessary to define prior distributions for the model parameters, and to define the probability distribution of the model errors (i.e., variances and covariances of the residual error terms if these terms are assumed normally distributed). In some particular cases, these parameters may be known but this is not the case in general, and they then need to be estimated. With frequentist estimation methods, variances and covariances of residual model errors are estimated using various techniques, e.g., by maximum likelihood (see Chapter 6). With Bayesian methods, they are estimated by combining prior distributions and data in the same way as the other parameters. Prior distributions of variances and covariances must be defined with special care because these parameters need to be restricted to some ranges of values. Variances only take positive values and covariances need to be defined in order to lead to correlations between -1 and 1 (the variance-covariance matrix needs to be positive-definite). In practice, prior distributions of variances are defined using uniform, gamma, or Jeffreys probability distributions (Carlin and Louis, 2008; Gelman et al., 2008). When covariances need to be considered (i.e., when the residual model errors cannot be considered as independent), a common practice is to define the prior distribution of the variance-covariance matrix of residual model errors using a Wishart probability distribution (Gelman et al., 2004).

2.1.2 Example 1: Yield Estimation

Suppose we want to estimate crop yield for a given agricultural field. The unknown true yield value is noted θ. We assume that an expert is able to provide information about yield value based on her/his past experience. According to this expert, the yield value is probably near $\mu = 5\,\text{t}\,\text{ha}^{-1}$ and the uncertainty about yield value is $\tau = 2\,\text{t}\,\text{ha}^{-1}$. μ and τ can be used to define a Gaussian prior distribution, $\theta \sim N(\mu, \tau^2)$. With this assumption, the yield value follows a Gaussian distribution. Although this choice is very convenient from a computational point of view (see Section 2.3), it may lead to unrealistic yield values as the Gaussian distribution gives non-zero probability to negative values (Figure 7.1). Another option is to represent the expert knowledge

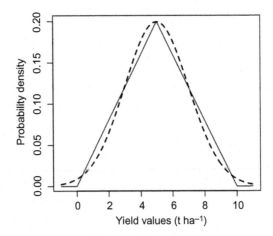

FIGURE 7.1 Prior probability distribution of yield values assuming a Gaussian distribution (thick dashed line) and a triangle distribution (thin continuous line). The Gaussian distribution has a mean value equal to 5 and a standard deviation equal to 2. The triangle distribution has minimum, most likely, and maximum values equal to 0, 5, and 10 t ha^{-1}, respectively.

about yield using a probability distribution which is non-zero for positive values only. An example is a triangle distribution with minimum, most likely, and maximum values equal to 0, 5, and $10 \, t \, ha^{-1}$, respectively (Figure 7.1). This distribution assumes that it is impossible to get negative yield values. It also assumes that possible yield values are lower than $10 \, t \, ha^{-1}$.

2.1.3 Example 2: Soil Carbon Model

This model has one state variable, the mass of carbon (C_t, $kg \, ha^{-1}$) in the top 20 cm of soil in a single field. Changes in soil C are simulated dynamically on a yearly basis. The model again has one unknown parameter, R, the fraction of soil C that is decomposed per year (yr^{-1}). The equation that describes the dynamics of this system is dynamic and is adapted from Jones et al. (2004) (see Appendix 1 for more details):

$$C_t = C_{t-1} - R \times C_{t-1} + b \times U_{t-1}$$

where t is time (in years) from an arbitrary starting year when initial values of soil C are known, C_t is the true soil C in year t, R is the soil C decomposition rate (yr^{-1}), U_t is the amount of C in crop biomass added to the soil in year t ($kg[C] \, ha^{-1} \, yr^{-1}$), and b is the fraction of crop biomass C that is added to the soil in year t that remains after one year (yr^{-1}). We assume here that U_t is not random and is perfectly known at each time step. We also assume that b is known.

According to Jones et al. (2004), the decomposition rate R is likely to be equal to 0.02, but the authors considered a standard deviation of 0.01 in order to describe the uncertainty about the parameter value. This knowledge about parameter values can be described using a Gaussian distribution with a mean equal to 0.02 and a standard deviation equal to 0.01. However, this distribution gives non-zero probabilities to negative decomposition rates. In order to avoid this problem, in a second study we took as the prior distribution a triangle distribution with minimum, most likely, and maximum values equal to 0, 0.02, and 0.04. Both distributions are shown in Figure 7.2.

2.2 Likelihood Function

2.2.1 Definition

A likelihood function is a function giving the probability of the observations for given values of parameters (see Chapter 6). This type of function is used by the maximum likelihood method for estimating parameter values. It is also used by Bayesian methods for relating parameter values to data. Likelihood functions can be based on different types of probability distributions such as Poisson, Binomial, and Gaussian. Suppose that N observations, $Y = (y_1, \ldots, y_N)^T$, are available for estimating parameters. Let E denote the vector of the N model error terms, $E = (\varepsilon_1, \ldots, \varepsilon_N)^T$, where $\varepsilon_i = y_i - f(x_i; \theta)$,

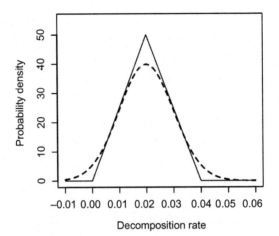

FIGURE 7.2 Gaussian (thick dashed line) and triangle (thin continuous line) prior probability distributions for the soil carbon decomposition rate. The Gaussian distribution has a mean value equal to 0.02 and a standard deviation equal to 0.01. The triangle distribution has minimum, most likely, and maximum values equal to 0, 0.02, and 0.04, respectively.

$i = 1, \ldots, N$, and $f(x_i; \theta)$ is the ith crop model prediction obtained with input variable x_i. We assume that E is normally distributed, $E \sim N(0, V)$ where V is $(N \times N)$ variance-covariance matrix of the model errors. Under this assumption, the likelihood of θ and V is defined by

$$L(y_1, \ldots, y_N; \theta, V) = P(y_1, \ldots, y_N | \theta, V) =$$
$$(2\pi)^{-N/2} |V|^{-1/2} \exp\left\{ -\frac{1}{2}[Y - F(\theta)]^T V^{-1} [Y - F(\theta)] \right\} \tag{1}$$

where $F(\theta)$ is a vector containing the N model predictions, $F(\theta) = [f(x_1; \theta), \ldots, f(x_N; \theta)]^T$, and $P(.)$ is a probability density function. The likelihood represents the probability of the observations for given values of θ and V.

2.2.2 Example 1: Yield Estimation

We assume that an imperfect yield measurement was performed in a small plot within the field. We assume that this measurement is normally distributed and is unbiased, i.e., the expected value of the measurement is equal to θ. Under this assumption, the distribution of the measurement is defined by $Y|\theta \sim N(\theta, \sigma^2)$ where Y is the measurement and σ^2 is the variance of the measurement. The probability density of $Y|\theta$ represents the likelihood of θ and is equal to

$$\frac{1}{\sqrt{2\pi\sigma^2}} \exp\left[-\frac{1}{2} \frac{(Y-\theta)^2}{\sigma^2} \right].$$

Note that, in this example, the likelihood function takes its maximum value when the parameter value θ is equal to the measured yield value Y. The measurement Y is the maximum likelihood estimate of the true crop yield θ. We will further assume that σ is known with $\sigma = 1\ \mathrm{t\ ha^{-1}}$.

2.2.3 Example 2: Soil Carbon Model

If we assume that N soil carbon measurements are available and that the model errors $\varepsilon_i = y_i - f(x_i; \theta)$, $i = 1, \dots, N$, are normally distributed, the likelihood function is defined by Equation (1). If we further assume that the model errors are independent with equal variance, the likelihood function becomes:

$$P(y_1, \dots, y_N | \theta, \sigma) = \prod_{i=1}^{N} \frac{1}{\sqrt{2\pi\sigma^2}} \exp\left[-\frac{1}{2} \frac{[y_i - f(x_i; \theta)]^2}{\sigma^2} \right]$$

$$\frac{1}{\sqrt{2\pi\sigma^2}} \exp\left[-\frac{1}{2} \frac{[y_1 - f(x_1; \theta)]^2}{\sigma^2} \right] \times \dots \times \frac{1}{\sqrt{2\pi\sigma^2}} \exp\left[-\frac{1}{2} \frac{[y_N - f(x_N; \theta)]^2}{\sigma^2} \right]$$

$$\tag{2}$$

In this example, θ includes a single element, the soil carbon decomposition rate and the vector of input variable x includes two elements, namely the time (in years) and U_t, which is the amount of C in crop biomass added to the soil in year t. The residual standard deviation σ is an additional parameter that also needs to be estimated from the data.

2.3 Posterior Probability Distribution

2.3.1 Analytical Expression for Posterior Probability Distribution

The posterior probability distribution is related to the prior distribution and to the likelihood function by Bayes' theorem: $P(\theta|Y) = \frac{P(Y|\theta)P(\theta)}{P(Y)}$. It is possible to determine the exact analytical expression of the posterior probability distribution $P(\theta|Y)$ in only a few special cases. In general, this expression is unknown and iterative algorithms need to be used to approximate the posterior distribution.

The yield estimation example with a Gaussian prior presented above is one of the few cases where the analytical expression of $P(\theta|Y)$ is known. In this example, the prior distribution is defined by $P(\theta) = N(\mu, \tau^2)$ and the likelihood is defined by $P(Y|\theta) = N(\theta, \sigma^2)$. Under these assumptions, the posterior distribution can be expressed as:

$$P(\theta|Y) = N\left[B\mu + (1 - B)Y, (1 - B)\sigma^2 \right] \tag{3}$$

where $B = \frac{\sigma^2}{\sigma^2 + \tau^2}$. According to Equation (3), $E(\theta|Y)$ is a weighted sum of the prior mean and of the measurement, $E(\theta|Y) = \mu + \frac{\tau^2}{\tau^2 + \sigma^2}(Y - \mu)$. The weight B depends on the prior variance τ^2 and on the variance of the error of measurement σ^2. As $\tau = 2\ \mathrm{t\ ha^{-1}}$ and $\sigma = 1\ \mathrm{t\ ha^{-1}}$, we have here

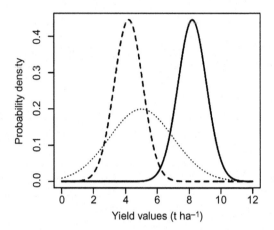

FIGURE 7.3 Posterior probability of the yield value when $Y = 4$ (dashed line) and when $Y = 9\,t\,ha^{-1}$ (continuous line). The dotted line shows the prior distribution.

$B = 1/5$, $E(\theta|Y) = \frac{1}{5}\mu + \frac{4}{5}Y = 1 + \frac{4}{5}Y$, and $var(\theta|Y) = \frac{4}{5}$. The posterior variance (equal to 4/5) is lower than the prior variance (equal to 4). The use of a measurement Y considerably reduced the uncertainty about yield value. Two examples of posterior distributions corresponding to two different measurements Y (4 and 9 t ha^{-1}) are shown in Figure 7.3.

Let us compare the maximum likelihood estimate of θ and the posterior mean $E(\theta|Y)$. The maximum likelihood estimate of θ is $\hat{\theta}_{ML} = Y$. The maximum likelihood estimate depends only on the observation whereas $E(\theta|Y)$ depends on both the data and the prior parameter distribution. It is interesting to look at the frequentist properties of these two estimators of yield, i.e., on the bias and variance of $\hat{\theta}_{ML}$ and of $E(\theta|Y)$ over all possible measured yield value Y. Contrary to $\hat{\theta}_{ML}$, $E(\theta|Y)$ is biased because its expected value is equal to $E(\theta|Y) = 1 + \frac{4}{5}\theta$ and, so, is different than the true value θ when $\theta \neq 5$. The advantage of using $E(\theta|Y)$ is that its variance over measured yield values Y (i.e., the frequentist variance) is only equal to $\frac{16}{25}\sigma^2 = \frac{16}{25}$ and, thus, is lower than the variance of $\hat{\theta}_{ML}$, $var_{Y|\theta}(\hat{\theta}_{ML}) = 1$.

2.3.2 Algorithms for Estimating Posterior Probability Distributions

Different algorithms have been proposed for approximating posterior probability distributions or, at least, some aspects of these distributions. A first approach is to calculate the mode of the posterior parameter distribution. This approach returns only a single value for each parameter, the value maximizing $P(\theta|Y)$. The drawback of this approach is that it gives information only on a very restricted aspect of the posterior distribution (its mode) and does not give information about the uncertainty of the parameter values. This approach is presented in detail in Section 3.

A much more popular approach is to use an algorithm to generate a large sample of parameter values whose distribution approximates the posterior probability distribution. The basic idea is to generate thousands of parameter values and to approximate key features of the posterior probability distribution (e.g., mean, median, quantiles, variance, covariance) from these values. Different algorithms were developed in the 1990s to do that, and they are now widely used in many areas. They are presented in Section 4.

3. COMPUTATION OF POSTERIOR MODE

3.1 Definition

Because crop models are complex, it is usually impossible to derive an analytical expression for $P(\theta|Y)$ but, under some assumptions, it is possible to calculate its mode. This method returns only a single value for each parameter, the value maximizing $P(\theta|Y)$. Suppose that p parameters, $\theta = (\theta_1, \ldots, \theta_p)^T$, are to be estimated. Suppose that the prior distribution is a normal distribution:

$$P(\theta) = (2\pi)^{-p/2}|\Omega|^{-1/2}\exp\left\{-\frac{1}{2}[\theta-\mu]^T\Omega^{-1}[\theta-\mu]\right\},$$

where

$$\mu = (\mu_1, \ldots, \mu_p)^T.$$

is the $(p \times 1)$ vector of prior means and Ω is the $(p \times p)$ variance-covariance matrix. Suppose that N observations, $Y = (y_1, \ldots, y_N)^T$, are available for estimating parameters and that these observations are normally distributed. The likelihood is then

$$P(Y|\theta) = (2\pi)^{-N/2}|V|^{-1/2}\exp\left\{-\frac{1}{2}[Y-F(\theta)]^T V^{-1}[Y-F(\theta)]\right\}$$

where $F(\theta)$ is a vector containing the N model predictions, $F(\theta) = [f(x_1; \theta), \ldots, f(x_N; \theta)]^T$, and V is $(N \times N)$ variance-covariance matrix of the model errors. According to Bayes' theorem, the posterior distribution $P(\theta|Y)$ is related with $P(Y|\theta)$ and to $P(\theta)$ as follows:

$$P(\theta|Y) = \frac{P(Y|\theta)P(\theta)}{P(Y)}$$

where $P(Y)$ is the distribution of the observations and does not depend on the parameters. Plugging likelihood and prior equations into Bayes' theorem gives:

$$P(\theta|Y) = K_1\exp\left\{-\frac{1}{2}[Y-F(\theta)]^T V^{-1}[Y-F(\theta)]\right\}\exp\left\{-\frac{1}{2}[\theta-\mu]^T\Omega^{-1}[\theta-\mu]\right\}$$

where K_1 is a constant independent of θ. The posterior mode is the value of θ that maximizes $P(\theta|Y)$ or equivalently that maximizes $\log P(\theta|Y)$, which is usually more convenient to work with. $\log P(\theta|Y)$ is

$$\log P(\theta|Y) = K_2 - [Y - F(\theta)]^T V^{-1}[Y - F(\theta)] - [\theta - \mu]^T \Omega^{-1}[\theta - \mu]$$

where K_2 is a constant independent of θ. Consequently, the posterior mode is the value of θ that minimizes

$$[Y - F(\theta)]^T V^{-1}[Y - F(\theta)] + [\theta - \mu]^T \Omega^{-1}[\theta - \mu] \qquad (4)$$

Equation (4) includes two terms. The first term, $[Y - F(\theta)]^T V^{-1}[Y - F(\theta)]$, is equal to the function minimized by the generalized least squares estimate $(Z_{GLS}(\theta))$ (see Chapter 6). The second term, $[\theta - \mu]^T \Omega^{-1}[\theta - \mu]$, is a penalty term that penalizes the parameter values that differ strongly from the prior mean μ. When the observations are mutually, independent and so are the parameters, the matrices V and Ω are diagonal and Equation (4) is equal to

$$\sum_{i=1}^{N} \frac{[y_i - f(x_i; \theta)]^2}{\sigma_i^2} + \sum_{j=1}^{p} \frac{\left[\theta_j - \mu_j\right]^2}{\omega_i^2} \qquad (5)$$

where σ_i^2, $i = 1, \ldots, N$, and ω_j^2, $j = 1, \ldots, p$ are the diagonal elements of V and Ω. If the ω_j^2, $j = 1, \ldots, p$, take very small values, meaning that the prior information has little uncertainty, then the parameter values minimizing Equation (5) won't differ much from the prior mean μ. On the contrary, if the prior variances are large, the parameter estimates will be very different from the prior means, and closer to the least squares estimates.

The minimization of Equation (4) or Equation (5) can be performed with the same algorithms as those used to apply generalized least squares (see Chapter 6). The trick is to consider the prior mean μ of the p parameters as p additional data and then to implement the generalized least squares method. In practice, the user needs to add the values of μ_j, $j = 1, \ldots, p$, to the list of the data and to include the θ_j, $j = 1, \ldots, p$, as additional outputs in the model function. The main drawback of this method is that it provides only the posterior mode and not the whole posterior parameter distribution.

In a case study, Tremblay and Wallach (2004) studied the use of the posterior mode as an estimator. The authors considered a model that is a part of the STICS model (Brisson et al., 1998), which we shall refer to as Mini-STICS. Mini-STICS includes 14 parameters and simulates sunflower development over a period of 20 days, starting at the stage Maximal Acceleration of Leaf growth (AMF). Tremblay and Wallach (2004) compared generalized least squares and a Bayesian approach that consists of minimizing Equation (4). Generalized least squares was applied to estimate a small number of parameters (1 to 7). The other parameters were fixed at their initial values. With the Bayesian approach, all 14 parameters were estimated simultaneously. The authors applied the two types of estimation method to several training data sets each with 14 observations and calculated *MSEP* values for different model output variables (*LAI* and soil water content, each at two dates). The results showed that the *MSEP* values were lower with the Bayesian approach than with generalized least squares.

In practice, it is often difficult to give a value to the variance-covariance matrix of the model errors V. Then, it is useful to estimate the elements of V at the same time as the model parameters θ. Different types of prior distribution can be used for V but, when no information about V is available, it is convenient to define a weakly informative prior density function for V, for example, the Jeffreys distribution $P(V) = K|V|^{-(N+1)/2}$, where $|V|$ is the determinant of V and K is a constant.

The posterior mode is then calculated by maximizing

$$P(\theta, V|Y) = \frac{P(Y|\theta, V)P(\theta, V)}{P(Y)} = \frac{P(Y|\theta, V)P(\theta)P(V)}{P(Y)}$$

or, equivalently, by minimizing

$$-\log P(\theta, V|Y) =$$

$$R + \left(\frac{N}{2} + 1\right)\log|V| + [Y - F(\theta)]^T V^{-1}[Y - F(\theta)] + [\theta - \mu]^T \Omega^{-1}[\theta - \mu] \quad (6)$$

where R does not depend on V or θ.

As already explained in Chapter 6, the number of non-zero elements in V can be large when the model errors are correlated. In such cases, the minimization of Equation (6) is often difficult and the parameter estimates may be inaccurate.

When the observations are mutually independent and so are the parameters, the matrices V and Ω are diagonal and the Jeffreys prior density function is

$$P(V) = K \frac{1}{\sigma_1^2 \times \ldots \times \sigma_i^2 \times \ldots \times \sigma_N^2}.$$

The posterior mode is then obtained by minimizing $-\log P(\theta, V|Y)$ with

$$-\log P(\theta, V|Y) = R + \frac{3}{2}\sum_{i=1}^{N}\log(\sigma_i^2) + \sum_{i=1}^{N}\frac{[y_i - f(x_i; \theta)]^2}{\sigma_i^2} + \sum_{j=1}^{p}\frac{[\theta_j - \mu_j]^2}{\omega_j^2}.$$

3.2 Example: Soil Carbon Model

Calculation of the posterior mode is illustrated here with the soil carbon model. Soil carbon data were collected in a plot over a period of 40 years (Figure 7.4) and these data were used to estimate the parameter R of the soil

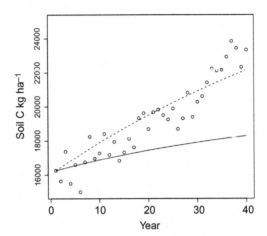

FIGURE 7.4 Soil carbon data (points), carbon model simulations when the parameter R is set equal to its prior mean, i.e., $R = 0.02$ (continuous line), and simulations when the parameter R is set equal to its ordinary least squares estimate (dashed line).

carbon model described above. The parameter R was first estimated using a frequentist method, the ordinary least squares method (see Chapter 6) with the R function nls, and the estimated value was $R = 0.01281$. This value is lower than the prior mean defined from expert knowledge, i.e., 0.02 (see Section 2.2.3 of this chapter) and leads to a better model fit compared to the simulated soil carbon values obtained with the prior mean (Figure 7.4).

The posterior mode was then calculated using a Gaussian prior distribution with mean 0.02 and standard deviation 0.01, and the likelihood function defined by Equation (2). In this example, the standard deviation of the model residuals σ was assumed to be known, and was set equal to 707.1 (Jones et al., 2004). Equation (5) was minimized with $\omega = 0.01$ and $\mu = 0.02$ again using the R function nls (see Chapter 6). The estimated parameter value was equal to 0.01282 and, thus, was almost equal to the ordinary least squares estimate. This is due to the high prior standard deviation (equal to 0.01 for a prior mean of 0.02) chosen for the computation. This high value gives a small weight to the prior mean compared to the data and, for this reason, the posterior mode does not differ much from the ordinary least squares estimate.

In order to illustrate the influence of the prior standard deviation ω on the result, the posterior mode was computed again using a series of prior standard deviations ranging from 0.0001 to 0.01. The posterior mode was computed by minimizing Equation (5) setting the quantity ω to each standard deviation value successively. Results are shown in Figure 7.5. The higher the prior standard deviation, the lower the posterior mode. When the prior standard deviation is set to a low value, a large weight is given to the prior mean and the posterior mode is close to the prior mean value. On the other hand, when the prior standard

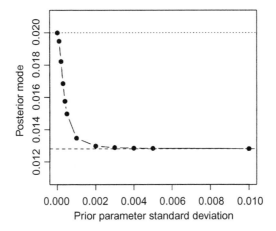

FIGURE 7.5   Posterior mode of the soil carbon model parameter R (points) estimated using a prior mean equal to 0.02 and prior standard deviation ranging from 0.00001 to 0.01. Horizontal dashed and dotted lines indicate ordinary least squares estimate and prior parameter mean, respectively.

deviation is set to a high value, the posterior mode is almost equal to the ordinary least squares estimate. This example shows that the choice of the prior distribution is critical when using a Bayesian method for parameter estimation.

4. ALGORITHMS FOR ESTIMATING POSTERIOR PROBABILITY DISTRIBUTION

We are interested in the posterior distribution,

$$P(\theta|Y) = P(Y|\theta)P(\theta)/P(Y).$$

In general we can calculate the numerator on the right hand side, since this is just the likelihood times the prior probability. However, in general we cannot calculate the denominator. This is the normalization constant, meaning that it is the value that makes the integral of the probability equal to 1. This is just a constant (it doesn't depend on θ), but it determines the height of the probability density at each point. Without it, we cannot directly calculate values of the probability density function. The problem then is to approximate the posterior distribution while knowing its probability density only relative to a constant.

The principle of the methods discussed here is to generate a sample of parameter values whose distribution approximates the posterior distribution. The interest of this approach is that it can be applied to complex nonlinear models with a large number of parameters. For example, Harmon and Challenor (1997) used a Monte Carlo method for estimating 10 parameters of a complex ecological model. Wallach et al. (2012) estimated the parameters of a maize model using the same approach.

Monte Carlo methods are useful for estimating parameters of complex non-linear models. According to an article in *Science* (Malakoff, 1999), these methods explain the new popularity of the Bayesian methods. A detailed description of Monte Carlo methods can be found, for example, in Geyer (1992), Smith and Gelfand (1992), Gilks et al. (1995), Robert and Casella (2004), and Carlin and Louis (2000, 2008). Here, we present two of these methods, importance sampling and the Metropolis-Hastings algorithm.

4.1 Importance Sampling

4.1.1 Definition

If we could sample directly from the posterior distribution, then we could just take a large sample of size Q and use that as an empirical approximation to the posterior density. Giving each value a probability of $1/Q$ ensures that the sum of the probabilities is 1. In general, however, we cannot do this, since the posterior distribution does not usually have some simple known form. However, we can create an approximation to the posterior distribution by taking a random sample from some other distribution, sufficiently simple so that we can sample from it, say $g(\theta)$, and then weighting each sampled value by $P(\theta|Y)/g(\theta)$.

It is easy to see then that the expectation of any function h of θ using the weighted sample is

$$E[h(\theta)] = \int h(\theta)g(\theta)\frac{P(\theta|Y)}{g(\theta)}\,d\theta = \int h(\theta)P(\theta|Y)d\theta$$

That is, the expectation is equal to the expectation over the posterior distribution. Note that this result is very general. First of all, anything we need to calculate using the posterior distribution can be expressed as an expectation. Secondly, any distribution $g(\theta)$ can be used. In practice, however, we will approximate $P(\theta|Y)$ using a finite sample, and if $g(\theta)$ is very different than the posterior distribution, the approximation may be poor.

The importance sampling method requires the definition of a density function $g(\theta)$, called the importance function, from which realizations of θ can be sampled. This function is used to generate a sample of Q parameter vector values denoted θ_q, $q = 1, \ldots, Q$. A weight ω_q is calculated for each generated vector as

$$\omega_q = \frac{P(Y|\theta_q)P(\theta_q)}{g(\theta_q)}.$$

The weight depends on the likelihood function, on the prior density, and on the importance function. The normalized weights are calculated as

$$\omega_q^* = \frac{\omega_q}{\sum\limits_{q=1}^{Q}\omega_q}$$

The weight values are used to describe the posterior parameter distribution. The pairs (θ_q, ω_q^*), $q = 1, \ldots, Q$, define a weighted discrete probability distribution and give an approximation to the posterior probability distribution. With this discrete distribution, the expected value of the posterior distribution of θ is approximated by:

$$\hat{E}(\theta|Y) = \sum_{q=1}^{Q} \omega_q^* \theta_q.$$

Importance sampling gives good results if the importance function $g(\theta)$ is not too different than $KP(Y|\theta)P(\theta)$ where K is the normalization constant. A particular importance function is the prior density function. In this case, we have $g(\theta) = P(\theta)$ and $\omega_q = P(Y|\theta_q)$. The weight is thus equal to the likelihood with this importance function; values of θ_q close to the maximum likelihood parameter estimate will then get higher weights than values far from the maximum likelihood estimate.

In general, the direct use of discrete pairs (θ_q, ω_q), $q = 1, \ldots, Q$, is not recommended for practical reasons and also because, in many cases, several weight values are almost equal to zero and are thus useless. It is usually suggested to draw a sample of size Q from the discrete pairs (θ_q, ω_q), $q = 1, \ldots, Q$. The algorithm including this resampling step is frequently called importance sampling resampling. Note that after resampling, each element in the sample has an equal probability of $1/Q$. It is much more convenient to work with a sample like this, with equal weights, than to work with weighted probabilities.

Although this algorithm is very simple, it should be used with care. When the prior distribution is far from the posterior distribution, the use of the prior as an importance function leads to a poor approximation because only very few parameter values get non-negligible weights. It is then necessary to use a very large number of parameter values Q (and of model simulations) in order to obtain a good approximation to the posterior parameter distribution.

Note that the estimation method GLUE (generalized likelihood uncertainty estimation, e.g., Shulz et al., 1999; Makowski et al., 2002) can be seen as a version of importance sampling with prior density as the importance function and with a particular type of likelihood function.

4.1.2 Example: Soil Carbon Model

In this example the importance sampling resampling algorithm is implemented to estimate the parameter R of the soil carbon model. In the first study, the standard deviation of the model residuals σ is assumed to be known, and is set equal to 707.1 (Jones et al., 2004). The likelihood function is given by Equation (2) and the data on variations in observed soil carbon

TABLE 7.1 Minimum, First Quartile, Median, Mean, Third Quartile, and Maximum Values Obtained for the Parameter R Using an Importance Sampling Resampling Algorithm with Q = 10,000 and Two Different Prior Distributions

Prior	Min	1st Quartile	Median	Mean	3rd Quartile	Max
Gaussian	0.01167	0.01261	0.01286	0.01283	0.01306	0.01396
Triangle	0.01165	0.01260	0.01283	0.01306	0.01306	0.01401

vs. time are shown in Figure 7.4. Two different prior distributions for R are used; a Gaussian and then a triangle probability distribution (Figure 7.2). The prior is used as the importance function in both cases. R provides functions for sampling from both these distributions. Results of the importance sampling resampling algorithm are summarized in Table 7.1 for $Q = 10,000$. The differences in results obtained with the two priors are very small. This is due to the relatively large number of data used to calculate the posterior distribution. As a large amount of data was used to estimate R, the influence of the prior on the result is not very important, and the uncertainty about the value of R is very small as shown by the small differences between the minimum and maximum values of R reported in Table 7.1.

Figure 7.6 shows the weights calculated by the Importance Sampling algorithm with the Gaussian prior distribution when $Q = 10, 100, 1000$ and 10000 parameter values are generated. When the number of values is small ($Q = 10$, $Q = 100$), the number of parameter values getting a non-negligible weight is small, and the posterior distribution of R is poorly approximated. However, even in these cases, the parameter value getting the highest weight is close to the posterior mode computed with the method presented in Section 3 (Figure 7.6AB). As expected, the number of parameter values with non-negligible weight is higher when Q is set to 1000 or more (Figure 7.6CD), and the posterior distribution is more accurately estimated in these cases. Figure 7.7 shows the histogram of the values of R resampled from the discrete distribution defined by the weights displayed in Figure 7.6D.

The results shown in Table 7.1, Figure 7.6, and Figure 7.7 were derived under the assumption that the standard deviation of the residual model error is known. It is more realistic to treat this standard deviation as unknown. Then it must be estimated at the same time as the parameter R. In this case, it is necessary to define a prior distribution for σ. Here, the prior distribution of σ is defined as a uniform distribution with lower and upper bounds set equal to 0 and 2000, respectively, and the prior distribution of R is defined as a Gaussian distribution with mean equal to 0.02 and standard deviation equal to 0.01 (as in Figure 7.6 and Figure 7.7). The two prior distributions

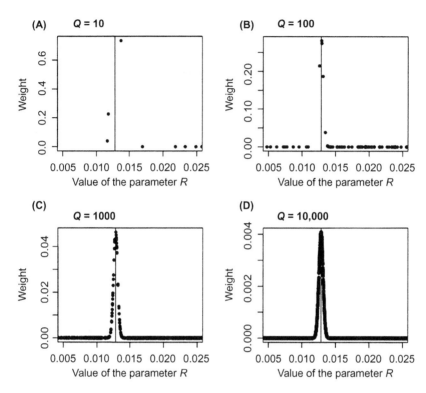

FIGURE 7.6   Weights calculated by importance sampling with $Q = 10$, (**A**) 100, (**B**) 1000, (**C**) and 10,000 (**D**) values of the parameter R of the soil carbon model using a Gaussian prior distribution (mean = 0.02 and standard deviation = 0.01). The vertical line corresponds to the posterior mode calculated with the same prior distribution.

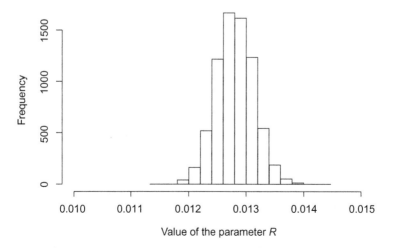

FIGURE 7.7   Histogram of the values of R after resampling when $Q = 10,000$.

are treated as independent, so the joint prior distribution is the product of the prior distributions for R and σ. The importance function is the same as the prior. The importance sampling resampling algorithm was run again under these assumptions with $Q = 10,000$, and the results are shown in Figure 7.8 and Table 7.2.

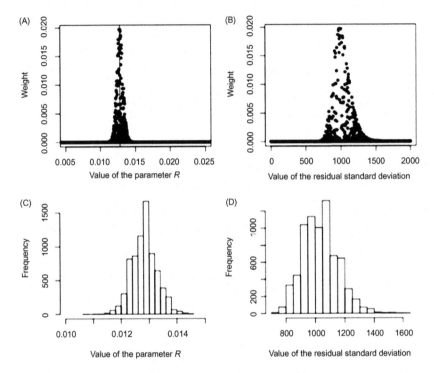

FIGURE 7.8 Weights calculated by importance sampling with $Q = 10,000$ values of the parameters R and σ (A, B), and histograms of the values of R and σ after resampling (C, D). The vertical line corresponds to the posterior mode of R (A).

TABLE 7.2 Minimum, First Quartile, Median, Mean, Third Quartile, and Maximum Values of the Posterior Distributions of the Parameters R and σ Obtained Using an Importance Sampling Resampling Algorithm with $Q = 10,000$

Parameter	Min	1st Quartile	Median	Mean	3rd Quartile	Max
R	0.01076	0.01253	0.01284	0.01283	0.01310	0.01452
σ	711.7	943.8	1030	1033	1103	1602

The importance sampling algorithm used to compute the weights shown in Figure 7.8 is presented below:

```
#Number of parameter values
Npara <- 10000
#Sampling from the prior distribution of the parameter R
Rsample <- rnorm(Npara,0.02,0.01)
#Sampling from the prior distribution of the residual model
 standard deviation
Ssample <- runif(Npara,0,2000)
#Initialisation of the vector of log likelihood
LogLike_vec <- rep(NA,Npara)
#Loop running the model and calculating log likelihood values
for (i in 1:Npara) {
    #Soil carbon simulations with the function CarbonMod for
    dates Sata$Time, parameter R value Rsample[i], and initial
    soil carbon Data$ScO
    Simul.i <- CarbonMod(Data$Time,Rsample[i],Data$ScO)
    #Calculation of log likelihood
    LogLikelihood.i <- sum(log(dnorm(Ym,Simul.i,Ssample[i])))
    #Addition of a constant to avoid very small values leading
     to computation errors
    LogLike_vec[i] <- LogLikelihood.i + 1000
}
#Weight calculation
Weight <- exp(LogLike_vec)/sum(exp(LogLike_vec))
```

The number of generated parameter values is $Npara = Q = 10,000$ in the above R code. It is important to check whether this number is sufficiently high or not. In order to study the influence of Q on the performance of the algorithm, the mean values of R and σ were calculated successively with $Q = 10, 11, 12, \ldots, 9999, 10,000$. The results are displayed graphically in Figure 7.9. The figure shows that mean values are highly instable when the number of generated parameter values is less than 2000, but become much more stable with a larger number of parameter values. Based on these results, the use of $Q = 10,000$ leads to accurate estimations of the posterior means of R and σ. Results obtained for small and large percentiles of parameter R values also indicate that $Q = 10,000$ is sufficient (results not shown).

4.2 The Metropolis-Hastings Algorithm

4.2.1 Definition

The Metropolis-Hastings algorithm is a Markov chain Monte Carlo algorithm (MCMC) (see, for example, Geyer, 1992 and Gilks et al., 1995, for more detail). The idea is that the algorithm builds up a chain of parameter vectors by taking random steps in parameter space. This is a Monte Carlo method in that new parameter values are generated using a probability distribution.

(A)   (B)

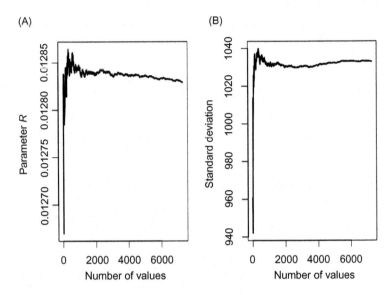

FIGURE 7.9 Mean values of R (A) and σ (B) obtained with the importance sampling resampling algorithm when $Q = 10, 11, 12, \ldots, 10,000$.

Each step depends on the previous parameter value, but not on values before that, according to the Markov property.

At each step a new value of θ is added to the chain of values generated. There are two possibilities; the new value in the chain can be the θ that results from the new step, or this value can be rejected and in that case the new value is the same as the previous value. The decision between these two possibilities depends on $P(Y|\theta)P(\theta)$, which is proportional to the posterior density. It doesn't depend on the normalization constant. The exact decision rule is given in the algorithm below. It is quite easy to prove that under quite reasonable assumptions, the chain of values generated by this algorithm tends toward the posterior distribution.

The algorithm is iterative and starts with an initial parameter vector θ_0. A series of Q vectors θ_q, $q = 1, \ldots, Q$, is then generated sequentially as follows:

i. Generate a candidate vector θ^* from a proposal distribution denoted as $P(\theta^*|\theta_{q-1})$. For instance, if the proposal distribution is a normal distribution with mean equal to θ_{q-1}, and the parameter vector has only a single component, then

$$P(\theta^*|\theta_{q-1}) = \frac{1}{(2\pi\sigma^{*2})^{1/2}} e^{-(\theta^* - \theta_{q-1})^2/(2\sigma^{*2})}$$

The average step size between θ_{q-1} and θ^* will depend on σ^{*2}, the variance of this normal distribution. A smaller variance will on average lead to smaller steps. For vector θ, there will be a variance-covariance matrix Σ in place of σ^{*2}. The diagonal elements of Σ will determine the step size for each element of θ, and the off-diagonal elements will determine how changes in one parameter are related to changes in another parameter. If Σ is diagonal, then the steps for different elements of θ are independent.

ii. Calculate

$$T = \frac{P(Y|\theta^*)P(\theta^*)P(\theta_{q-1}|\theta^*)}{P(Y|\theta_{q-1})P(\theta_{q-1})P(\theta^*|\theta_{q-1})}.$$

Note that if the proposal distribution is a normal distribution, then

$$P(\theta_{q-1}|\theta^*) = P(\theta^*|\theta_{q-1})$$

so the terms on the right of the numerator and denominator cancel. Then T just depends on the ratio of the posterior probabilities. T will be larger than 1 if the posterior probability of the proposal point is larger than the posterior probability of the previous point. Since this is a ratio, the normalization constants cancel.

iii. If min $(1, T) > u$, where u is drawn from a uniform distribution on the interval (0,1) then $\theta_q = \theta^*$ otherwise $\theta_q = \theta_{q-1}$. If T is greater than or equal to 1, then min$(1, T) = 1$ and so it will be larger than u (which has 0 probability of being exactly equal to 1). In this case θ^* is added to the chain. If T is less than 1, then θ^* has a certain probability of being added anyway. For example, if $T = 0.6$, there is a 60% chance that u will be smaller than T, in which case θ^* will be added to the chain, and a 40% chance that u will be larger than T, in which case θ_{q-1} will be repeated in the chain.

After a phase of, say, M iterations, the chain of values $\theta_1, \theta_2\ldots$ thus constructed will converge to a chain with elements drawn from the posterior parameter distribution. The first M iterations are influenced by the starting point and should be discarded.

In order to use the Metropolis-Hastings algorithm, it is necessary to choose the starting value θ_0, the proposal distribution $P(\theta^*|\theta_{q-1})$, the total number of iterations Q, and the number of discarded iterations M. The definition of precise rules for choosing these elements is currently an area of active research.

According to Gilks et al. (1995), the choice of θ_0 is not very critical. On the other hand, the choice of the proposal distribution $P(\theta^*|\theta_{q-1})$ is an important issue. A poor choice can lead to very slow convergence toward the posterior distribution. A common practice is to use a normal distribution with mean θ_{q-1} and covariance matrix Σ, i.e., $\theta^*|\theta_{q-1} \sim N(\theta_{q-1}, \Sigma)$. Several authors (Campbell et al., 1999; Harmon and Challenor, 1997) suggest choosing Σ such that the acceptance rate of the test performed in Step iii of the algorithm is in the range

20−50%. If the acceptance rate is very small, then the same parameter values will be repeated very many times in the chain. A large acceptance rate often occurs because the steps are too small. The problem is that it then requires a very large number of steps to fully cover the parameter space.

Several methods for determining M (number of iterations to be discarded) and Q (total number of iterations) are presented in Geyer (1992), Gilks et al. (1995), and Carlin and Louis (2000). Their principle is to run several chains (three or four in general) with different starting points, and to compare key quantities calculated with the different chains (means, quantiles, variances) in order to determine if the number of parameter values Q-M is sufficiently high to lead to stable results. In particular, one can test whether the variance within a chain, parameter by parameter, is similar to the variance between chains (Gelman and Rubin, 1992). If not, this indicates that the chains are still affected by the starting point and have not converged to the same distribution. Tools for convergence diagnosis are available in the R package coda (Plummer et al., 2012).

The series of Q vectors θ_q, $q = 1, \ldots, Q$, generated by the Metropolis-Hastings algorithm is often auto-correlated. This is due to the fact that the parameter θ_q is generated from the previous value θ_{q-1}. The use of an auto-correlated series may lead to an underestimation of the posterior variance. When auto-correlations are strong, it is advisable to thin the series of generated parameter values (e.g., by keeping one value out of every ten) in order to get a nearly uncorrelated series.

Although the Metropolis-Hastings algorithm looks more complex than importance sampling, the former is more frequently used in practice because it can give a good approximation of the posterior distribution even when the prior is far from the posterior. Makowski et al. (2002) compared the GLUE method and the Metropolis-Hastings algorithm in a simulation study and showed that the latter gives slightly better results in terms of *MSEP* values. The main difficulty when implementing the Metropolis-Hastings algorithm is to find a proposal distribution leading to an acceptance rate of 20−50%. This is usually easy when the number of parameters is small (1, 2 or 3), but more difficult with complex models.

The Metropolis-Hastings algorithm is one of the MCMC algorithms developed to approximate posterior probability distributions. Gibbs sampling is another popular MCMC algorithm that consists of generating values for each parameter in turn using conditional probability distributions (Casella and George, 1992). For example, if two parameters $\theta^{(1)}$ and $\theta^{(2)}$ need to be estimated, values of $\theta^{(1)}$ are generated from $P(\theta^{(1)}|\theta^{(2)})$ and values of $\theta^{(2)}$ are generated from $P(\theta^{(2)}|\theta^{(1)})$. The sequential implementation of this procedure leads to a Gibbs sequence $\theta_0^{(1)}, \theta_0^{(2)}, \theta_1^{(1)}, \theta_1^{(2)}, \ldots, \theta_Q^{(1)}, \theta_Q^{(2)}$ that can be used to approximate the posterior distribution. To implement this algorithm, it is necessary to define a starting value $\theta_0^{(1)}$ but no proposal distribution needs to be specified. This is a clear advantage compared to Metropolis-Hastings. However, Gibbs sampling requires the knowledge of all the conditional distributions and this requirement is not always satisfied. Hybrid algorithms combining Metropolis-Hastings and Gibbs sampling have been proposed, especially the Metropolis-Hastings within Gibbs algorithm (Gelman et al., 2004, p. 202). This algorithm was recently used to estimate the parameters of a maize crop model (Wallach et al., 2012).

4.2.2 Example: Soil Carbon Model

The Metropolis-Hastings algorithm was implemented to estimate the parameter R of the soil carbon model. The standard deviation of the model residuals σ was first assumed to be known, and was set equal to 707.1 as before. The posterior distribution of R was estimated using the Gaussian prior probability distribution presented in Figure 7.2. The likelihood function defined by Equation (2) and the data shown in Figure 7.4 were used to calculate the posterior distribution of R. The initial value of R was set equal to 0.02 (prior mean). The proposal distribution used to generate candidate values for the parameter R was a Gaussian probability distribution $R_q \sim N(R_{q-1}, 0.0015^2)$. The variance of this distribution was chosen in order to get an acceptance rate of about 25%.

The first 100 values of the parameter R selected by the algorithm are displayed in Figure 7.10A. The graphic shows that the algorithm quickly moves from the initial value 0.02 and selects values ranging from 0.012 to 0.014. The reason for throwing out some number of initial values of the chain is clear. At the start of the chain, the values are strongly influenced by the initial value. The curve shown in Figure 7.10A includes a high number of plateaus. This is due to the fact that a large majority of the values of R

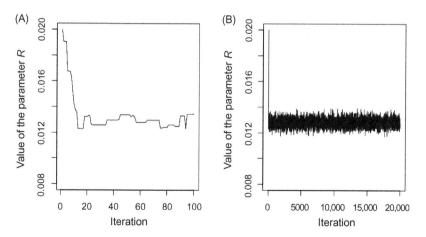

FIGURE 7.10   Results of 100 iterations (**A**) and of 20,000 iterations (**B**) of the Metropolis-Hastings algorithm for estimating the posterior distribution of the parameter R of the soil carbon model.

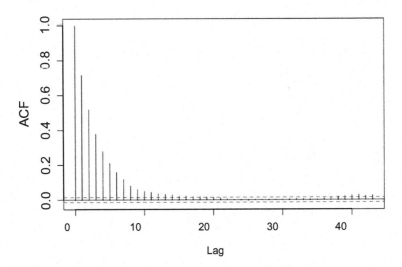

FIGURE 7.11   Auto-correlations of the last 15,000 values of the parameter R generated by the Metropolis-Hastings algorithm. Blue dashed lines indicate significant values.

generated by the proposal distribution is rejected (about 75%); the parameter value selected by the algorithm at a given iteration is thus kept unchanged in most cases. Figure 7.10B shows the chain of selected values of R after 20,000 iterations. The plateaus are still there, but are not visible at this scale. These results show that the algorithm quickly converged toward parameter values ranging from 0.012–0.014.

It is always recommended to check whether the chain of parameter values generated by the Metropolis-Hastings algorithm is strongly auto-correlated or not. Auto-correlations can be computed with the `acf` function of R. Figure 7.11 shows the auto-correlations of the chain of values displayed in Figure 7.10B. The x-axis shows the lag, i.e., the number of iterations separating two parameter values, and the y-axis shows the correlation between parameter values separated by a number of iterations equal to the lag. Results show that the auto-correlations are smaller than 0.1 for lags equal to or higher than 10. In this example, it is recommended to thin the distribution and to retain only one parameter value out of 10 in order to reduce the auto-correlation of the chain.

We now consider the more complex problem where two parameters are estimated at the same time; the parameter R and the standard deviation of the residual model error σ. As before, the prior distribution of σ is defined as a uniform distribution with lower and upper bounds set equal to 0 and 2000, respectively, and the prior distribution of R is defined as a Gaussian distribution with mean equal to 0.02 and standard deviation equal to 0.01.

As two parameters are considered, two proposal distributions need to be defined. The initial value of R was set equal to 0.02, and the initial value of σ was set equal to 1500. Several short MCMC chains were run with different variances for the proposal distributions. Based on these preliminary runs, the proposal distribution of R was defined as

$$R_q \sim N(R_{q-1}, 0.0015^2)$$

and the proposal distribution of σ was set equal to

$$\log(\sigma_q) \sim N\left[\log(\sigma_{q-1}), 0.2^2\right]$$

in order to get an acceptance rate of about 25%. A log-transformation was used for generating only positive values for σ. The R code used to implement the algorithm is shown below:

```
#Number of iterations
Npara <-20000
#Standard deviations of the proposal Gaussian distributions
SigMH.R <-0.0015
SigMH.S <-0.2
#Intial value of the parameter R and of the residual model
  error SD
R.1 <-0.02
lS.1 <-log(1500)
#Initialisation of the vectors including selected values of R
  and SD
Rsample <-rep(NA, Npara)
Rsample[1] <-R.1
lSsample <- rep(NA, Npara)
lSsample[1] <-lS.1
#Initialisation of the vector including log likelihood values
LogLike_vec <-rep(NA,Npara)
Simul.1 <-CarbonMod(Data$Time,Rsample[1],Data$Sc0)
LogLikelihood.1 <-sum(log(dnorm(Ym,Simul.1,exp(lS.1))))
LogLike_vec[1] <-LogLikelihood.1
Count <-1
#Loop over the iterations of the algorithm
for (i in 2:Npara) {
  Rsample[i] <-Rsample[i-1]
  lSsample[i] <-lSsample[i-1]
  LogLike_vec[i] <-LogLike_vec[i-1]
#Sampling of candidate parameter values
  R.i <-rnorm(1,Rsample[i-1],SigMH.R)
  lS.i <-rnorm(1,lSsample[i-1],SigMH.S)
#Simulation and log likelihood with the candidate parameter
  values
  Simul.i <-CarbonMod(Data$Time,R.i,Data$Sc0)
  LogLikelihood.i <-sum(log(dnorm(Ym,Simul.i,exp(lS.i))))
```

```
#Test
Test.i <-
LogLikelihood.i + log(dnorm(R.i,0.02,0.01)) + log(dunif(exp(lS.i),
0, 2000)) - LogLike_vec[i - 1] - log(dnorm(Rsample[i - 1],0.02,0.01)) -
log(dunif(exp(lSsample[i - 1]), 0, 2000))
#Selection or rejection of the candidate parameter values
if (Test.i > 0 | Test.i > log(runif(1,0,1))) {
LogLike_vec[i] <- LogLikelihood.i
Rsample[i] <- R.i
lSsample[i] <- lS.i
Count <- Count + 1
                }

      }
```

Figure 7.12 shows chains of 20,000 iterations for R and σ. The algorithm quickly converges toward values of R ranging from 0.012−0.014. For σ, the algorithm selects values in the range 700−1600 after a few hundred

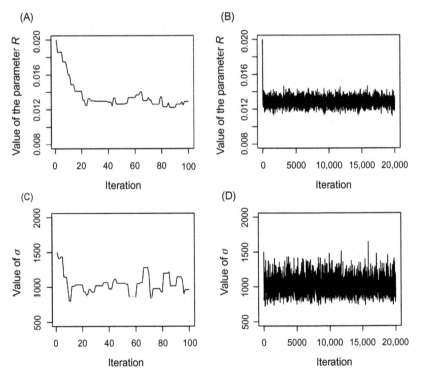

FIGURE 7.12 Results of 100 iterations (A) and of 20,000 iterations (B) of the Metropolis-Hastings algorithm for estimating the posterior distribution of the parameter R of the soil carbon model and of the residual standard deviation σ.

iterations. The auto-correlations of these chains are presented in Figure 7.13. Results show significant auto-correlations for lag values lower than 20. It is thus necessary to thin the series of values in order to decrease the auto-correlations. Figure 7.14 shows the distributions of the values of R and σ after thinning the chains (one value kept every 20). These distributions are approximations of the posterior parameter distributions. Note that, although 20,000 iterations were performed, the histograms displayed in Figure 7.14 include 751 values only. This is because the first 5000 values were discarded in order to deal with the transient phase before the convergence of the algorithm, and because the remaining 15,000 values were thinned in order to decrease the level of auto-correlation.

In order to study the stability of the results, the algorithm was run three times and the resulting parameter distributions were summarized by their 1st and 3rd quartiles, mean, median, and standard deviation. Table 7.3 shows that the results obtained with the three runs are similar. The posterior standard deviations are much lower than the prior standard deviations. Posterior standard deviations of R range from 0.00044 to 0.00046 over the three runs, and they are thus about 20 times lower than the prior standard deviation (0.01). The posterior standard deviations of σ range from 117 to 124 and are about 4−5 times lower than the prior standard deviation (577.35). These results show that the use of the soil carbon data has strongly decreased the uncertainty about the parameter values.

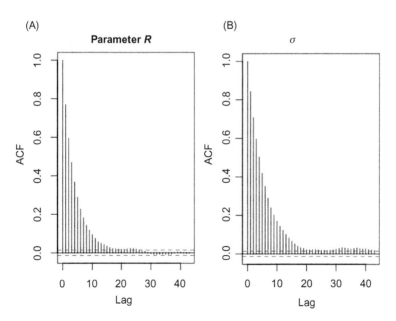

FIGURE 7.13   Auto-correlations of the last 15,000 values of the parameter R and of σ generated by the Metropolis-Hastings algorithm. Blue dashed lines indicate significant values.

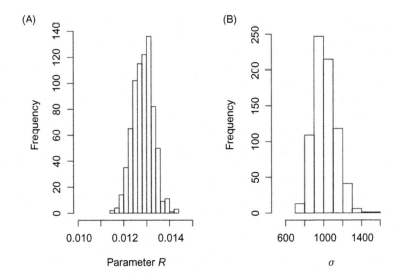

FIGURE 7.14 Histograms of 750 values of the parameter R and of σ generated by the Metropolis-Hastings algorithm. The total number of iterations was 20,000. The first 5,000 values were discarded and only one value in every 20 was kept.

5. CONCLUDING REMARKS

Bayesian methods allow modelers to combine two sources of information for estimating parameters: prior information based on expert knowledge and experimental data. With these methods, it is possible to treat model parameters as partially known, and to use the available experimental data to update their values.

Another advantage of the Bayesian approach is the information it gives on uncertainty. In a classical statistical framework, parameter estimation methods give standard deviations of the parameter estimators, but these standard deviations are based on a linear approximation to the model and may not be very good. Estimating the variances of other quantities may require further linear approximations. In a Bayesian approach, on the other hand, one can easily calculate the distribution of any quantity that depends on the parameters. Bayesian parameter estimation methods are thus very useful for uncertainty analysis (see Chapter 5).

Although the basic principles of Bayesian parameter estimation are simple, the calculation of a posterior distribution may be difficult in practice. Several algorithms have been developed to approximate posterior distributions and these algorithms can be implemented using freely available software such as R, as shown in this chapter. Chapter 10 presents a case study where the parameters of a maize model are estimated using an MCMC algorithm. However, the application of these algorithms to system models including large number of parameters remains a challenge.

TABLE 7.3 Summary of the Posterior Distributions of the Parameter R and of σ Obtained with Three Different Chains ($Q = 20,000$, First 5000 Values Discarded, and One Value Every 20 Selected)

Chain Number	Parameter	1st Quartile	Median	Post. Mean	3rd Quartile	Posterior Standard Deviation
1	R	0.01253	0.01286	0.01284	0.01314	0.000456
2	R	0.01254	0.01281	0.01283	0.01309	0.000440
3	R	0.01256	0.01286	0.01285	0.01313	0.000441
1	σ	924.1	1002.0	1012.0	1090.0	123.83
2	σ	929.5	1007.0	1015.0	1088.0	119.38
3	Σ	933.9	1003.0	1014.0	1086.0	116.75

EXERCISES

Easy

1. Suppose we want to estimate yield value for a given field. The unknown true yield value is denoted as θ. Two types of information are available for estimating θ. First, according to an expert, the yield value must be close to $\mu = 5$ t ha^{-1} and the uncertainty about yield value is $\tau = 2$ t ha^{-1}. μ and τ are used to define a prior yield distribution as follows:

$$\theta \sim N(\mu, \tau^2).$$

   Second, an imperfect yield measurement is performed in the field. We suppose that this measurement is distributed as:

$$Y \sim N(\theta, \sigma^2)$$

   where Y is the measurement and σ^2 is the variance of the measurement. σ^2 is calculated from replicates.
   a. Suppose that the measured yield value is $Y = 3.0$ t ha^{-1} and $\sigma = 0.75$ t ha^{-1}. Calculate the expected value and variance of the posterior distribution of θ when $\mu = 5$ t ha^{-1} and $\tau = 2$ t ha^{-1}.
   b. What is the effect of using less accurate measurement (e.g., $\sigma = 2$ t ha^{-1})?

Moderate

2. We consider a simple non-linear model predicting relative growth as a function of some factor x (e.g., x = nitrogen fertilizer) defined by:

$$f(x, \theta) = 1 - \exp(-\theta x)$$

   This model includes one parameter θ: the growth rate. Suppose that the prior probability distribution of this parameter is defined from expert knowledge as $P(\theta) = N(0.03, 0.015^2)$. We also assume that two measurements of relative growth y_1 and y_2 are available:
   $y_1 = 0.83$ for $x_1 = 100$,
   $y_2 = 0.95$ for $x_2 = 200$.
   a. The two measurements are assumed to be independent. We define a statistical model relating the function f to measurement:

$$y = f(x, \theta) + \varepsilon \quad \text{with} \quad \varepsilon \sim N(0, \sigma^2)$$

   with independance between the values of ε. Finally, we assume that σ is known and equal to 0.02.

   a. Write the likelihood function of θ.
   b. Which algorithms could be used to approximate the posterior distribution of θ?

   c. Look at the R code given below. Which algorithm is implemented by this code?

   d. Modify the program in order to approximate the posterior distribution with 1000.

   e. Run four times the program with 10,000 iterations.

   f. Is 10,000 iterations sufficient to approximate the posterior mean and the posterior standard deviation with a coefficient of variation of 10%?

   g. What is the consequence of using a lower number of iterations?

```
###Number of simulations###
N<-100
###Prior distribution parameters###
mu <-0.03
tau <-0.015
###Data###
Y <-c(0.83,0.95)
X <-c(100,200)
sigma <-0.02
###Simulations from prior###
thetaPrior <-rnorm(N,mu,tau)
Likeli <-matrix(nrow = length(thetaPrior),ncol = length(Y))
LikeliProd <-thetaPrior
###Weight calculation###
for (i in 1:N) {
    for(j in 1:length(Y)) {
            Model.i.j <-1-exp(-thetaPrior[i]*X[j])
            Likeli[i,j] <-dnorm(Y[j],Model.i.j,sigma)
                }
            LikeliProd[i] <-prod(Likeli[i,])
                }
Weight <-LikeliProd
###Weight normalization###
Weight <-LikeliProd/sum(LikeliProd)
###Plot of posterior###
par(mfrow = c(2,2))
plot(thetaPrior,Weight,xlab = "Theta",ylab = "Weight")
title("Plot of normalized weights")
###Resampling###
thetaPost <-sample(thetaPrior,replace = T,prob = Weight)
hist(thetaPost,labels = F,xlab = "Theta",main = "Parameter values
drawn from posterior")
```

Difficult

3. We consider a linear-plus-plateau model relating wheat yield to nitrogen fertilizer dose defined by:

$$y = f(x; \theta) + \varepsilon$$

where x is the applied nitrogen dose, y is the yield observation, $f(x; \theta)$ is a linear-plus-plateau function (see below), and ε is the model residual, i.e., the difference between the yield observation and the linear-plus-plateau function, assumed to be normally distributed, $\varepsilon \sim N(0, \sigma^2)$.

a. The linear-plus-plateau function is defined by:

$$f(x; \theta) = \theta_1 + \theta_2(Dose - \theta_3) \text{ if } Dose \leq \theta_3$$

$$f(x; \theta) = \theta_1 \text{ if } Dose > \theta_3$$

where $\theta = (\theta_1, \theta_2, \theta_3)^T$ is a vector of three parameters representing potential yield, yield per unit of N fertilizer, and dose threshold (minimum dose leading to maximum yield).

The objective of this paper is to estimate the posterior distribution of $\theta = (\theta_1, \theta_2, \theta_3)^T$ for a given wheat field using a prior distribution and a set of measurements obtained in the considered field. The prior distribution comes from the paper by Makowski and Lavielle, 2006 (JABES 11, 45 − 60) and is defined by:

$P(\theta_1) = N(9.18, 1.16^2)$ t ha^{-1} (maximal yield value)
$P(\theta_2) = N(0.026, 0.0065^2)$ t kg N^{-1} (slope of the linear part)
$P(\theta_1) = N(123.85, 46.7^2)$ kg N ha^{-1} (N dose threshold)

The three distributions are assumed independent. The same paper gives the value of the residual standard deviation: $\sigma = 0.3$ t ha^{-1}.

Four yield measurements are available in the considered site. They were obtained for four different N doses:

The tested doses are 0, 50, 100, and 200 kg/ha.

The corresponding yield measurements in the site are 2.50, 5.01, 7.45, and 7.51 t ha^{-1}.

a. Write the likelihood function of the model parameters.

b. Write an R function to simulate yield with the function $f(x; \theta)$ and to calculate the likelihood function.

c. Use the function rnorm to generate 1000 values of $\theta = (\theta_1, \theta_2, \theta_3)^T$ from the prior distributions.

d. Calculate the likelihood for each generated parameter value.

e. Calculate a normalized weight for each generated parameter value using the equation of the importance sampling algorithm.

f. Resample from the discrete distribution defined by the generated values and the calculated weights.

g. Summarize the approximated posterior distribution using key statistics (mean, median, quartile etc.).

h. Check whether the number of iterations is sufficiently high.

Very Difficult

4. Consider the model and data described in Exercise 3.

  a. Approximate the posterior distribution of the model parameters using the Metropolis-Hastings algorithm.

  b. Assume that the residual standard deviation σ is unknown. Define a uniform prior distribution for σ and approximate the posterior distributions of $\theta = (\theta_1, \theta_2, \theta_3)^T$ and σ using either the importance sampling or the Metropolis-Hastings algorithm.

REFERENCES

Campbell, E.P., Fox, D.R., Bates, B.C., 1999. A Bayesian approach to parameter estimation and pooling in nonlinear flood event models. Water Resource Research 35, 211−220.

Carlin, B.P., Louis, T.A., 2000. Bayes and empirical Bayes methods for data analysis. Chapman and Hall, London.

Carlin, B.P., Louis, T.A., 2008. Bayesian Methods for Data Analysis, Third Edition Chapman & Hall/CRC Texts in Statistical Science.

Casella, G., George, E.I., 1992. Explaining the Gibbs Sampler. Am. Stat. 46 (1992), 167−174.

Gelman, Carlin, Stern, Rubin, 2004. Bayesian Data Analysis. CRC Press.

Gelman, A., Rubin, D., 1992. Inference from Iterative Simulation Using Multiple Sequences (with discussion). Stat. Sci. 7, 457−511.

Geyer, C.J., 1992. Practical Markov chain Monte Carlo. Stat. Sci. 7, 473−511.

Gilks, W.R., Richardson, S., Spiegelhalter, D.J., 1995. Markov Chain Monte Carlo in practice. Chapman and Hall, London.

Harmon, R., Challenor, P., 1997. A Markov chain Monte Carlo method for estimation and assimilation into models. Ecol. Modell. 101, 41−59.

Makowski, D., Wallach, D., Tremblay, M., 2002. Using a Bayesian approach to parameter estimation; comparison of the GLUE and MCMC methods. Agronomie 22, 191−203.

Malakoff, D., 1999. Bayes offers a 'New' way to make sense of numbers. Science 286, 1460−1464.

Plummer M., Best N., Cowles K., Vines K., Sarkar D., Almond R. (2012). coda: Output analysis and diagnostics for MCMC. < http://cran.r-project.org/web/packages/coda/index.html > .

Robert, C.P., Casella, G., 2004. Monte Carlo Statistical Methods. Springer, New York.

Smith, A.F.M., Gelfand, A.E., 1992. Bayesian statistics without tears: A sampling-resampling perspective. Am. Stat. 46, 84−88.

Wallach, D., Brun, F., Keussayan, N., Lacroix, B., Bergez, J.-E., 2012. Assessing the uncertainty when using a model to compare irrigation strategies. Agron. J. 104, 1274−1283.

## Chapter 8

Data Assimilation for Dynamic Models

1. INTRODUCTION

In this chapter, we consider modifying a dynamic model to make it a better predictor for a specific situation (e.g., for a specific agricultural field). We assume that we have some information that allows us to modify the model specifically for the situation of interest. In general, this information takes the form of one or several measurements of model state variables during the crop growth period. The model outputs can then be modified based on the measurement, and the modified model used to make predictions for the future growth of the crop. If, for example, the measurement shows that the value of leaf area index (LAI) is smaller than the model predicts, then we can change LAI in the model to a smaller value, and use the corrected model for predicting the future evolution of the system.

Model modification based on measurements is called data assimilation because the data are incorporated into or assimilated into the model. Various methods can be used for data assimilation. Methods of parameter estimation can be used to adjust the values of the model parameters to the data obtained for the situation of interest. This approach was described in detail in the chapters on parameter estimation. In this chapter, we consider another family of methods often referred to as filtering. A *filter* is an algorithm that is applied to a time series to improve it in some way (like filtering out noise). Here the time series is the successive values for the model state variables, and the improvement comes from using measured values to update the model state variables. The state variables are updated sequentially, i.e., each time an observation is available. The best known algorithm for doing this is the *Kalman filter*, which applies if the model is linear and the errors have a normal distribution (Kalman, 1961). This chapter is mostly devoted to that algorithm and its extensions (Welch and Bishop, 2002; Burgers et al., 1998; Anderson and Anderson, 1999; Pastres et al., 2003).

Filtering takes into account the fact that one has two sources of information about model state variables; observations and the model equations. It is

Working with Dynamic Crop Models. DOI: http://dx.doi.org/10.1016/B978-0-12-397008-4.00008-3

assumed that both have errors, so they are combined to give an improved prediction. Taking into account an observation improves the estimation of the state variables, and therefore reduces the uncertainty in future model predictions.

Measurements of output variables are in fact more and more commonly available, with increases in detection and transmission capability. Satellite systems give information about plant biomass, LAI, or leaf chlorophyll content. Tensiometers can be used to give information about soil moisture. Several different methods are available for giving information about plant nitrogen status. In each case, the results of the measurements can be compared to model predictions, and the model can be adjusted in the light of those measurements.

Potentially, such measurements could lead to a large improvement in model predictions. In Chapter 9 on model evaluation we show that MSEP, the mean squared error of prediction, cannot be smaller than a lower bound, which is a measure of how much variability remains that is not accounted for by the explanatory variables in the model. That lower limit to prediction error no longer applies, however, if measurements are used to correct the model. In this case, we are using extra information in addition to the explanatory variables. How good will predictions be after correction with measurements? That depends on a number of factors, in particular, how closely the measured variables and the predicted variables are related. A measurement of LAI early in the season may not improve yield prediction much, while a late measurement of biomass may lead to substantial improvement in yield prediction. We emphasize again that this discussion only concerns the situation where the measurements were made. The correction will not be used when applying the model in other situations.

Section 2 introduces notation and presents four simple examples. Section 3 treats the case of models whose dynamic equations are linear in the state variables and parameters to be modified. We first treat simple special cases, which help introduce the methods and the consequences of data assimilation. Then we treat a fairly general case that covers all the special cases. Of course, essentially, all crop models are non-linear. Nonetheless it is important to treat the linear case. First of all, it is easier to get a basic understanding of how assimilation works from the linear case, where we can derive analytical equations. Secondly, the methods for non-linear models draw more or less on the theory developed for linear models. In Section 4 we consider the problem of data assimilation for non-linear models. We discuss two different approaches to data assimilation for such models. The first one consists of working with a linearized model, and the second approach uses Monte Carlo simulations performed with the original non-linear model.

2. MODEL SPECIFICATION

2.1 Observation and System Equations

Before implementing the Kalman filter or other filtering techniques, it is necessary to express the dynamic model as a combination of two equations, namely an observation equation and a system equation.

2.1.1 Observation Equation

The observation equation relates an observation collected at time t, Y_t, to the model state variables, Z_t as follows:

$$Y_t = f(Z_t, X_t, \delta, \varepsilon_t) \qquad (1)$$

Y_t corresponds to the measurements available at time t. Y_t can include one measurement or several. In the latter case, Y_t is a vector. The state variables Z_t correspond to the variables that need to be estimated from the available measurements, e.g., crop biomass, LAI, soil mineral N, soil C, N_2O gas emission. The function f relates the state variables to the observation; it can be linear or non-linear. A simple special case is $f = $ identity. In this case, Y_t corresponds to a direct measurement of the state variable, Z_i, for example, a biomass measurement. Another special case is

$$f(Z_t, \delta) = \delta_0 + \delta_1 Z_t.$$

In this case, Y_t is not a direct measurement of a state variable but can be related to Z_t using a linear regression model including two parameters $\delta = (\delta_0, \delta_1)$. For example, Naud et al. (2009) used a linear regression model to relate leaf transmittance observations to wheat nitrogen uptake. More complex functions can be used when needed. Some of them may include covariables noted X_t in Equation (1). An example of a function including a covariable is presented in Section 2.4. The residual term ε_t in Equation (1) is a random variable that accounts for the imperfection of the relationship between Y_t and Z_t due to the error of measurement and to the inability of f to relate observations to state variables perfectly. In most cases, ε_t is additive and assumed normally distributed but, in some cases, other probability distributions can be used, especially when the measurements are count data (Makowski et al., 2010). It is usually assumed that the ε_t for different t are all mutually independent and identically distributed (iid).

2.1.2 System Equation

The system equation describes the dynamic behavior of the state variables. It relates the values of the vector of the state variables Z_t at time t to the values at time $t - 1$ as follows

$$Z_t = g\left(Z_{t-1}, X_t, \theta, \eta_{t-1}\right) \qquad (2)$$

where g is a function describing the dynamic of the system, X_t is a vector including input variables (e.g., climate), θ is a vector of parameters, and η_t is a random error term. Depending on the situations, the function g can correspond to a simple function or to a complex one. When g − identity and when the random error is additive, Equation (2) describes a random walk defined by $Z_t = Z_{t-1} + \eta_{t-1}$. In this case, the state variable at time t is then equal to its value at time $t - 1$ plus a random noise. In some cases, Equation (2) will correspond to a complex process-based function (see Chapter 1) calculating several state variables, such as biomass, LAI, and plant nitrogen, at a daily time step as a function of several input variables (e.g., temperature, radiation, farmer practices). It is usually assumed that the η_t for different t are iid.

2.2 Model Parameters

The model defined by Equations (1)−(2) includes four types of parameters: the parameters of the function f (δ), the parameters of the function g (θ), the parameters of the probability distribution of ε_t, and the parameters of the probability distribution of η_t. The number of parameters of the probability distributions of ε_t and η_t depends on several factors, notably on the type of measurements, the number of state variables, and on the family of the probability distributions. For example, if the system equation includes two state variables (i.e., a vector Z_t including two elements) and if η_t is assumed to follow a zero-mean normal distribution, the distribution of η_t is defined by three parameters i.e., two variances and one covariance. All these parameters need to be estimated. Parameter estimation will be discussed in detail in Sections 3 and 4.

2.3 Examples

We present below examples of models including observation and system equations. The first two examples are simple and can be used to analyze time series data. The last two examples concern dynamic system models and are more complex. We only present the equations of the models in this section, and details on the application of the Kalman filter and other filtering techniques to these models will be given in Sections 3 and 4.

2.3.1 Models for Analyzing Yield Time Series

Analysis of crop yield trends constitutes an important component of prospective studies on food security (Paillard et al., 2010). They are used to estimate future levels of production in different areas in the world and to assess the capabilities of cropping systems to satisfy the population's food demand. It is thus necessary to estimate crop yield trends in a rigorous manner and to quantify uncertainties about yield trend estimates. Yield trends can be estimated from yield time series data including yearly yield data, such as those

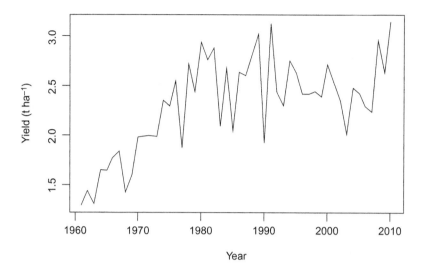

FIGURE 8.1 Wheat yield in Greece from 1961 to 2010 (Source: FAO).

presented in the FAO-stat database. An example of yield time series obtained for wheat in Greece is presented in Figure 8.1. Two simple dynamic models are presented below for analyzing such yield time series. The first model is a random-walk model and the second one is a polynomial dynamic linear model.

Observation Equation

The random-walk model is defined by the following equations:

$$Y_t = Z_t + \varepsilon_t \tag{3}$$

with Y_t the yield data obtained at time t (FAO data), Z_t the true expected yield value at time t, and ε_t a random term describing the between-year yield variability, $\varepsilon_t \sim N(0, \sigma_\varepsilon^2)$.

System Equation

$$Z_t = Z_{t-1} + \eta_{t-1} \tag{4}$$

with $\eta_{t-1} \sim N(0, \sigma_\eta^2)$. The system equation describes the evolution of the expected yield value, i.e., the evolution of the yield value that can be expected across time.

Note that here the state variable at time t, Z_t, is treated as a random variable. There are two sources of information about it. First, there is the dynamic model. If this was the only information available (no measurements), then our knowledge of Z_t would become less and less precise from year to year. Suppose that Z_0 is perfectly well known. Then we would have

$var(Z_1) = \sigma_\eta^2$, $var(Z_2) = var(Z_1) + \sigma_\eta^2 = 2\sigma_\eta^2$, etc. However, the additional information provided by the measurements reduces the variance of Z_t at each measurement time.

An advantage of this model is that it includes only one parameter, namely σ_η^2. A limitation of the random-walk model is that it cannot be used to predict future yield values as the result of a possible trend. If yield is expected to increase linearly across time, it will be useful to take this linear trend into account for predicting future yield values. This is not possible with the random-walk model.

The yield time series presented in Figure 8.1 can be analyzed using a slightly more complex model corresponding to a polynomial dynamic linear model defined as follows:

Observation Equation

$$Y_t = \alpha_{0t} + \alpha_{1t} \times \Delta t + \varepsilon_t \tag{5}$$

with Δt = time duration since the last measurement, and $\varepsilon_t \sim N(0, \sigma_\varepsilon^2)$. Here, we will use $\Delta t = 1$ year because we will consider that a yield measurement is available every year. In Equation (5), the true expected yield value at time t is expressed as a linear function of time. Contrary to the classical linear regression, the intercept and slope are assumed to vary from year to year.

System Equation

$$Z_t = Z_{t-1} + \eta_{t-1} \tag{6}$$

with $Z_t = \begin{pmatrix} \alpha_{0t} \\ \alpha_{1t} \end{pmatrix}$, $\eta_{t-1} \sim N(0, \Sigma)$, and $\Sigma = \begin{pmatrix} \sigma_{\alpha 0}^2 & 0 \\ 0 & \sigma_{\alpha 1}^2 \end{pmatrix}$.

In this model, the quantities that play the role of state variables are the intercept and slope of a linear regression equation. They are assumed to vary from year to year according to a stochastic process defined by the system equation. The slope α_{1t} corresponds to the yearly yield increase rate. The observation equation relates the yield data Y_t, $t = 1, \ldots, N$, to the two state variables

$$Z_t = \begin{pmatrix} \alpha_{0t} \\ \alpha_{1t} \end{pmatrix},$$

and the system equation relates the values of the two state variables at time t to the values at time $t - 1$. The model includes three unknown parameters, namely σ_ε^2, $\sigma_{\alpha 0}^2$, and $\sigma_{\alpha 1}^2$. The variance σ_ε^2 quantifies the variability of yield around the trend. The variances $\sigma_{\alpha 0}^2$ and $\sigma_{\alpha 1}^2$ quantify the variability of the intercept and slope of the linear trend and define their evolution across time.

In these two examples, the function g of Equation (2) is the identity function, but the function f used by the two models is different. In the first model, f = identity but, in the second one, f is a linear regression function. In Section 3, we will show how to use these models for estimating yield trends and predicting future yield values.

2.3.2 Soil Carbon Model

The soil C model is presented in detail in Appendix 1. It simulates the mass of carbon in the top 20 cm of soil in a field. In order to apply the Kalman filter to this model, we define below a stochastic version of this model relating soil C measurements Y_t to C_t, the true soil C in year t, and to U_t, the amount of C in crop biomass added to the soil in year t.

Observation Equation

Soil C measurements (Y_t) may be made yearly or less frequently. It is assumed that soil C measurement error is normally distributed. Thus,

$$Y_t = C_t + \varepsilon_t \tag{7}$$

with ε_t a random term describing measurement error,

$$\varepsilon_t \sim N\left(0, \sigma_\varepsilon^2\right).$$

System Equation

The main model output is the mass of carbon (C_t, kg ha^{-1}) in the top 20 cm of soil in a single field. Changes in soil C are simulated dynamically on a yearly basis. The model also has one unknown parameter, R, the fraction of soil C that is decomposed per year (yr^{-1}). The equation that describes the behavior of this system is dynamic and is adapted from Jones et al. (2004):

$$C_t = C_{t-1} - R \times C_{t-1} + b \times U_{t-1} + \eta_{t-1} \tag{8}$$

where t is time (in years) from an arbitrary starting year when initial values of soil C are known, C_t is the true soil C in year t, R is the true soil C decomposition rate (yr^{-1}), U_t is the amount of C in crop biomass added to the soil in year t (kg[C] ha^{-1} yr^{-1}), b is the fraction of crop biomass C that is added to the soil in year t that remains after one year (yr^{-1}), and

$$\eta_{t-1} \sim N\left(0, \sigma_\eta^2\right)$$

is a random term representing model error (kg[C] ha^{-1}). In principle, this model includes two state variables,

$$Z_t = \begin{pmatrix} C_t \\ U_t \end{pmatrix},$$

but we consider here that U_t is not random and is perfectly known at each time step. Thus, only one state variable needs to be estimated here (soil carbon C_t).

2.3.3 Water Balance

We consider the water balance model described in Appendix 1. This model is a budget-based soil water balance developed for grassland. The water balance has a single state variable, the amount of water at the beginning of the day. We assume that the amount of water in the soil can be measured several times every year. Measurements are related to the state variables using an observation equation, and the dynamics of the state variable are described using a non-linear system equation.

Observation Equation

It is assumed that soil water content measurements are related to the true soil water content as follows:

$$Y_t = Z_t + \varepsilon_t \tag{9}$$

with ε_t a random term describing measurement error, $\varepsilon_t \sim N(0, \sigma_\varepsilon^2)$, independent in time and independent of Z_t.

System Equation

The dynamics of the state variable are described using a non-linear model presented in Appendix 1 that can be expressed as

$$Z_t = g(Z_{t-1}, X_t, \theta) + \eta_{t-1} \tag{10}$$

where X_t is a vector of two input variables, precipitation and evapotranspiration, and θ is a vector of five parameters (Appendix 1).

2.3.4 Weed Population Model

We consider here the population demography model developed for *A. myo suroides* by Munier-Jolain et al. (2002). This model includes four state variables: the early seedling density before post-emergence weed control at year t noted d_t, the soil seed bank in the upper soil layer after tillage at year t noted SSB_t, the deep seed bank density after tillage at year t noted DSB_t, and the number of weed seeds produced in year t noted by S_t. All the state variables are simulated by the model for a number of years in one or several plots of interest. Each year is supposed to begin at the harvest of the previous crop. The vector including the values of the four state variables at year t is denoted further as

$$Z_t = \begin{pmatrix} d_t \\ SSB_t \\ DSB_t \\ S_t \end{pmatrix}$$

As shown by Munier-Jolain et al. (2002), the four state variables defined above can be used to compute other variables of interest like seedbank density before tillage, weed density after weed control, and crop yield.

Observation Equation

Weed counts were related to the weed densities simulated by the model using a probability density function. As the measurements correspond here to count data (number of weed plants), the measurements collected in a given plot were related to the weed density d_t simulated by the model for the same plot using a Poisson probability density. Let Y_t denote the number of weeds measured in year t in n subplots of the studied plot. The conditional probability of Y_t given d_t is given by a Poisson probability distribution defined by:

$$P(Y_t|d_t) = \frac{\exp(-d_t \times s \times n) \times (d_t \times s \times n)^{Y_t}}{Y_t!}$$ (11)

where

$$Y_t = \sum_{k=1}^{n} Y_{tk}$$

is the total number of weed plants counted in year t in the n subplots taken in a given plot, Y_{tk} is the weed count in the kth subplot, $k = 1, \ldots, n$, s is the surface area of each subplot, and d_t is the weed density before weed control computed by the model. The conditional distribution is in fact all that is needed to do the assimilation calculations because it defines the relationship between measurements and model state variables.

System Equation

The values of the four state variables at year t are related to their values at year $t - 1$ using a complex non-linear function expressed here as:

$$Z_t = g(Z_{t-1}, X_{t-1}, \theta) \times \eta_t$$ (12)

where X_t is a vector including input variables (type of crop, soil tillage, type of herbicide treatment),

$$\eta_t = \begin{pmatrix} \eta_t^{(1)} \\ \eta_t^{(2)} \\ \eta_t^{(3)} \\ \eta_t^{(4)} \end{pmatrix}$$

is the vector of the random term errors, $\eta_t^{(k)}$ is a random term representing the model error associated with the kth state variable, $k = 1, 2, 3, 4$. A Gamma probability distribution is used for $\eta_t^{(k)}$ in order to obtain positive state variable values. The random term errors are thus assumed identically and independently distributed as,

$$\eta_t^{(k)} \sim \text{Gamma}(\lambda, \lambda), \qquad k = 1, 2, 3, 4$$

This probability distribution depends on a single parameter λ, is defined for positive values of $\eta_t^{(k)}$ only, and has an expectation equal to one. The Gamma probability distribution is a natural choice here because this distribution is frequently used for positive random variables (e.g., Carlin and Louis, 2000). The same parameter λ is used for the four state variables but, because of the multiplicative relationship between the state variables and the errors, the variances of the four state variables are different.

3. FILTER AND SMOOTHER FOR GAUSSIAN DYNAMIC LINEAR MODELS

3.1 Definitions

In a Gaussian dynamic linear model, the functions f and g of Equations (1)–(2) are linear in the state variables and the random error terms ε_t and η_{t-1} are normally distributed. More formally, a Gaussian dynamic linear model is defined as follows:

Observation Equation

$$Y_t = FZ_t + \varepsilon_t \tag{13}$$

where F is a matrix and ε_t is a Gaussian random term. If Y_t includes N measurements and if Z_t includes m state variables, F is a $(N \times m)$ matrix, $\varepsilon_t \sim N(0, V)$, and V is a $(N \times N)$ variance-covariance matrix.

System Equation

$$Z_t = GZ_{t-1} + \eta_{t-1} \tag{14}$$

where G is a $(m \times m)$ matrix, $\eta_t \sim N(0, W)$, and W is a $(m \times m)$ variance-covariance matrix. Examples of matrices F, G, V, and W are given in Section 3.2.

The model defined by Equations (13)–(14) is a special case of space state model (Petris et al., 2009). The two yield models defined in Section 2.3.1 and the soil carbon model defined in Section 2.3.2 are Gaussian linear models, but the water balance model presented in Section 2.3.3 and in Appendix 1 is not. For example, in the random-walk yield model, both F and G are equal to a (1×1) identity matrix.

The Kalman filter aims at calculating the conditional expected value and variance of the state variables Z_t given the measurements collected up to time t, denoted by $E(Z_t|y_{1:t})$ and $Var(Z_t|y_{1:t})$, respectively. The vector $y_{1:t}$ includes the measurements collected up to time t. $E(Z_t|y_{1:t})$ is the conditional expected value of the state variable at time t given the data available up to and including time t. It can be used as a point estimate of Z_t. $Var(Z_t|y_{1:t})$ quantifies how much the value of Z_t remains uncertain.

The smoother aims at calculating the conditional expected value and variance of Z_t given all available measurements. The expected value and variance computed by the smoother are denoted by $E(Z_t|y_{1:N})$ and $Var(Z_t|y_{1:N})$, where $y_{1:N}$ is the set of the N available data, including those obtained after time t. The filter and the smoother differ in the number of data points used to compute the conditional expected values and variances. The filter is useful to estimate the state variables sequentially at each time step and its results obtained at the time of the last measurement can be used for predicting future state variable values. Contrary to the filter, the smoother uses all available data and this technique is useful to go back in time, i.e., to estimate past values of the state variable. Results of the smoother and of the filter are identical at the time of the last measurement.

When the model is a Gaussian linear model defined by Equations (13)–(14) and when the matrices F, G, V, and W are known, conditional expected values and variances of the state variables can be calculated analytically. It is thus possible to calculate the exact values of the conditional expected values and variances. With a model defined by Equations (13)–(14), the value of Z_t conditioned to the measurements follows a normal distribution whose expected values and variances are given by the Kalman filter and the Kalman smoother.

A basic feature of the Kalman filter is that the values of $E(Z_t|y_{1:t})$ and $Var(Z_t|y_{1:t})$ can be computed sequentially at each time. The Kalman filter includes two steps: a propagation step and a correction step. The propagation step calculates $E(Z_t|y_{1:t-1})$ and $Var(Z_t|y_{1:t-1})$ from $E(Z_{t-1}|y_{1:t-1})$ and $Var(Z_{t-1}|y_{1:t-1})$. The correction step calculates $E(Z_t|y_{1:t})$ and $Var(Z_t|y_{1:t})$ from $E(Z_t|y_{1:t-1})$ and $Var(Z_t|y_{1:t-1})$. When no new measurement is available at time t, only the propagation step is applied.

The equations of the correction step are presented below (Petris et al., 2009):

$$E(Z_t|y_{1:t}) = E(Z_t|y_{1:t-1}) + K\left[y_t - FE(Z_t|y_{1:t-1})\right] \tag{15}$$

$$Var(Z_t|y_{1:t}) = (I - KF)Var(Z_t|y_{1:t-1}) \tag{16}$$

with

$$K = Var(Z_t|y_{1:t-1})F^T\left[F\,Var(Z_t|y_{1:t-1})F^T + V\right]^{-1} \tag{17}$$

where I is the identity matrix and K is the Kalman gain matrix. Equations (15)–(17) are used to update the expected value and the variable of the state variables at each time step when a new observation is available. Equations (15)–(17) are simply the standard equations for the conditional means and variances of variables with a joint normal distribution (see Chapter 2). In Equation (15), the term $y_t - FE(Z_t|y_{1:t-1})$ is the difference between the measurement at time t and the predicted value of this measurement derived from the model state variables before the correction step. If the difference between the measurement at time t and its prediction is large, $E(Z_t|y_{1:t})$ will differ strongly from $E(Z_t|y_{1:t-1})$. On the contrary, $E(Z_t|y_{1:t})$ will remain close to $E(Z_t|y_{1:t-1})$ if $y_t - FE(Z_t|y_{1:t-1})$ is small.

When the measurement error has a large variance (i.e., a high value of V), the Kalman gain tends toward zero. In this case, the use of the new observation does not improve our knowledge of the state variables and $E(Z_t|y_{1:t}) \approx E(Z_t|y_{1:t-1})$ and $Var(Z_t|y_{1:t}) \approx Var(Z_t|y_{1:t-1})$. On the other hand, when the measurement error has a small variance V, the expected value and variance of the state variable are strongly influenced by the new measurement and could then differ strongly from the values obtained at the previous time step.

The *smoother* is a backward recursive algorithm to compute the conditional distribution of Z_t given $y_1, \ldots, y_t, \ldots, y_N$ (i.e., given all the measurements, including those after time t). This algorithm is useful when the objective is to retrospectively reconstruct the behavior of the system. The algorithm is based on the following equations (Petris et al., 2009):

$$
\begin{aligned}
E(Z_t|y_{1:N}) = {} & E(Z_t|y_{1:t}) + Var(Z_t|y_{1:t})G^T Var(Z_{t+1}|y_{1:t})^{-1} \\
& \times \left[E(Z_{t+1}|y_{1:N}) - E(Z_{t+1}|y_{1:t}) \right]
\end{aligned}
\tag{18}
$$

$$
\begin{aligned}
Var(Z_t|y_{1:N}) = {} & Var(Z_t|y_{1:t}) - Var(Z_t|y_{1:t})G^T Var(Z_{t+1}|y_{1:t})^{-1} \\
& \times \left[Var(Z_{t+1}|y_{1:t}) - Var(Z_{t+1}|y_{1:N}) \right] \\
& \times Var(Z_{t+1}|y_{1:t})^{-1} G\, Var(Z_t|y_{1:t})
\end{aligned}
\tag{19}
$$

3.2 Gaussian Dynamic Linear Models with Known Parameters

When the model parameters are known, the Kalman filter and the smoother can directly be used to update the model state variables using a set of available measurements. We show below how these techniques are implemented in the R package dlm (dynamic linear model) (Petris, 2010) using several examples of dynamic linear models.

3.2.1 Example 1: Yield Models

The random-walk model in Equations (3)−(4) is specified with the function dlmModPoly of the package dlm as follows:

```
MyModel <-function(x) {
return(dlmModPoly(1, dV = exp(x[1]), dW = exp(x[2])))}
```

The first input of dlmModPoly defines the number of terms of the linear function used in the observation equation (i.e., only one term here). The *R* function MyModel defines a random-walk model including two parameters, namely σ_ε^2 and σ_η^2. In MyModel, these two parameters are called dV and dW, respectively. dV and dW are keywords for dlm. They correspond to V and W in Equations (13)−(14). Values of dV and dW are specified using the vector x. For example, values of 0 and -5 are allocated to x[1] and x[2] using the following *R* code:

```
FittedModel <-MyModel(c(0,-5))
```

The vector $c(0, -5)$ includes two elements, x[1] and x[2], related to dV and dW using an exponential function in order to constrain the variances to take positive values only. The vector is passed to the function MyModel to define a random-walk model with specific value of σ_ε^2 and σ_η^2. This model is called FittedModel. Once the parameter values are specified, the Kalman filter and smoother are implemented as follows:

```
YieldFilter<-dlmFilter(Yield, FittedModel)
YieldSmooth<-dlmSmooth(Yield, FittedModel)
```

where `Yield` is a vector including all the yield data. The first element corresponds to the first date of measurement, and the last element corresponds to the last date of measurement. If data are missing at a given date (missing relative to the dates at which model outputs are produced), an NA should be included in the vector. By default, the filter and smoother are implemented using a pre-sample expected state value equal to zero and a very large pre-sample state variance (10^7). If necessary, these values can be modified by the user if they have some prior knowledge about the initial state of the system. The conditional expected values and variances of the model state variable Z_t are calculated using dlmFilter and dlmSmooth at each time step using Equations (15)−(19).

Results can then be plotted using the usual R functions (Figure 8.2):

```
plot(Year,Yield,ylab = "Yield (t ha-1)", type = "l",lwd = 1)
lines(Year,YieldFilter$m[-1],lwd = 2, lty = 3)
lines(Year,YieldSmooth$s[-1],lwd = 2)
```

The Kalman filter gives values of $E(Z_t|y_{1:t})$ at each time step, whereas the smoother gives values of $E(Z_t|y_{1:N})$ for all t, where N is the total number of yield measurements. Figure 8.2 shows that the results obtained with the Kalman filter and smoother are quite different. The yield values estimated with the Kalman filter are lower and much less smooth compared to the yield values estimated with the Kalman smoother. These differences are due to the

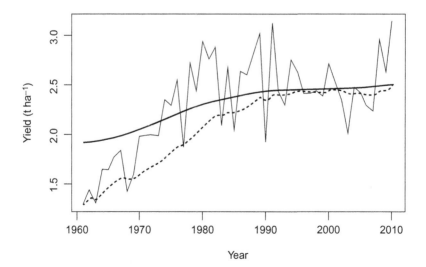

FIGURE 8.2   Results of the Kalman filter (dotted line) and of the Kalman smoother (thick continuous line) obtained with the random-walk model (Equations (3)−(4)) for the Greek wheat yield data (thin continuous line). Parameter values were not adjusted by the data.

fact that the yield trend estimated for a given year with the Kalman filter is derived from the data collected up to this year, whereas the yield trend estimated with the Kalman smoother is always derived using the whole time series. The difference between the two techniques is much smaller for the most recent years and is even equal to zero for the last year.

The dynamic linear model from Equations (5)–(6) is defined below using the function dlmModPoly:

```
MyModel <-function(x) {
return(dlmModPoly(2, dV = exp(x[1]), dW = c(exp(x[2]), exp(x[3]))))}
```

The first input of dlmModPoly indicates that the linear function used in the observation equation includes two terms. The variances σ_ε^2, $\sigma_{\alpha0}^2$, and $\sigma_{\alpha1}^2$ are calculated from x[1], x[2], and x[3]. Their values are specified using the vector x. For example, the values 0, −5, and −5 are allocated to x[1], x[2], and x[3] using the following R code:

```
FittedModel <- MyModel(c(0,-5,-5))
```

FittedModel is an *R* function defining a dynamic linear model with specific values of σ_ε^2, $\sigma_{\alpha0}^2$, and $\sigma_{\alpha1}^2$. Once the parameter values are specified, the intercept α_{0t} and slope α_{1t}, $t = 1, \ldots, N$, can be estimated using the Kalman filter and smoother. With the smoother, the slope at time t (i.e., the yield increase rate) is estimated by the expected value of α_{1t} conditionally to all available measurements, $E(\alpha_{1t}|Y_{1:N})$, and the yield trend at time t is estimated by $E(\alpha_{0t}|Y_{1:N}) + E(\alpha_{1t}|Y_{1:N}) \times \Delta t$ (here, we use $\Delta t = 1$ because a yield measurement is available at each time step). It is possible to define credibility ranges by computing some quantiles of the state variables from the variances of α_{0t} and α_{1t} conditionally to all the available measurements. The filter proceeds in a similar way but, with this method, only the measurements available up to time t are used. By default, the filter and smoother are implemented using a pre-sample expected state vector value equal to zero and a very large pre-sample state vector variance (10^7).

The Kalman filter and smoother are implemented with dlm as follows:

```
YieldFilter <-dlmFilter(Yield, FittedModel)
YieldSmooth <-dlmSmooth(Yield, FittedModel)
plot(Year,Yield,ylab = "Yield (t ha-1)", type = "l",lwd = 1)
lines(Year,YieldFilter$m[,1][-1]+YieldFilter$m[,2][-1],lwd = 2,
lty = 3)
lines(Year,YieldSmooth$s[,1][-1]+YieldSmooth$s[,2][-1],lwd = 2)
```

Results are shown in Figure 8.3. Here also, the yield values estimated with the Kalman filter are much less smooth than the yield values estimated with the Kalman smoother. When one is interested by the past yield trend, it is more appropriate to use the results given by the smoother because these results are based on a larger number of data.

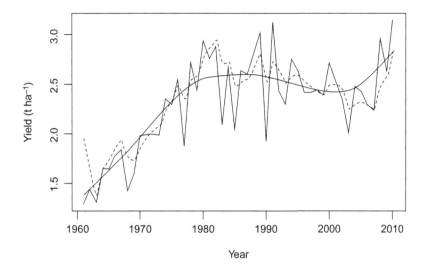

FIGURE 8.3   Results of the Kalman filter (dotted line) and of the Kalman smoother (thick continuous line) obtained with the dynamic linear model (Equations (5)−(6)) for the Greek wheat yield data (thin continuous line). Parameter values were not adjusted to the data.

The parameter values considered in Figures 8.2 and 8.3 were only used for illustration and were not adjusted to the yield data. We will show how to estimate these parameters by maximum likelihood in Section 3.3.

3.2.2 Example 2: Soil Carbon Model

The soil carbon model shown in Equations (7)−(8) corresponds to a Gaussian linear model (Equations (13−14)) with

$$Z_t = \begin{pmatrix} C_t \\ U_t \end{pmatrix}, \quad \varepsilon_t \sim N(0, V),$$

and with the following matrices G, F, and W:

$$G = \begin{pmatrix} 1 - R & b \\ 0 & 1 \end{pmatrix}, \quad F = \begin{pmatrix} 1 & 0 \end{pmatrix}, \quad W = \begin{pmatrix} \sigma_C^2 & 0 \\ 0 & 0 \end{pmatrix}.$$

The matrix W defines the year to year variability of the model state variables. The model includes two state variables but, as mentioned above, we consider here that the second state variable, the amount of C in crop biomass added to the soil in year t U_t, is not random and is always equal to $U_t = U_0 = 2000$ kg ha^{-1}. For this reason, the second element of the diagonal of matrix W is set equal to zero. This model includes four parameters, namely R, b, V, and σ_C^2. We set $b = 0.2$ and $V = 500{,}000$ (Jones et al., 2004). The other two parameters are discussed below.

The carbon model is specified using the `dlm` function as follows:

```
ModelC <-function(x) {
mGG <-matrix(0,2,2)
mGG[1,1] <-x[1]
mGG[2,2] <-1
mGG[1,2] <-0.2
mW <-matrix(0,2,2)
mW[1,1] <-exp(x[2])
mW[2,2] <-0
mCO <-matrix(0,2,2)
mCO[1,1] <-500000
mCO[2,2] <-0
return(dlm(FF = matrix(c(1,0),1,2),  V = 500000,  GG = mGG,  W = mW,
m0 = c(16000,2000),CO = mCO))
}
```

The `dlm` function uses the following arguments: `FF` specifies the matrix F, V specifies the variance-covariance matrix of the observation equation (reduced here to one variance), `GG` specifies the matrix G, `W` specifies the matrix W, `m0` specifies the initial values of the state variables Z, and `CO` specifies the initial variance-covariance matrix of the state variables. The parameter R and σ_C^2 are obtained from the function argument `x`; R is equal to $1 - x[1]$ and σ_C^2 is equal to `exp(x[2])`. As before, we use the exponential function in order to restrict the variance to positive values. For this example, the values 0.999 and 13 are allocated to `x[1]` and `x[2]`, respectively, using the following R code:

```
FittedModel <-ModelC(c(0.999, 13))
```

As shown here, the parameters can either be defined within the function that calls `dlm`, as done here for the parameters b and V, or they can be passed as arguments to that function (the case here for `x[1]` and `x[2]` used, respectively, to calculate R and σ_C^2). As we will see below, parameters that are to be estimated by maximum likelihood must be passed as arguments.

Once the parameter values are specified, the Kalman filter and smoother are implemented using a series of soil carbon measurements (listed in a vector noted by `Ym` that may include missing data) as follows:

```
CarbFilter <-dlmFilter(Ym, FittedModel)
CarbSmooth <-dlmSmooth(Ym, FittedModel)
Smooth <-CarbSmooth$s
```

Soil carbon can be predicted for the next 10 years using the `dlmForecast` function:

```
foreCarb <-dlmForecast(CarbFilter,nAhead = 10)
Pred <-foreCarb$f
```

The forecasted values are recovered in the *R* object forCarb$f and their variances are recovered in foreCarb$Q.

The variance of the state variable is computed using the function dlmSvd2var. An example is given below where the variance of $Z_t|y_1, \ldots, y_N$ is computed from the smoother as follows:

```
Var <-dlmSvd2var(CarbSmooth$U.S,CarbSmooth$D.S)
VarC <-1:length(Ym)
for (i in 1:length(VarC)) {
VarC[i] <-Var[[i]][1,1]
}
```

where CarbSmooth$U.S and CarbSmooth$D.S are two matrices used for calculating the variances of the state variables.

The vector VarC includes the variances of the soil carbon state variable for the period of time where data were collected. Results are plotted using the plot and lines functions:

```
plot(1:40,Ym,xlab = "Year",ylab = "Soil C kg ha−1",xlim = c(1,50),
ylim = c(15000,27000))
lines(1:N,Smooth[,1][−1],lwd = 2)
lines(1:N,Smooth[,1][−1] + qnorm(0.95)*sqrt(VarC),lty = 2)
lines(1:N,Smooth[,1][−1]−qnorm(0.95)*sqrt(VarC),lty = 2)
lines(41:50,Pred)
lines(41:50,Pred + qnorm(0.95)*sqrt(unlist(foreCarb$Q)),lty = 2)
lines(41:50,Pred−qnorm(0.95)*sqrt(unlist(foreCarb$Q)),lty = 2)
```

Figure 8.4 shows the smoothed and predicted soil carbon values and their 5% and 95% percentiles obtained with x[1] = 0.999 and x[2] = 13 (arbitrary values considered for illustration). The 5% to 95% interval is much larger for predicted values than for smoothed values because the use of soil C measurements led to a large reduction in the variance of the state variable.

Note that after the last measurement date the range between the 5 and 95 percentiles is increasing with time. This reflects the fact that, in the absence of measurements, the model predictions become more and more uncertain with time. Note also that the results of the smoother are not very 'smooth'. As we will see in the next section, this is because the values of R and σ_C^2 were not well chosen.

3.3 Gaussian Linear Model with Unknown Parameters

3.3.1 Principles

In practice, at least some of the model parameters are unknown. Usually, it is thus necessary to estimate some of the elements of the matrices F, G, V, and W. A first approach is to guess the parameter values and, thus, to fix the

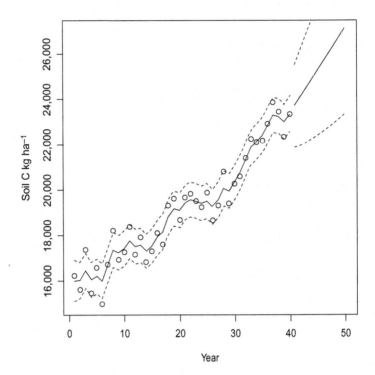

FIGURE 8.4 Smoothed and predicted soil carbon values using the model defined by Equations (7)–(8). Points correspond to soil carbon measurements, continuous lines correspond to smoothed (bold) and predicted (thin) values, dashed lines correspond to 5% and 95% percentiles. Parameter values were not adjusted to the data.

parameters to some arbitrary values. The risk is then one of obtaining poor results with the Kalman filter and smoother. Another approach is to estimate the model parameters from data.

Several methods of parameter estimation were presented in Chapter 6. In this section, we consider one of these methods, namely maximum likelihood. We show how model parameters can be estimated by maximum likelihood from the series of measurements $y_1, \ldots, y_t, \ldots, y_N$ using the R package dlm. The likelihood of the model defined in Equations (13)–(14) is expressed as:

$$P(y_1, \ldots, y_N | \varphi) = \prod_{t=1}^{N} P(y_t | y_{1:t-1}, \varphi)$$

where φ is the vector including all the unknown elements that need to be estimated.

The maximum likelihood estimation method aims at finding the values of the unknown elements of F, G, V, and W maximizing this likelihood function.

Maximum likelihood estimation is implemented in the function dlmMLE (Petris, 2010). This function requires starting values for the model parameters, and estimates these parameters from a set of available measurements using an iterative algorithm implemented with the R function optim already described in Chapter 6. Other methods can be used for parameter estimation, especially a Bayesian method (Petris et al., 2009), but their implementation for a dynamic linear model is more complex and they are thus not presented here.

Before running dlmMLE, it is necessary to define the model using, for example, the function dlm or the function dlmModPoly, as shown above. The estimated parameters must be arguments to the model function. Model parameters are then iteratively estimated from the data. Once the parameters are estimated, the filtering and smoothing algorithms can be applied as shown above.

3.3.2 Yield Models

The two parameters of the random-walk model, σ_ε^2 and σ_η^2, are estimated with the following R code dlmMLE(Yield,parm = c(0,0), build = MyModel). The algorithm implemented by dlmMLE uses $\sigma_\varepsilon^2 = \sigma_\eta^2 = \exp(0) = 1$ as starting values and, after a few iterations, returns two estimated parameter values: -2.65 and -4.26. These two values correspond to the estimated values of x[1] and x[2]. The estimated values for σ_ε^2 and σ_η^2 are equal to $\exp(-2.65)$ and $\exp(-4.26)$, respectively.

Similarly, the three parameters of the polynomial model Equations (5)–(6) are estimated using dlmMLE(Yield,parm = c(0,0,0), build = MyModel), and the estimated values of σ_ε^2, $\sigma_{\alpha0}^2$, and $\sigma_{\alpha1}^2$ are equal to $\exp(-2.56)$, $\exp(-20.38)$, and $\exp(-8.34)$, respectively.

With both models, the estimated parameter values are stored in an R object called fitMyModel$par. The Kalman filter and smoother can be implemented using the estimated parameter values as follows:

```
FittedModel <-MyModel(fitMyModel$par)
YieldFilter <-dlmFilter(Ym, FittedModel)
YieldSmooth <-dlmSmooth(Ym, FittedModel)
```

Filtered and smoothed yield values can then be presented graphically as before. Figure 8.5 shows the results. Compared to the results obtained without parameter estimation (Figures 8.2–8.3), the yield trends obtained after parameter estimation fit the data better (Figure 8.5). Results obtained with the two models are much more similar, although the results obtained with the polynomial model Equations (5)–(6) are smoother.

3.3.3 Soil Carbon Model

The parameters R and σ_C^2 of the soil carbon model can be estimated by maximum likelihood using soil carbon measurements using dlmMLE:

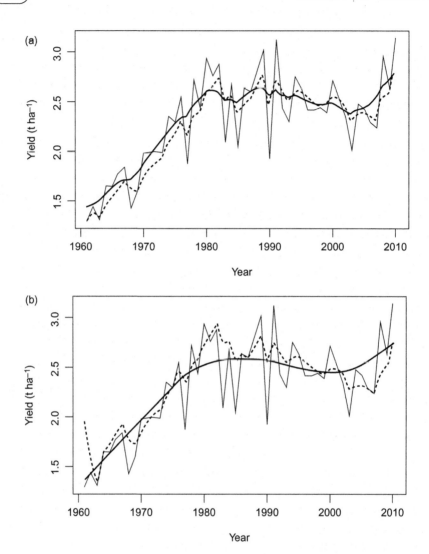

FIGURE 8.5 Results of the Kalman filter (dotted line) and of the Kalman smoother (thick continuous line) obtained with the random-walk model, Equations (3)–(4) **(A)** and with the polynomial model, Equations (5)–(6) **(B)** for the Greek wheat yield data (thin continuous line). Parameter values were estimated from the data by maximum likelihood

```
fitTemp <-dlmMLE(Ym, parm = c(0.5,30), build = ModelC, lower = c(0,
1), upper = c(1, 100))
```

As mentioned in Section 3.2.2, the parameter values to be estimated are passed as arguments to the function ModelC. Here, the arguments are the two components of the vector x. From these, the two parameters to be estimated are calculated as $R = 1 - x[1]$ and $\sigma_C^2 = \exp(x[2])$. The starting values are

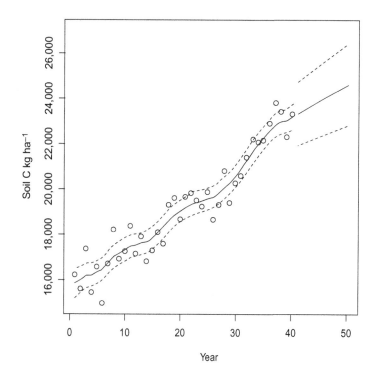

FIGURE 8.6 Smoothed and predicted soil carbon values using the model defined by Equations (7)–(8). Points correspond to soil carbon measurements, continuous lines correspond to smoothed (bold) and predicted (thin) values, dashed lines correspond to 5% and 95% percentiles. Parameter values were estimated by maximum likelihood from the soil carbon data.

set to $x[1] = 0.5$ and $x[2] = 30$. In this example, we illustrate the use of lower and upper bounds in dlmMLE. We constrain the parameter values to be in the range $0-1$ for R and to be in the range $1-100$ for $x[2]$ using lower $= c(0,1)$ and upper $= c(1,100)$. The estimated parameter values are $x[1] = 0.9891$ (i.e., $R = 0.0109$), and $\sigma_C^2 = 65{,}512.75$. Results obtained with the estimated parameter values are shown in Figure 8.6. Compared to the results obtained with arbitrary parameter values (Figure 8.4), the soil carbon trend obtained with the maximum likelihood estimates is much smoother and variations in predicted values are smaller (Figure 8.5).

4. FILTER AND SMOOTHER FOR NON-LINEAR MODELS

In Section 3, we treated the case where the functions f and g of Equations (1)–(2) are linear and the random error terms ε_t and η_{t-1} are normally distributed. In this section, we consider the case where the function g is not linear and we show several extensions of the Kalman filters that can be implemented with this type of model. Some of these extensions assume that the random error terms ε_t and η_{t-1} are normally distributed. However, we will also present a method called particle filter that can be implemented with non-Gaussian error terms.

4.1 Working with a Linearized Model

In this section, we present two approaches where the original non-linear model $Z_t = g(Z_{t-1}, X_t, \theta) + \eta_{t-1}$ is replaced by a linear model. The standard Kalman filter is then applied as explained above.

4.1.1 Extended Kalman Filter

Observation Equation

We consider a dynamic model defined by:

$$Y_t = FZ_t + \varepsilon_t$$

where F is a matrix and ε_t is a Gaussian random term. If Y_t includes n measurements and if Z_t includes m state variables, F is an $(n \times m)$ matrix, $\varepsilon_t \sim N(0, V)$, and V is an $(n \times n)$ variance-covariance matrix.

System Equation

$$Z_t = g(Z_{t-1}, X_t, \theta) + \eta_{t-1}$$

where g is a function describing the dynamic of the system, X_t is a vector including input variables (e.g., climate), θ is a vector of parameters, and η_t is a random error term. In the previous section, we considered a particular case with $g(Z_{t-1}, X_t, \theta) = GZ_{t-1}$. Here, we consider a more general case where $g(Z_{t-1}, X_t, \theta)$ can be non-linear.

The principle of the Extended Kalman filter (e.g., Welch and Bishop, 2002; Pastres et al., 2003) is to linearize $g(Z_{t-1}, X_t, \theta)$ and to apply the standard Kalman filter method described above to the following model:

$$Z_t \approx g(\hat{Z}_{t-1}, X_t, \theta) + H_{t-1}(Z_{t-1} - \hat{Z}_{t-1}) + w_{t-1}$$

where H_t is an $(m \times m)$ matrix of partial derivatives of g with respect to the m elements of Z_t, $\hat{Z}_t$ is the predicted state variable at time t,

$$\hat{Z}_t = \hat{E}(Z_t | y_1, \ldots, y_t),$$

and w_t is an m-error term vector assumed to be normally distributed. The main drawback of this method is that the linearization has been shown to be a poor approximation in a number of applications. The linear approximation may not give a good description of how the model errors evolve over time.

4.1.2 Dynamic Regression Model

We consider here a different model defined as follows:

Observation Equation

$$Y_t = \alpha_{0t} + \alpha_{1t}O_t + \varepsilon_t$$

where O_t is the output of the original non-linear model,

$$O_t = g(X_t, \theta),$$

at time t, and

$$\varepsilon_t \sim N\left(0, \sigma_\varepsilon^2\right).$$

System Equation

$$Z_t = Z_{t-1} + \eta_{t-1}$$

with $Z_t = \begin{pmatrix} \alpha_{0t} \\ \alpha_{1t} \end{pmatrix}$, $\eta_{t-1} \sim N(0, \Sigma)$, and $\Sigma = \begin{pmatrix} \sigma_{\alpha0}^2 & 0 \\ 0 & \sigma_{\alpha1}^2 \end{pmatrix}$.

This model is a dynamic regression model (Petris, 2009) where the measurements Y_t are related linearly to the output of the original model O_t, $t = 1, \ldots, N$. The intercept and slope of the regression equation are treated as the two state variables,

$$Z_t = \begin{pmatrix} \alpha_{0t} \\ \alpha_{1t} \end{pmatrix}.$$

They are assumed to vary over time according to a stochastic process defined by the system equation. It means that the relationship between measurements and state variables is allowed to vary across time.

The system equation relates the values of the two state variables at time t to the values at time $t-1$. The model includes three parameters, namely σ_ε^2, $\sigma_{\alpha0}^2$, and $\sigma_{\alpha1}^2$. The variance σ_ε^2 quantifies the variability of the measurements around the linear trend. The variances $\sigma_{\alpha0}^2$ and $\sigma_{\alpha1}^2$ quantify the variability of the intercept and slope of the linear trend and define their evolution over time.

The use of this approach is illustrated here with the water balance non-linear model presented in Appendix 1. In this case study, O_t is the water balance simulated at day t by the process-based water balance model, and Y_t is the measured soil water content on day t. The model is defined with the package dlm using the function dlmModReg as shown below:

```
buildFun <- function(Para) {
modWAT <- dlmModReg(x, dV = Vemp, dW = c(exp(Para[1]), exp(Para[2])))
return(modWAT)}
```

The variable x is a variable that corresponds to the variable O_t defined above and Para is a vector including the parameters that need to be estimated from the data. As explained above, the dynamic regression model includes three parameters, namely σ_ε^2, $\sigma_{\alpha0}^2$, and $\sigma_{\alpha1}^2$. We consider here that

$\sigma_{\alpha 0}^2$ and $\sigma_{\alpha 1}^2$ are unknown and need to be estimated from the data by maximum likelihood. The parameter σ_{ε}^2 is assumed to be known and is fixed to the empirical variance of the soil water content measurements (variances calculated from the replicates). The two parameters, $\sigma_{\alpha 0}^2$ and $\sigma_{\alpha 1}^2$, are estimated from the vector of measurements Y using the dlmMLE function, and the Kalman filter is applied with the dlmFilter:

```
fit<-dlmMLE(Y,parm=c(0,0),build=buildFun)
fitted.modWAT<-buildFun(fit$par)
modFilter<-dlmFilter(Y,mod=fitted.modWAT)
plot(1:length(x),100*Y,ylim=c(0,50),pch=19, xlab="Time (days)",
ylab="Soil water content (%)")
lines(1:length(x),100*x,lty=2)
lines(1:length(x),100*(modFilter$m[-1,1]+modFilter$m[-1,2]*x))
```

The results are shown in Figure 8.7. The soil water contents estimated by the Kalman filter are closer to the measurements than the initial simulations of the water-balance model.

4.2 Methods Based on Monte Carlo Simulations

In this section, we present two methods based on Monte Carlo simulation. These two methods can be applied to the non-linear dynamic model without linearization of the system equation. The first method is the Ensemble Kalman

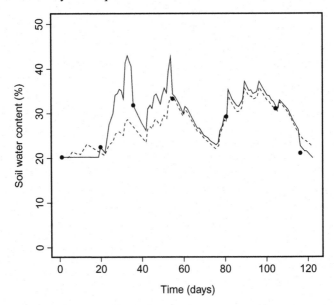

FIGURE 8.7 Soil water content simulated with the process-based water balance model (dashed line) and by using the Kalman filter (continuous line). Black points indicate the water content measurements used for both parameter estimation and filtering.

filter. This method assumes that the probability distribution of the Z_t conditionally to the set of measurements $y_1, \ldots, y_t$, denoted by $P(Z_t | y_1, \ldots, y_t)$, is Gaussian. The second method, called particle filter, is more recent, but it is also more general; it does not assume that $P(Z_t | y_1, \ldots, y_t)$ is Gaussian.

4.2.1 Ensemble Kalman Filter

The Ensemble Kalman filter is a method described by Burgers et al. (1998). The principle is to approximate the probability distribution $P(Z_t | y_1, \ldots, y_t)$ using random samples of state variables. First, an ensemble of M values of $Z_t = g(Z_{t-1}, X_t, \theta) + \eta_{t-1}$, noted $Z_t^1, \ldots, Z_t^j, \ldots, Z_t^M$, and an ensemble of M values of $Y_t + \varepsilon_t$, noted $Y_t^1, \ldots, Y_t^j, \ldots, Y_t^M$, are randomly generated as follows (Allen et al., 2002; Margulis et al., 2002):

$$Z_t^j = g\left(Z_{t-1}^j, X_t, \theta\right) + \eta_{t-1}^j \text{ with } \eta_{t-1}^j \sim N(0, W)$$

$$Y_t^j = Y_t + \varepsilon_t^j \text{ with } \varepsilon_t^j \sim N(0, V)$$

Second, the Kalman filter equation is applied to each ensemble element as follows:

$$Z_{t,K}^j = Z_t^j + K_t^e \left(Y_t^j - F Z_t^j\right)$$

where K_t^e is a matrix defined by

$$K_t^e = \Sigma_t^e F^T \left[F \Sigma_t^e F^T + \text{var}(\varepsilon_t)\right]^{-1},$$

$\text{var}(\varepsilon_t)$ is the variance-covariance matrix of the measurement error, and Σ_t^e is the $m \times m$ variance-covariance matrix of M vectors Z_t^j, $j = 1, \ldots, M$. The ensemble of state variables Z_t^j, $j = 1, \ldots, M$, describes the uncertainty in the state variable and parameter values before using the measurement Y_t. In this approach, the updated model prediction is set equal to the average value of $Z_{t,K}^j$, $j = 1, \ldots, M$, noted $\overline{Z}_{t,K}$. Note that $\overline{Z}_{t,K}$ is related to the average value, $\overline{Z}_t$, of the initial ensemble Z_t^j, $j = 1, \ldots, M$, by

$$\overline{Z}_{t,K} = \overline{Z}_t + K_t^e \left(\overline{Y}_t - F \overline{Z}_t\right)$$

where $\overline{Y}_t$ is the average value of Y_t^j, $j = 1, \ldots, M$. The attractive feature of this method is that its implementation does not require a linear approximation of the system equation

$$Z_t = g(Z_{t-1}, X_t, \theta) + \eta_{t-1}.$$

The value of M must be chosen carefully. Too small an ensemble can give a very poor approximation. Moreover, according to Burgers et al. (1998), the matrix Σ_t^e tends to underestimate the true error variance-covariance matrix when M is too small. Another issue is that this method assumes that $P(Z_t | y_1, \ldots, y_t)$ is Gaussian. This assumption is usually wrong when the system equation is non-linear and, in practice, it is recommended to use the particle filter.

4.2.2 Particle Filter

The particle filter (Gordon et al., 1993; Doucet et al., 2000; Doucet et al., 2001) approximates the distribution

$$P(Z_t|y_1,\ldots,y_t)$$

using the importance sampling algorithm described in Chapter 7 (on Bayesian parameter estimation). This method is applied sequentially in time in order to approximate

$$P(Z_1|y_1),\ P(Z_2|y_1,y_2),\ P(Z_t|y_1,\ldots,y_t).$$

An attractive feature of this method is that it can be applied to non-linear system models with non-Gaussian error terms.

The use of the particle filter is illustrated here with the weed model defined by Equations (11)–(12). A series of six yearly weed density measurements was used to estimate the four state variables. Each measurement consisted of weed counts obtained in 5 subplots of 0.04m^2 taken in a field located in Burgundy, France, between 1996 and 2001 (Makowski et al., 2010). The parameter λ of the Gamma distribution of η_t in Equation (12) was unknown and its value was estimated at the same time as the model state variables from the measurements. As the model errors may vary from year to year, the value of λ was not assumed constant between years. Note that the parameters θ of the deterministic model were not re-estimated here, but were fixed to the values published by Munier-Jolain et al. (2002). However, they could be estimated at the same time as the state variables. The particle filter was applied as follows:

1. Initialization step. $t = 0$ (t is time expressed in years from the first year of simulation). Let N be the number of Monte Carlo simulations. Randomly generate M values of the initial state vector, $Z_0^{(i)} = \left(d_0^{(i)}, SSB_0^{(i)}, DSB_0^{(i)}, S_0^{(i)}\right)^T$, and of the initial parameter value $\lambda_0^{(i)}$, $i = 1,\ldots,M$. All values are generated from independent uniform distributions, $d_0^{(i)} \sim \text{Unif}(300, 500)$, $SSB_0^{(i)} \sim \text{Unif}(2700, 4000)$, $DSB_0^{(i)} \sim \text{Unif}(60, 500)$, $S_0^{(i)} \sim \text{Unif}(2000, 6000)$, $\lambda_0^{(i)} \sim \text{Unif}(0.01, 1)$. These distributions were chosen independently from the available data in order to cover large but realistic ranges of values. They represent the uncertainty about the initial state of the system. Each generated value of $Z_0^{(i)}$ and $\lambda_0^{(i)}$ is called a 'particle' and $\hat{w}_0^{(i)} = \frac{1}{M}$ is the weight of the ith particle.

2. Propagation step. The M particles are propagated by using the model equations for the next year. For $i = 1, \ldots, M$,

   i. Generate $\tilde{\lambda}_t^{(i)}$ where $log\left(\tilde{\lambda}_t^{(i)}\right) = log\left(\lambda_{t-1}^{(i)}\right) + \delta_{i,t}^{(\lambda)}$ and $\delta_{i,t}^{(\lambda)}$ is randomly drawn from $\delta_t^{(\lambda)} \sim N\left(0, \sigma^2\right)$. The standard deviation σ was set equal to 0.5 in order to generate a large range of values of $\tilde{\lambda}_t^{(i)}$ because no precise knowledge about this parameter was available.

   ii. Compute $\tilde{Z}_{tt}^{i} = g(Z_{t-1}, X_{t-1}, \theta) \times \eta_{i,t}$ where $\eta_{i,t}$ is randomly generated from a Gamma distribution with $\lambda = \tilde{\lambda}_t^{(i)}$.

   iii. Define $\hat{w}_0^{(i)} = \frac{1}{M}$.

3. Correction step. The weights of the particles are updated by using the measurement y_t. For $i = 1, \ldots, M$, compute

$$w_t^{(i)} = P\left(y_t \middle| \tilde{Z}_t^{(i)}\right)$$

where

$$P\left(y_t \middle| \tilde{Z}_t^{(i)}\right)$$

is the Poisson probability density defined by Equation (11). Compute normalized weights,

$$\hat{w}_t^{(i)} = \frac{w_t^{(i)}}{\sum_{l=1,\ldots,N} w_t^{(l)}} \quad , i = 1, \ldots, M.$$

4. Re-sampling step. Generate

$$Z_t^{(i)} = \left(d_t^{(i)}, SSB_t^{(i)}, DSB_t^{(i)}, S_t^{(i)}\right)^T$$

and $\lambda_t^{(i)}$, $i = 1, \ldots, M$, by sampling with replacement from the weighted discrete probability distribution defined by the

$$\left[\left(\tilde{Z}_t^{(i)}, \tilde{\lambda}_t^{(i)}\right), \hat{w}_t^{(i)}\right]$$

couples. $x_t^{(i)}$ and $\lambda_t^{(i)}$, $i = 1, \ldots, M$, become the new particles.

5. Define

$$\hat{w}_t^{(i)} = \frac{1}{M}.$$

If the number of simulated years t is lower than 6, return to step 2. Stop otherwise.

Steps 1 and 2 are used to approximate the prior probability

$$P(Z_t)$$

of the state variables. Steps 3 and 4 are used to approximate the posterior distribution

$$P\left(Z_t \middle| y_t\right)$$

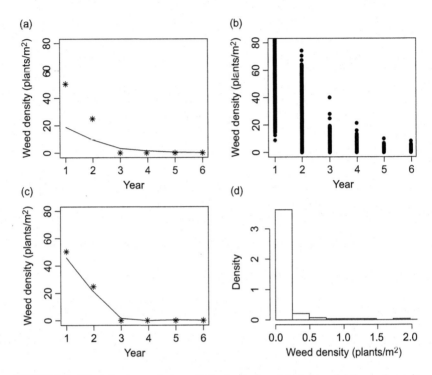

FIGURE 8.8 Weed density predictions obtained with the original deterministic weed population model (continuous curve) (a), $M = 10{,}000$ weed density values simulated at step 4 of the
particle filter algorithm (b), average values of the $M = 10{,}000$ weed density values (c), distribution of the $M = 10{,}000$ values on year 6 (d). The stars indicate the weed density measurements.

of the state variables from the prior distribution and the likelihood function
defined by Equation (11).

Results obtained with $M = 10{,}000$ simulations are shown in Figure 8.8. The
use of a higher number of simulations has no effect on the results. The discrepancy between the available measurements and the weed density simulated by
the original deterministic model was large (Figure 8.8a). The use of the particle
improved the quality of fit of the weed population model (Figure 8.8c).
Figure 8.8b and Figure 8.8d show the distributions of the $M = 10{,}000$ weed
density values and describe the uncertainty in weed density predictions.

5. CONCLUDING REMARKS

Filtering and smoothing allow one to combine a model and measurements in
useful ways, taking into account the uncertainties in each. Filtering is useful
for estimating sequentially in time the values of one or several state

variables, whereas smoothing can be used to estimate past values of state variables using all available measurements.

To implement these methods, it is necessary to define the system models using two different equations; an observation equation (relating observation to state variables) and a system equation (describing the dynamic of the state variables). Filtering and smoothing use these equations to calculate the expected values and variances of the state variables conditionally to one or several measurements. For linear Gaussian models, the expected values and variances can be computed analytically and the dlm R package makes the calculations very accessible.

For non-linear models, a first option is to approximate the original model by a linear one and then to apply the standard Kalman filter and smoother. Another option is to use the original non-linear model and to approximate the state variable expected values and variances with special algorithms based on Monte Carlo simulations. The particle filter appears to be a promising algorithm for filtering and smoothing with non-linear models.

Although the use of filtering and smoothing is currently limited in agriculture and environmental science, the increase in detection and transmission capability makes this approach attractive for improving predictions of system models. Due to the recent development of new algorithms and of efficient tools, such as those available in R, the application of filtering and smoothing is now relatively easy.

EXERCISES

Easy

1. Use the rnorm R function to simulate the output of a random-walk model defined for $t = 1$ to $t = 100$ by $Z_t = Z_{t-1} + \eta_{t-1}$ with $\eta_{t-1} \sim N\left(0, \sigma_\eta^2\right)$ and $\sigma_\eta = 1$. Repeat the procedure with different seeds.

2. We now add an observation equation to the random-walk model defined above: $Y_t = Z_t + \varepsilon_t$ with ε_t, a random term describing measurement error, $\varepsilon_t \sim N\left(0, \sigma_\varepsilon^2\right)$, and $\sigma_\varepsilon = 0.5$. Use the rnorm R function to simulate values of Y_t for $t = 1, \ldots, 100$.

3. Suppose that the following series of measurements is available for $t = 1, \ldots, 30$:

    ```
    30.0 32.2 31.1 30.2 30.6 32.7 32.0 31.5 31.1 30.7 29.8 30.6 29.9
    29.3 27.7 28.1 26.6 26.7 26.6 25.7 25.5 24.8 22.1 24.4 24.5 22.4 22.5
    22.9 22.4 22.1
    ```

    Use the package dlm to estimate the values of Z_t with both the Kalman filter and the smoother, assuming that $\sigma_\eta = 1$ and that $\sigma_\varepsilon = 0.5$.

    Plot the results as a function of time.

4. Apply the Kalman filter and smoother with the data given above now assuming that σ_η and that σ_ε are unknown.

Moderate

5. The objective of this exercise is to estimate the nitrogen content of a crop, noted by Z, by combining expert knowledge and a measurement designated by M. Suppose that, according to an expert, the possible values of the nitrogen content can be described by a normal distribution,

$$Z \sim N\left(\mu_Z, \sigma_Z^2\right).$$

We assume that a measurement M is performed in the field and that this measurement represents an index in the range $0-1$ based on reflectance values measured for different wavelengths of light. The measurement M is related to the nitrogen content as follows:

$$M = aZ + b + \tau$$

where a and b are two parameters and

$$\tau \sim N\left(0, \sigma_M^2\right)$$

is a measurement error independent of Z. We assume that μ_Z, σ_Z^2, a, b, and σ_M^2 are known.

   a. What is the joint distribution of the vector of random variables $\theta = \begin{pmatrix} Z \\ M \end{pmatrix}$? Express the expected value of θ and its variance-covariance matrix in terms of μ_Z, σ_Z^2, and σ_M^2.

   b. Determine the conditional distribution of Z given the measurement M. Express $E(Z|M)$ and var($Z|M$) in terms of M, μ_Z, σ_Z^2, and σ_M^2. See Chapter 2.

   c. What is the effect of σ_Z^2 and σ_M^2 on $E(Z|M)$ and var($Z|M$)?

   d. Numerical application. Calculate $E(Z|M)$ and var($Z|M$) with $M = 0.9$, $\mu_Z = 0.25$, $\sigma_Z^2 = 0.01$, $a = 1.2$, $b = 0$, and $\sigma_M^2 = 0.001$ from the equations obtained in (b).

   e. Suppose that the expert is not very confident in the value of μ_Z. Perform a sensitivity analysis of $E(Z|M)$ to the value of μ_Z when μ_Z varies in the range $0.15-0.35$.

Difficult

6. In this exercise, the expert knowledge is replaced by the prediction of a dynamic model defined as:

$$Z_t = Z_{t-1} + B_{t-1} + \varepsilon_{t-1}$$

where Z_t is the crop nitrogen content at time t, B_{t-1} is a known input variable, and ε_{t-1} is the model error. ε_{t-1} is normally distributed and has zero mean and constant variance, $\varepsilon_{t-1} \sim N(0, Q)$. We also assume that $Z_1 = 0$ and $\mathrm{var}(Z_1) = 0$.

Suppose that a measurement M is performed at $t = 10$ and that M is related to the nitrogen content as $M = aZ_{10} + b + \tau$ where a and b are two parameters and $\tau \sim N\left(0, \sigma_M^2\right)$ is a measurement error independent of Z.

a. What is the distribution for Z_{10} before using the measurement? Express $E(Z_{10})$ and $\mathrm{var}(Z_{10})$ in terms of the input variable values and Q.

b. Determine the joint distribution for the vector of random variables $\theta_{10} = \begin{pmatrix} Z_{10} \\ M \end{pmatrix}$. Express the expected value of θ and its variance-covariance matrix in terms of $B_1 \ldots B_9$, Q, and σ_M^2.

c. Give the distribution for Z_{10} conditionally to the measurement M. Express $E(Z_{10}|M)$ and $\mathrm{var}(Z_{10}|M)$ in terms of M, $B_1 \ldots B_9$, Q, and σ_M^2.

d. Compute the Kalman gain.

e. What are the effects of Q and σ_M^2 on $E(Z_{10}|M)$ and $\mathrm{var}(Z_{10}|M)$?

f. Numerical application. Calculate $E(Z_{10}|M)$, $\mathrm{var}(Z_{10}|M)$, and the Kalman gain with $B_1 = 0.01$, $B_2 = 0.015$, $B_3 = 0.02$, $B_4 = 0.035$, $B_5 = 0.02$, $B_6 = 0.025$, $B_7 = 0.025$, $B_8 = 0.018$, $B_9 = 0.02$, $Q = 0.001$, $a = 1.2$, $b = 0$, $M = 0.9$, and $\sigma_M^2 = 0.001$.

g. Is the uncertainty higher with the expert knowledge or with the model?

7. In this exercise, we consider a more sophisticated dynamic model simulating two state variables, noted $Z_t^{(1)}$ and $Z_t^{(2)}$. $Z_t^{(1)}$ is the crop nitrogen content at time t and $Z_t^{(2)}$ is the crop biomass at time t. The model is defined by:

$$\begin{pmatrix} Z_t^{(1)} \\ Z_t^{(2)} \end{pmatrix} = \begin{pmatrix} 1 & c \\ 0 & 1 \end{pmatrix} \begin{pmatrix} Z_{t-1}^{(1)} \\ Z_{t-1}^{(2)} \end{pmatrix} + \begin{pmatrix} B_{t-1}^{(1)} \\ B_{t-1}^{(2)} \end{pmatrix} + \begin{pmatrix} \varepsilon_{t-1}^{(1)} \\ \varepsilon_{t-1}^{(2)} \end{pmatrix}$$

where c is a known parameter, $B_{t-1}^{(1)}$ and $B_{t-1}^{(2)}$ are two known input variables, and $\varepsilon_{t-1}^{(1)}$ and $\varepsilon_{t-1}^{(2)}$ are two errors. We assume that $\varepsilon_{t-1}^{(1)}$ and $\varepsilon_{t-1}^{(2)}$ are normally distributed, independent, have zero means, and that their variances are equal to Q_1 and Q_2. We also assume that $Z_1^{(1)} = Z_1^{(2)} = 0$ and that their variances are equal to zero.

As before, we suppose that a measurement M is performed at $t = 10$ and that M is related to the nitrogen content as $M = aZ_{10}^{(1)} + b + \tau$ where a and b are two parameters and $\tau \sim N\left(0, \sigma_M^2\right)$ is a measurement error independent of $\varepsilon_{t-1}^{(1)}$ and $\varepsilon_{t-1}^{(2)}$.

a. Determine the distribution for $\begin{pmatrix} Z_{10}^{(1)} \\ Z_{10}^{(2)} \end{pmatrix}$ before using the measurement.

Give the expected value and the variance-covariance matrix for this random vector.

b. What is the correlation between $Z_{10}^{(1)}$ and $Z_{10}^{(2)}$?

   c. What is the sensitivity of this correlation to the values of c, Q_1, and Q_2?

   d. What is the joint distribution of $\begin{pmatrix} Z_{10}^{(1)} \\ Z_{10}^{(2)} \\ M \end{pmatrix}$?

   e. Determine the distribution for $\begin{pmatrix} Z_{10}^{(1)} \\ Z_{10}^{(2)} \end{pmatrix}$ conditionally to M.

   f. What are the effects of Q_1, Q_2, and σ_M^2 on $E\begin{pmatrix} Z_{10}^{(1)} \\ Z_{10}^{(2)} \end{pmatrix} \bigg| M\end{pmatrix}$ and $\mathrm{var}\begin{pmatrix} Z_{10}^{(1)} \\ Z_{10}^{(2)} \end{pmatrix} \bigg| M\end{pmatrix}$?

   g. Develop the expression of the Kalman gain for the two state variables.

   h. Numerical application. Calculate

$$E\left(\begin{matrix} Z_{10}^{(1)} \\ Z_{10}^{(2)} \end{matrix} \bigg| M \right), \mathrm{var}\left(\begin{matrix} Z_{10}^{(1)} \\ Z_{10}^{(2)} \end{matrix} \bigg| M \right),$$

and the Kalman gains with $B_1^{(1)} = 0.001$, $B_2^{(1)} = 0.0015$, $B_3^{(1)} = 0.002$, $B_4^{(1)} = 0.0035$, $B_5^{(1)} = 0.002$, $B_6^{(1)} = 0.0025$, $B_7^{(1)} = 0.0025$, $B_8^{(1)} = 0.0018$, $B_9^{(1)} = 0.002$, $B_1^{(2)} = 10$, $B_2^{(2)} = 50$, $B_3^{(2)} = 65$, $B_4^{(2)} = 64$, $B_5^{(2)} = 70$, $B_6^{(2)} = 35$, $B_7^{(2)} = 38$, $B_8^{(2)} = 51$, $B_9^{(2)} = 25$, $Q_1 = 0.0005$, $Q_2 = 0.09$, $a = 1.2$, $b = 0$, $c = 0.001$, and $\sigma_M^2 = 0.001$.

REFERENCES

Allen, J.I., Eknes, M., Evensen, G., 2002. An Ensemble Kalman Filter with a complex marine eco-system model: hindcasting phytoplankton in the Cretan sea. Annales Geophysicae 20, 1–13.

Anderson, J.L., Anderson, S.L., 1999. A Monte Carlo implementation of the non-linear filtering problem to produce ensemble assimilations and forecasts. Monthly weather review 127, 2741–2758.

Burgers, G., van Leeuwen, P.J., Evensen, G., 1998. Analysis scheme in the ensemble Kalman fil-ter. Monthly weather review 126, 1719–1724.

Doucet, A., de Freitas, N., Gordon, N., 2001. Sequential Monte Carlo in practice. Springer.

Doucet, A., Godsill, S., Andrieu, C., 2000. On sequential Monte Carlo sampling methods for Bayesian filtering. Statistics and Computing 10, 197–208.

Kalman, R.E., 1961. A new approach to linear filtering and prediction theory. J. Basic Eng 83, 95–108.

Jones, J.W., Graham, W.D., Wallach, D., Bostick, W.M., Koo, J., 2004. Estimating soil carbon levels using an ensemble Kalman filter. Transactions of the ASAE 47, 331–339.

Makowski, D., Chauvel, B., Munier-Jolain, N., 2010. Improving weed population model using a sequential Monte Carlo method. Weed Research 50, 373–382.

Margulis, S.A., McLaughlin, D., Entekhabi, D., Dunne, S., 2002. Land data assimilation and estimation of soil moisture using measurements from the Southern Great Plains 1997 field experiment. Water. Resour. Res. 38, 1299–1318.

Munier-Jolain, N.M., Chauvel, B., Gasquez, J., 2002. Long-term modelling of weed control strategies: analysis of threshold-based options for weed species with contrasted competitive abilities. Weed Research 42, 107–122.

Naud, C., Makowski, D., Jeuffroy, M.-H., 2009. Transmittance measurements can improve predictions of the nitrogen status for winter wheat crop. Field Crop Research 110, 27–34.

Paillard S., Treyer S., Dorin (coord.). 2010. Agrimonde, scénarios et défis pour nourrir le monde en 2050. Quae, Paris.

Pastres, R., Ciavatta, S., Solidoro, C., 2003. The Extended Kalman Filter as a toll for the assimilation of high frequency water quality data. Ecol. Modell. 170, 227–235.

Petris, G., 2010. An R package for dynamic linear models. J.Stat. softw. 36, 1–16.

Petris, G., Petrone, S., Campagnoli, P., 2009. Dynamic linear models with R. Springer, Dordrecht.

Welch, G., Bishop, G., 2002. An introduction to the Kalman Filter. Department of Computer Science, University of North Carolina: Report TR, 95–104.

## Chapter 9

Model Evaluation

1. INTRODUCTION

1.1 Definition

The dictionary definition of evaluation is to 'ascertain the value of ', and that is the meaning that we use here. The goal of evaluation is to determine the value of a crop model, with respect to the proposed use of the model. The results of an evaluation study can include qualitative conclusions about the quality of a model, graphs comparing observed and predicted values or quantitative measures of quality.

In the literature, one often encounters the term 'validation' rather than 'evaluation'. A rather common definition is that validation concerns determining whether a model is adequate for its intended purpose or not. This emphasizes the important fact that a model should be judged with reference to an objective. On the other hand, this definition seems to indicate that the result of a validation exercise is 'yes' (the model is valid) or 'no' (not valid). In practice, it is rarely the case that one makes, or even wishes to make, such a categorical decision. Rather one seeks a diversity of indications about how well the model represents system responses. We therefore prefer the term 'evaluation'.

General discussions and reviews of evaluation for ecological or agronomic models can be found in Loehle (1987), Mayer and Butler (1993), Rykiel (1996), Swartzman and Kalzuny (1987) and Tedeschi (2006).

1.2 The Role of Evaluation in a Modeling Project

Evaluation should not be envisioned as an activity that is undertaken just once, at the end of a modeling project. It is rather an activity that, in its different forms, accompanies the project throughout its lifetime.

Evaluation should begin at the start of a modeling project. At that time it is necessary to identify the goals of the project, and consequently the criteria for model evaluation. It is at the start of the project that the range of conditions of interest is specified, that the output variables of interest are identified, and that an acceptable level of error is defined.

A second evaluation step involves the model equations, which will be compared with results in the literature or with expert opinion. In general, the

Working with Dynamic Crop Models. DOI: http://dx.doi.org/10.1016/B978-0-12-397008-4.00009-5

model is then embodied in a computer program and there is the essential step of evaluation of the computer program. Testing a computer program, to ensure that it performs the intended calculations, is an important field with its own literature and we will not consider it here.

Once the computer program exists, one often continues with sensitivity analysis (Chapter 5) and parameter estimation (Chapters 6 and 7). Both of these activities include elements of evaluation. Sensitivity analysis allows one to evaluate whether model response to input factors is reasonable, and also to see if the input factors of most importance are sufficiently well known. Parameter estimation normally includes indications of the quality of the parameter estimators, which is an important aspect of model quality, as we shall see.

Once the parameter values are determined, one can proceed to evaluation of the model results. This is the aspect of evaluation that is treated in the rest of this chapter. That is, we suppose that the model is fully defined, including the parameter values. We also suppose that one has data with which the model results can be compared.

Model evaluation is essential. Firstly, the simple fact of deciding to evaluate a model obliges one to answer some basic questions, including what is the objective of the model, under what range of conditions will the model be used, and what level of quality will be acceptable.

Secondly, model improvement is impossible without evaluation. In the absence of a measure of model quality, how can one decide whether improvement is called for, and how can one know if a modified model, or perhaps an alternative model, is an improvement?

Finally, evaluation is important for potential users of a model. The user needs information about the quality of the model in order to decide how much credence to give to model results. Evaluation is a major activity in all modeling work, but it is particularly important for system modeling in agronomy. The systems in question are very complex, and as a consequence, model errors can be quite large. Rather than simply saying that one is furnishing predictions with errors that may be large, it is important to give some quantitative measure of how large those errors are.

1.3 Validation or Evaluation?

A system model consists of equations that describe the various processes that determine the way the system functions. It has a set of system inputs (explanatory variables) and system outputs (response variables). A schematic diagram for the simple maize model presented in Chapter 1 is shown in Figure 9.1.

Each process is itself a model that describes the output of the process as a function of the inputs of that process. For example, in the simple maize model daily increase in biomass is a function of thermal time, leaf area index (LAI), and daily solar radiation.

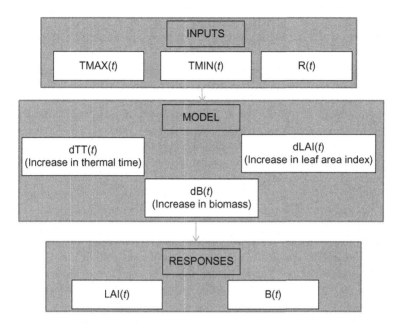

FIGURE 9.1 Schematic diagram of the simple maize model.

There are two quite different types of question we can ask about the quality of a system model. First, one can ask whether the representation of the processes, and their interaction, is correct. We will refer to this as treating the model as a scientific hypothesis, the hypothesis being that the processes behave in the same way in the real world as in the model. It is tempting in this case to speak of model 'validation'. Is the model description of the processes correct ('valid') or not? This question concerns the inner workings of the system model.

A rather different approach is to treat the system model as an engineering tool. Here the question is, how well does the model predict responses as a function of the explanatory variables? This is not a question about the processes within the model. The model is treated as a black box. The question is not how it works, but how well it works. This is a question of model evaluation.

That having been said, we must now retreat a bit from that extreme position. While in principle we would be satisfied with a model that predicts well even if the processes are seriously misrepresented, we would usually want to have some confidence in the descriptions of the processes. A practical reason for this is that often we don't have enough data to be really confident in the estimated quality of prediction of the model. It is important then to also ensure that the model equations are sound (based on current knowledge) and that the model meets 'behavioral tests' that indicate that the process equations have expected qualitative behavior (Sinclair and Seligman, 2000). That is, even when the model is treated as an engineering tool, it is important to check that the inner workings of the model are in accord with current knowledge.

1.4 In This Chapter

We distinguish two quite different objectives of system modeling: to test hypotheses about how the system functions, or to make predictions about how the system responds to different values of the explanatory variables.

The first step in evaluating model predictions is the comparison of the model with experimental data. Several different measures of goodness-of-fit are presented. We also consider two rather particular cases: evaluation when the objective is to classify results between two mutually exclusive categories, and evaluation when one has probabilistic rather than deterministic predictions. We then consider the conditions for extrapolating model quality as based on a sample, to model quality for a target population. We consider particularly the evaluation of predictive quality for a target population. This is the main topic in this chapter, since prediction is often the main objective in modeling. Our criterion here is mean squared error of prediction (MSEP). We show that when the same data are used for parameter estimation and to estimate error, the error estimate is in general unrealistically optimistic. We show several approaches (data splitting, cross validation, and bootstrap) that provide more realistic estimates of MSEP. We show that MSEP can be expressed as the sum of two contributions, one that decreases as extra explanatory variables are added, and the other that may initially decrease but finally increases. This can provide some insight into the question of how complex a model should be.

The R code for carrying out the analyses discussed in the chapter is shown in Section 7 and is available in the R package ZeBook.

2. A MODEL AS A SCIENTIFIC HYPOTHESIS

The objective in this subsection is to consider the use of measurements on the system to draw conclusions about the processes that are described in the model. We observe the overall behavior of the system, and want to draw conclusions from that about the underlying processes in the system. Our question is, does the model correctly describe those processes?

2.1 One Can Invalidate, but Never Definitively Validate, a Model

We begin with a general result about the possibility of validating a scientific hypothesis. To formalize the argument, we define three logical statements which can each have the value TRUE or FALSE:

A: The data come from situations where the model is meant to apply.
B: The model processes have the same behavior as the true world for situations where the model is meant to apply. This is the question

underlying validation. If this statement is true, the model is valid. If false, the model is invalid.

C: The observed output of the real system and the outcomes simulated using the model are identical. This is what we can check using our system data, assuming for the moment that there is no experimental error.

The basis of our argument is the syllogism

$$\text{IF A AND B, THEN C} \tag{1}$$

This says that if our data come from situations where the model is meant to apply (if A is TRUE), and if the model processes have the same behavior as the real world in those situations (if B is TRUE), then true and simulated results will be identical (C will be TRUE).

Logic allows us to conclude from Equation (1) that

IF (NOT C) THEN (NOT A) OR (NOT B)

That is, if observed and simulated results are not identical (NOT C is TRUE, i.e., C is FALSE), then either we are in a situation where the model is not meant to apply (A is FALSE), and/or the model processes do not have the same behavior as the real world (B is FALSE). Assuming that we only use data for situations where the model is meant to apply, then A is TRUE so it must be that B is FALSE.

This shows that we can invalidate a model, i.e., conclude that the processes according to the model do not have the same behavior as in the real world.

Suppose now that the model and the data do agree (C is TRUE). We continue to assume that the data represent situations where the model is meant to apply, so A is TRUE. It is not correct to conclude in this case that the model processes behave in the same way as the processes in the real world. That is, Equation (1) does not imply that IF C THEN A AND B. This is intuitively obvious. First of all, at best we have only tested the model in a limited number of situations. It may be that while simulated and observed results agree for these situations, they will not agree for other situations where the model is meant to apply. We cannot generalize our conclusions based on a sample of data to all the situations where the model is meant to apply. Secondly, it is possible that the model processes are incorrect, but that the errors cancel. The model could correctly describe the way the overall system behaves even if the descriptions of the processes are wrong.

This shows that one can never definitively validate a model. In other words, it is possible to prove a model incorrect but it is not possible to prove that it is correct.

If there is measurement error, then we don't observe the true output in the real world, but only an approximation to it. Then categorical statements such as 'C is FALSE' or 'C is TRUE' would be replaced by probabilistic statements like 'it is unlikely that C is TRUE' or 'it is not unlikely that C is TRUE'. More specifically, one would calculate the probability p of observing a difference between measured and simulated values at least as large as that actually observed, if C were indeed TRUE. If p is small (often less than 0.05 for example), one concludes that 'it is unlikely that C is TRUE'. If p is large (say larger than 0.05), then one concludes that 'it is not unlikely that C is TRUE'. This would not basically change the fact that one can invalidate, but not validate, a model, because we still have the situation that even if the model gives results reasonably close to the observed results, it is still for only a limited range of conditions and could still be the result of compensation of errors.

The above arguments apply to system models but also to models of the individual processes or to any model. In general one can invalidate a model but one can never definitively validate a model. (Some basic models in physics have been tested very thoroughly, so that one is quite confident that they apply over a wide range of conditions. On the other hand, even here there are surprises. It was thought for several centuries that Newton's equations of motion were universally true, until it was found that they must be replaced by quantum mechanics at the scale of very small particles. Despite all the verifications showing the agreement between Newtonian mechanics and the real world, it would have been a mistake to conclude that the model was definitively validated).

We have just shown that one can never definitely validate a model. However, if the agreement between the model and the data is 'acceptable' (close compared to experimental error), then at least one would not conclude that the model is invalid. That is already of some comfort, right?

Not really. In fact, system models in agronomy are, by design, a simplification of reality, not meant to be identical to the real world. There are few, if any, cases where we seriously entertain the hypothesis that the model outputs are strictly identical to real world outputs. It may occur that in a case with measurement error we have a fairly large p value, so that we do not conclude that the model is incorrect. However, this will often simply be the result of poor data. Data with large measurement errors will tend to lead to large p values. That is, we do not really need to test the model against data to conclude that it is invalid. We know that by construction the model is invalid.

2.2 But Models Can Be Useful as Scientific Hypotheses Nonetheless

Does this mean that models have nothing to say about how the system functions? They do, but not in the sense of providing a hypothesis of exactly how the system functions. A more reasonable question is which description of the processes in a system, among a small number of clearly expressed alternatives, is most compatible with observed data or has the smallest prediction error. This can be useful as a way forward in understanding the behavior of the system, even if none of the models is exactly correct. An example is presented below.

Another use of system models is to propose a simple description of the system. This can allow one to deduce new aspects of the behavior of the system, even if the quantitative predictions of the model are incorrect. It can also be the basis for more detailed models. For example, the Lotka-Volterra model for predator-prey interactions (Haefner, 2005) is a very simple model (just two state variables, predator population and prey population). In general, it does not give quantitative agreement with data. Nevertheless, it predicts cyclic behavior in both populations with a phase lag between them, which cannot be deduced from the behavior of the individual populations and which is often observed. It is also the basis for many more complex models.

Example 1

Gent (1994) considers three different hypotheses concerning the use of reserve photosynthate in wheat. The hypotheses are shown schematically in Figure 9.2 for the period of late grain filling. According to hypothesis 1, there is no

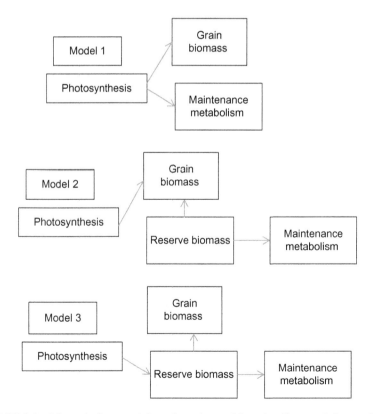

FIGURE 9.2 Schematic diagram of three alternative models to describe use of photosynthates during grain filling in wheat.

reserve photosynthate compartment. In hypothesis 2, current photosynthesis goes for grain filling. Reserve photosynthate is used preferentially for maintenance metabolism. In hypothesis 3, current photosynthate goes to reserve biomass, and reserve biomass then supports both maintenance metabolism and grain filling. Three different models were developed, embodying the three hypotheses.

The experimental data were daily gross photosynthesis and dark respiration, shoot biomass at various times, harvest index at various times, and values for the distribution of ^{14}C label after photosynthetic assimilation of $^{14}CO_2$.

It was found that the model based on hypothesis 3 gave the closest agreement with the observed data. The conclusion is not that this model is exactly correct, but rather that it provides a closer approximation to the system behavior than the competing hypotheses.

2.3 Conclusions

If by 'valid' we mean that the processes described in a system model have behavior identical to the processes in the real world, then we can never conclude that a model is valid. In fact we are sure that system models in agronomy are not valid. Testing them against data is not necessary to reach this conclusion.

This does not imply that if the goal is better understanding of the system, testing models against data is useless. Such tests can be useful in choosing between alternative descriptions of the system. Even if no description is perfectly correct, it is important to know which description more closely conforms to the behavior in the real world.

The usefulness of models for drawing conclusions about real-world behavior is probably limited to very simple models. If the model is complex, like most crop models for example, then in fact it represents a large number of hypotheses. Even if one model gives better agreement with data than another, it is often impossible to ascribe that to specific processes.

3. COMPARING SIMULATED AND OBSERVED VALUES

In the rest of this chapter, we treat system models as engineering tools. We want to evaluate how well the model imitates the input-output relationships in a sample of observations or in a larger population. We begin by comparisons between the observed and simulated values for the sample. We will illustrate using the data of example 2.

Example 2

Mohanty et al. (2012) present results for the evaluation of the APSIM model. Table 9.1 shows their observed and predicted soybean N uptake values, from 3 treatments in each of 3 years, at one location in India.

TABLE 9.1 N Uptake by Soybean[1] Kg/Ha

Year	Treatment	Observed Value	Simulated Value
2002–2003	Control	78	126
2002–2003	Inorganic	110	126
2002–2003	Organic	92	126
2003–2004	Control	75	105
2003–2004	Inorganic	110	105
2003–2004	Organic	108	105
2005–2006	Control	113	147
2005–2006	Inorganic	155	147
2005–2006	Organic	150	147

[1] Based on (Mohanty et al., 2012).

3.1 Graphical Representations of Model Error

3.1.1 Graph of Simulated Versus Measured Values

Probably the most widespread graphical presentation of the agreement between measured and calculated values for system models is a plot as shown in Figure 9.3. For each measurement, the y value is the measured value and the x value is the corresponding value simulated with the model. It is also usual to draw the 1:1 line on such a graph. If there is no model error the calculated values and measured values are identical and then each point will be exactly on the 1:1 line. Model error is given by the vertical or equivalently the horizontal distance from the point to the line.

The advantage of this type of graph is that one can see at a glance how well model calculations and measurements agree.

A graph of measured versus calculated values can be used not only for output variables that have a single value for each situation (for example yield), but also for output variables that are functions of time (for example LAI). In this case, there may be several points for each situation (for example, in the same field there may be several measurements of LAI at different dates).

It can be useful to distinguish groups of points. For example, if there are several measurements of LAI per field, the use of different symbols for each field will make it easier to see if there is a field effect on error. Another example would be where different fertilizer doses are studied. The use of different symbols for different levels will allow one to see if model error tends to be different depending on fertilizer level.

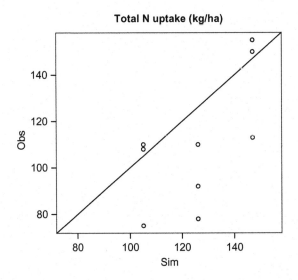

FIGURE 9.3   Graph of observed versus simulated values of N uptake, with 1:1 line.

3.1.2 Measured and Simulated Values Versus Time or Other Variable

For an output variable which is a function of time, it is fairly common practice to graph measured and calculated values versus time for each situation separately, as in Figure 9.4. In general the model produces calculated values every day while the measurements are much sparser.

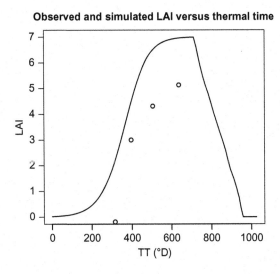

FIGURE 9.4   Graph of simulated (solid line) and measured (circles) LAI values.

Mayer and Butler (1993) discuss the difficulty of visual evaluation of model error based on graphs like this when the output variable fluctuates rather rapidly with time. This is often the case for example for soil moisture. They present an artificial example where the 'measurements' are generated by a random mechanism, with no relation to the model. Because of the rapid fluctuations there is always some part of the simulated curve that is fairly close horizontally to the measured values, and this gives the impression that the agreement between the data and the simulated values is reasonable. The fact that such graphs may be difficult to interpret is an additional reason for using several different types of graph for assessing model agreement with data.

It may also be of interest to graph measured and calculated values versus some variable other than time. For example, Pang (1997) presents observed and calculated values of total absorbed N as a function of applied N. The objective is to see how model error varies with the dependent variable.

3.1.3 Graphing the Residuals

The classical method of examining model error in statistics is to plot model residuals (measured value minus simulated value) on the y-axis, against measured values on the x-axis. This type of graph is rarely used to present crop model results, which is unfortunate, since it brings out other aspects of the residuals. For example, graphs of residuals are very useful for bringing attention to systematic patterns in the errors. The data of Table 9.1 are plotted in this way in Figure 9.5.

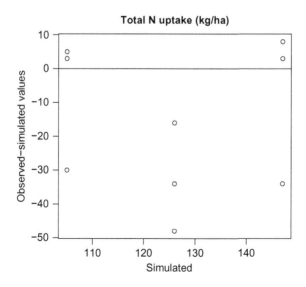

FIGURE 9.5 Graph of model residuals (observed − simulated) versus simulated values.

It is also of interest to plot model error versus variables such as total water input (rainfall plus irrigation), date of sowing, total applied nitrogen etc. to see if there is any trend in the errors.

3.2 Evaluating How Well the Model Fits the Data

3.2.1 Quantitative Measures of Distance Between Measured and Simulated Values

Graphs are valuable visual aids in model evaluation, but one also wants a quantitative summary of goodness-of-fit. Arguably the most widely used is mean squared error (MSE), defined as

$$MSE = (1/n) \sum_{i=1}^{n} (y_i - \hat{y}_i)^2 \tag{2}$$

where y_i is the measured value, $\hat{y}_i$ the corresponding simulated value, and n the number of measurements.

Because MSE is an average of squared differences, large differences are heavily weighted. It is worthwhile to verify if MSE is not essentially due to one or two large differences. If this is the case, one should examine those specific cases. There might be a problem with the data or perhaps there are exceptional circumstances such as extreme stress.

Often it is more convenient to work with the square root of MSE, called the root mean squared error (RMSE), which has the same units as y:

$$RMSE = \sqrt{MSE}$$

Another evaluation measure is relative root mean squared error (RRMSE), defined as

$$RRMSE = RMSE / \bar{y}$$

where $\bar{y}$ is the average of the measured values. This measure is dimensionless; it expresses error as a fraction of average measured value. For the data of example 2 RRMSE = 0.23. This gives a quick idea of the level of error in this case, even if one has no idea of typical soybean N uptake values. Also, RRMSE may be more meaningful than RMSE for comparing errors based on different data sets with different average responses.

An alternative summary measure of model error, which like MSE avoids compensation between under- and over-prediction, is the mean absolute error (MAE):

$$MAE = \frac{1}{N} \sum_{i=1}^{N} |y_i - \hat{y}_i|$$

The units of MAE are the same as for y. Furthermore, there is no over-weighting of large differences here. Thus MAE has advantages over MSE or RMSE, if the objective is simply to examine overall model error (Willmott and

Matsuura, 2005). An advantage of MSE is that it can be decomposed into separate contributions, as we shall see below. It is always the case that MAE $\leq$ RMSE, and equality only occurs if all the errors being averaged are equal.

One can also look at relative mean absolute error (RMAE), defined by

$$RMAE = \frac{MAE}{(1/n)\sum |y_i|}.$$

A variant would be

$$RMAE' = (1/n)\sum_{i=1}^{n} |y_i - \bar{y}_i|/|y_i|$$

where each difference is divided by the measured value for that situation. Note that if y_i is small, the contribution to $RMAE'$ could be very large.

Example 2 cont.

The values of the above criteria for the data of this example are shown in Table 9.2.

TABLE 9.2 Goodness-of-Fit Measures for Data from Example 2

Goodness-of-fit Measure	Value
MSE	653.2 $(kg/ha)^2$
RMSE	25.6 (kg/ha)
RRMSE	0.23
MAE	201.1 kg/ha
RMAE	0.18
RMAE'	0.22

3.2.2 Skill Scores

A skill score is the ratio of some measure of model error to the same measure for some naïve simple predictor. It thus compares the skill of the model with the skill of the simple predictor. In meteorological studies, the naive predictor for short term predictions is often the same value as most recently observed (tomorrow's maximum temperature will be the same as today's), and for long-term predictions is often the climatological average (maximum temperature next June will be the same as long-term average maximum temperature in June). For models in agronomy the naïve predictor will often be the average of the observed values.

Probably the most widely used measure of this type is model efficiency, defined as

$$EF = 1 - \frac{\sum\limits_{i=1}^{n} \left(y_i - \hat{y}_i\right)^2}{\sum\limits_{i=1}^{n} (y_i - \bar{y})^2} \qquad (3)$$

If the model is perfect then $y_i = \hat{y}_i$ for each i and $EF = 1$. That is the upper bound on EF. If one uses the average of observed values as the predictor for every case (the naïve predictor here) so that $\hat{y}_i = \bar{y}$ for all i, then $EF = 0$. A model with EF = 0 is not a good model; it does no better than the average, which does not explain any of the variability in the data. There is no lower bound on EF. A model that is a worse predictor than the average of observed values will have EF < 0.

The relationship between EF and MSE is

$$EF = 1 - \frac{\sum\limits_{i=1}^{n} \left(y_i - \hat{y}_i\right)^2}{\sum\limits_{i=1}^{n} (y_i - \bar{y})^2} = 1 - \frac{MSE}{MSE_{\bar{y}}}$$

where

$$MSE_{\bar{y}} = \sum\limits_{i=1}^{n} (y_i - \bar{y})^2 = \text{var}(y)^*(n-1)/n$$

is the MSE value when $\bar{y}$ is used to predict for all cases. EF compares the MSE value for the model with the MSE value for the naïve predictor. That is, EF is the skill score based on MSE, where the average of measured values is the naïve predictor.

Finally, note that the numerator in Equation (3) measures the variability in observed data that is not explained by the model, while the denominator measures the total variability in the observed values. This is the same as one of the definitions of R^2, the coefficient of determination.

Any distance criterion can be converted to a skill score that is 1 for a perfect model and 0 for a model that has the same value for the criterion as the naïve predictor. The general formula is

$$CritSS = 1 - \frac{Crit_{model}}{Crit_{naive}}$$

where $CritSS$ is the skill score associated with the criterion, $Crit_{model}$ is the criterion value for the model being evaluated, and $Crit_{naive}$ is the criterion value for the naïve predictor. The skill score is 1 for a model which agrees perfectly with the data, because the distance between simulated and measured values in this case is 0.

For example, a skill measure related to mean absolute error, noted MAESS, would be

$$MAESS = 1 - \frac{\sum\limits_{i=1}^{n} |y_i - \hat{y}_i|}{\sum\limits_{i=1}^{n} |y_i - \bar{y}|} = 1 - \frac{MAE}{MAE_{\bar{y}}}$$

This is 1 for a model which gives perfect agreement with the observed data, and 0 for a model with the same MAE value as using $\bar{y}$ for all predictions.

Although skill scores typically have the form of Equation (4), one can develop skill scores which do not have that form. Willmott (1981) proposed an agreement index defined as

$$index = 1 - \frac{\sum\limits_{i=1}^{N} (y_i - \hat{y}_i)^2}{\sum\limits_{i=1}^{N} (|\hat{y}_i - \bar{y}| + |y_i - \bar{y}|)^2} \tag{4}$$

This is somewhat similar to EF, however, the denominator is related to the variability in both the measured and calculated values. If the model is perfect, then $y_i = \hat{y}_i$ and $index = 1$. If the model predictions are identical in all cases and equal to the average of the observed values, i.e., $\hat{y}_i = \bar{y}$, then $index = 0$. These special cases give the same values as EF, but for other cases, the two criteria will have different values. EF is the more often used.

Example 2 cont.

For the data of this example, EF = 0.053. The model is only slightly better than predicting that every result is equal to the average of the measured values. This also says that the model only explains about 5% of the variability in the observations.

The value of the criterion of Equation (4) is index = 0.70. There is a large difference in this case between EF and index, because the denominator in index is much larger. This fairly large value of index seems a bit misleading, since in fact the model explains only a small part (5%) of the variability in the observations.

3.2.3 Decomposition of MSE

MSE can be written as a sum of individual contributions, and this decomposition can help in understanding the origin of the errors. This possibility of delving into the components of MSE is the major advantage of this criterion. In fact, several different decompositions of MSE have been suggested.

Kobayashi and Salam (2000) show that MSE can be decomposed as

$$MSE = (Bias)^2 + SDSD + LCS \qquad (5)$$

with

$$Bias^2 = \left[(1/n)\sum_{i=1}^{n}(y_i - \hat{y}_i) \right]^2$$

$$SDSD = (\sigma_Y - \sigma_{\hat{y}})^2 *(n-1)/n$$

$$LCS = 2\sigma_Y\sigma_{\hat{y}}(1-r)*(n-1)/n$$

where $\hat{\sigma}_Y^2$, $\hat{\sigma}_{\hat{y}}^2$, $\hat{\sigma}_{Y\hat{y}}$, and r are respectively the sample estimates of the variance of y, of the variance of $\hat{y}$, of the covariance of y and $\hat{y}$, and of the correlation between y and $\hat{y}$. They are calculated as

$$\hat{\sigma}_y^2 = \frac{1}{n-1}\sum_{i=1}^{n}\left[(y_i - \bar{y})^2 \right]$$

$$\hat{\sigma}_{\hat{y}}^2 = \frac{1}{n-1}\sum_{i=1}^{n}\left[(\hat{y}_i - \bar{\hat{y}})^2 \right]$$

$$\hat{\sigma}_{y\hat{y}} = \frac{1}{n-1}\sum_{i=1}^{n}\left[(y_i - \bar{y})(\hat{y}_i - \bar{\hat{y}}) \right]$$

$$r = \frac{\hat{\sigma}_{y\hat{y}}}{\sigma_y\sigma_{\hat{y}}}$$

$$\bar{\hat{y}} = \frac{1}{n}\sum_{i=1}^{N}\hat{y}_i$$

The first term in Equation (5) is the bias squared. Bias measures the average difference between measured and simulated values. If on the average the model under-predicts, the bias is positive, and conversely if the model over-predicts on the average, the bias is negative. The cause of model bias is in many cases relatively easy to identify and perhaps to correct. For example, if yield is systematically under-predicted (positive bias), one might start by examining whether average final biomass or harvest index or both are under-predicted, and then by asking what would lead to systematically low values.

The second term in the decomposition of MSE is related to the difference between the standard deviation of the measurements and the standard deviation of the simulated values. It measures the extent to which observed and simulated values have similar ranges. Once again, the causes of the difference can sometimes be identified. For example, if the model predicts that yield for different situations varies only slightly, whereas the measurements show a larger variation, one might look at the effect of stress. Is the difference due to the fact that the simulated values are not sufficiently sensitive to water stress for example?

The last term in the decomposition is related to the correlation between observed and predicted values. This term depends in detail on how well the model mimics the observed variation of y from situation to situation. As such, it may often be the result of many small errors rather than a single major error, and thus be relatively difficult to analyze and correct.

Gauch et al. (2004) suggest that it would be advantageous to have a decomposition with terms explicitly related to the regression of y on $\hat{y}$. The decomposition they propose is

$$MSE = (Bias)^2 + NU + LC \qquad (6)$$

$$NU = (1 - b_{y\hat{y}})^2 \sigma_{\hat{y}}^2 {}^*(n-1)/n$$
$$LC = (1 - r^2)\sigma_y^2 {}^*(n-1)/n$$

$$b_{Y\hat{Y}} = \frac{\sigma_{y\hat{y}}^2}{\sigma_{\hat{y}}^2}$$

The term $b_{y\hat{y}}$ is the slope of the regression of y on $\hat{y}$. The decomposition of Equation (6) is quite similar to that of Equation (5). The first term is again the squared bias, and the last, LC, 'lack of correlation', depends in detail on how variation in y and $\hat{y}$ are correlated. However the second term here, NU (non-unity slope), depends on how close the slope of the regression of y on $\hat{y}$ is to 1.

Willmott (1981) proposed a decomposition of MSE based on the linear regression of $\hat{y}$ on y, $\hat{y} = a + by$. The estimated regression parameters are given by the standard formulas for linear regression,

$$\hat{b} = \frac{\sigma_{y\hat{y}}}{\sigma_y^2}$$

$$\hat{a} = \overline{\hat{y}} - b\overline{y}$$

The estimated value of $\hat{y}$ based on the regression is

$$\hat{y}^{reg} = \hat{a} + \hat{b}y$$

The decomposition is then

$$MSE = MSE_s + MSE_u$$
$$MSE_s = (1/n)\sum (\hat{y}_i^{reg} - y_i)^2 \qquad (7)$$
$$MSE_u = (1/n)\sum (\hat{y}_i - \hat{y}_i^{reg})^2$$

The term MSE_s is called the systematic part of MSE and MSE_u the unsystematic part.

Suppose that $\hat{a} = 0$ and $\hat{b} = 1$ so that the regression line is the 1:1 line. Then $\hat{y}^{reg} = y$ and $MSE_s = 0$. The systematic contribution is zero in this case. In general, MSE_s is a measure of how far the regression line deviates from the 1:1 line. The MSE_u term on the other hand measures the variability of $\hat{y}_i$ around the regression line. An example of the use of this decomposition is given in Ben Nouna et al. (2000).

Example 2 cont.

The results of the above decompositions for the data of Table 9.1 are shown in Table 9.3.

TABLE 9.3 Values of the Terms in the Decomposition of MSE

	Equation (5)	Equation (6)	Equation (7)
	Bias^2 = 252	Bias^2 = 252	MSEs = 482
	SDSD = 83	NU = 0.02	MSEu = 171
	LCS = 318	LC = 401	
Sum = MSE	653	653	653

In Equation (5) the Bias^2 term is about 40% of total MSE; an appreciable contribution. (The bias is negative, so the model over-predicts). The SDSD term is relatively small, which indicates that the dispersion of the simulated values is similar to that of the measured values.

In Equation (6) the Bias^2 term is of course the same as in Equation (5). The NU term is negligible, because the regression slope is very close to 1 (0.99).

For the last decomposition, the linear regression of simulated on observed values is

$$\hat{y}_i^{reg} = 79 + 0.42^* y_i$$

The systematic part of MSE makes the larger contribution.

In each case the sum of the terms in the decomposition is 653 (kg/ha)^2 = MSE, as it must be.

3.2.4 Statistical Tests

Several authors propose using statistical tests for evaluating a model. Two tests in particular have been used in the literature. The first is related to the regression model

$$y = a + b\hat{y}$$

where y is the observed value and $\hat{y}$ the model prediction. A perfect model would have $y = \hat{y}$, i.e., $a = 0$ and $b = 1$. The idea then is to estimate the regression coefficients a and b, using ordinary least squares (OLS), and to test the two hypotheses

$$H_0^{(a)}:a = 0$$

and

$$H_0^{(b)}:b = 1.$$

The second test concerns the hypothesis

$$H_0:y_i = \hat{y}_i$$

for all i. A paired t test can then be used to test equality. This is equivalent to testing that the means of the measured and simulated values are equal.

Although one sees these tests in the literature, they are less appropriate than distance measures for evaluating system models, and several authors (for example, Mitchell, 1997) have criticized them. One problem is that the tests are based on assumptions which are often not satisfied. Another objection is that acceptance depends on the variability in the data and the amount of data. Having few data with lots of variability makes it more likely that the hypotheses will be accepted. Thus one is testing not only the model, which is our main interest, but also how good the data are, which is not our main interest. A third objection is that the choice of a level for the tests is rather arbitrary. Perhaps the major objection relates to the nature of tests. We have already made this argument above. We know that mathematical models are simplifications of the complex reality of agronomic systems, and are not exactly correct. That is, we know that if we had enough data, we would reject H_0 in every case. So we are not really interested in the hypotheses. We are interested in the quality of the model as a predictor, and the tests do not directly give us information on this.

3.2.5 Measures Based on a Threshold of Model Accuracy

In some cases one might accept a few large differences between measured and calculated values as long as the agreement is good for the majority of situations. The two measures presented in this section are adapted to this viewpoint.

In the first measure, one fixes the percentage p of situations which should show acceptable agreement. The measure is $TDI(p)$ = smallest value of d such that

$$|y_i - \hat{y}_i| \leq |d|$$

for at least p% of the observed situations.

The second measure in this group fixes a maximum error $|d|$ and measures the percentage p of situations with

$$|y_i - \hat{y}_i| \leq |d|.$$

This is called the coverage probability.

Example 2 cont.

The errors in increasing order of absolute value are shown in Table 9.4. For each error, the table also shows the percentage of values with errors that small or smaller. Consider first TDI(80%). That is, what is the smallest error such that at least 80% of the errors are at that level or smaller? From the table, this is seen to be TDI(80%) = 34 kg/ha. Consider next CP(10), the percentage of cases that have an error of 10 kg/ha or less. From the table, CP(10) = 44%.

TABLE 9.4 Measures Based on Threshold of Accuracy for Data of Example 2

d (kg/ha)	Situations with $\|y_i - \hat{y}_i\| \leq d$ (%)
3	22 (2/9)
5	33 (3/9)
8	44 (4/9)
16	56 (5/9)
30	67 (6/9)
34	89 (8/9)
48	100 (9/9)

3.3 Evaluating a Model Used for Binary Classification

3.3.1 Introduction

In certain cases it is not the overall fit of the model to the data that is of interest, but rather how useful the model is in discriminating between two mutually exclusive alternatives. Here we consider how to evaluate a model in this case.

A common binary decision problem in agronomy is whether or not to apply some treatment (fungicide, herbicide, insecticide, fertilizer, etc.) to a field. Many other problems in agronomy could be usefully formulated as binary decision problems (Makowski et al., 2009). For example, one might want to predict whether or not the amount of mineral N left in the soil at harvest will exceed some threshold for different management decisions, since this is closely related to environmental risk of N leaching over winter. Rather than predicting the actual value of mineral N, it might be sufficient to determine whether the risk is high or low (a binary decision). Predicting whether or not grain protein content (GPC) will exceed some threshold is another example of a binary decision problem.

This type of problem is extremely widespread in medicine (Zweig and Campbell, 1993). The typical problem is that one has the result of some diagnostic test (a level of some chemical in the blood, an image of some organ) and one has to decide if the patient does or does not have some pathological condition. In ecology, a widespread binary decision problem is deciding whether or not some particular species, in particular an endangered species, is present or not in some particular habitat.

We now express the problem more formally. We note the two possible outcomes of interest $X = 0$ and $X = 1$ (not to be confused with explanatory variables).We suppose that we have some variable or vector of variables that could be used to predict which of the two outcomes will occur. We will refer to this as the decision variable Y. In medicine this could be the result of a diagnostic test. In agronomy, this could be the value of some indicator variable or, more relevant here, a value simulated by a system model.

In general, Y will be a continuous variable. To establish a decision rule, we need not only to identify Y, but also to fix a decision threshold t. The use of the threshold is as follows: if $Y < t$, then we predict $X = 0$, and if $Y \geq t$ we predict $X = 1$. In some cases there may be a pre-defined threshold that we want to test, but in many cases the value of the decision threshold t has to be chosen.

The evaluation problem then has two aspects. First, for a specific pair (Y,t) of decision variable and threshold, one wants to evaluate how well that pair discriminates between $X = 0$ and $X = 1$. Secondly, one would like to remove the threshold t from the problem and ask more generally how well Y can discriminate. We discuss both of these evaluation questions below. We will also be led to discuss the choice of threshold.

We assume that we have a sample of individuals from the target population and that for each individual we know the value of the decision variable Y and the true value of X (0 or 1).

Example 3

Barbottin et al. (2008) compared the capacity of various models and indicators to predict whether GPC in wheat would be above or below some threshold (we use 11.5%). In some cases a premium is paid for higher protein content; it is therefore important to be able to predict protein content, or more specifically whether or not protein content will be sufficiently high to earn the premium. In this case $X = 0$ if GPC is below 11.5% and $X = 1$ if GPC is at or above 11.5%.

The authors considered various decision variables including measurements of leaf chlorophyll at a particular wheat growth stage using a SPAD meter, GPC simulated using the crop model AZODYN (Jeuffroy and Recous, 1999), and various simple models that relate GPC to measurable variables like applied N. As this example shows, one can often imagine many different ways of predicting the output of interest, in this case grain protein content. Evaluation is important in choosing which decision method is most effective.

The sample consisted of 43 different fields. For each field the value of each decision variable is known. Also, for each field GPC was measured, so X is known. We illustrate using just eight of these fields and two decision variables. The data for these fields are shown in Table 9.5.

It is expected that higher SPAD meter readings correspond to higher levels of GPC. Thus the decision rule is that GPC = 0 if SPAD < t and GPC = 0 if SPAD ≥ t for some threshold t. However, there is no obvious choice for the threshold t. This will have to be determined from the sample.

The AZODYN model directly predicts GPC. It seems natural to set the threshold here at 11.5%. If simulated GPC < 11.5%, then we decide that $X = 0$. If simulated GPC ≥ 11.5%, we decide $X = 1$. However, it was observed in this case that the model seems to be quite seriously biased. On the average, the model under-predicts GPC by almost 3%. So a threshold below 11.5% for predicted GPC would no doubt be better for deciding whether or not actual GPC will be above or below 11.5%. Thus, in this case, we still need to decide on an appropriate threshold.

TABLE 9.5 Values of Two Possible Decision Variables (SPAD and AZODYN) and Observed GPC (%)[1]

SPAD Measurement	AZODYN Prediction of GPC	GPC	X[2]
38.0	7.9	14.6	1
38.3	8.0	10.6	0
39.7	7.8	11.3	0
42.6	11.3	10.3	0
45.4	12.0	11.9	1
45.5	6.0	11.4	0
47.8	10.9	12.4	1
48.0	6.9	14.1	1

[1] Extracted from Barbottin et al. (2008)
[2] $X = 0$ if GPC < 11.5, $X = 1$ if GPC ≥ 11.5

We consider first the criteria for evaluating a couple (Y,t) and the choice of threshold. Then we consider the question of the intrinsic quality of a decision variable Y for discriminating between $X = 0$ and $X = 1$. Finally, we present the ROC analysis, which allows one to evaluate decision variables without reference to a specific threshold.

3.3.2 Evaluation Criteria

Given Y and t, we can summarize the information in the sample in a 2-way contingency table (Table 9.6).

TABLE 9.6 Form of Contingency Table for a Binary Decision Problem

Prediction\True	$X = 1$	$X = 0$	Total
$Y \geq t$	a True positive	b False positive	$N_{Y \geq t}$
$Y < t$	c False negative	d True negative	$N_{Y < t}$
Total	N_1	N_0	N

For fixed $N_{Y \geq t}$ and $N_{Y < t}$ there are only two independent pieces of information in this table, so there are two independent criteria that can be defined. However, there are many possible formulations for these two criteria. We will concentrate on the true positive rate and the true negative rate. The true positive rate is a/N_1, the fraction of cases with $X = 1$ that are correctly classified. This is also called the sensitivity. The true negative rate is d/N_0. This is the fraction of cases with $X = 0$ that are correctly classified. It is also called the specificity. The specificity can also be expressed as $d/N_0 = (N_0 - b)/N_0 = 1 - b/N_0 = 1 -$ false positive rate. Sensitivity and specificity are in the range 0 to 1.

To summarize, our two criteria for evaluating how well the couple (Y,t) predicts a binary outcome are

sensitivity = true positive rate = a/N_1 = $1 -$ false negative rate
specificity = true negative rate = d/N_0 = $1 -$ false positive rate

The larger these two values, the better the discrimination.

If one wants to choose an optimal threshold, it is necessary to define the relative weights to give to different types of misclassification. If the criterion is to maximize the total number of correctly classified individuals, then one simply wants to maximize (sensitivity + specificity). In other cases, different weights may be appropriate.

Sensitivity and specificity vary with the threshold. We could test very many thresholds, but in fact we can be more efficient. If the observed

Y values are arranged in ascending order, all thresholds between successive pairs of values will give the same sensitivity and specificity values. Therefore, it is sufficient to test a single threshold between each pair of successive Y values.

Example 3 cont.

Figure 9.6 shows how sensitivity and specificity of the SPAD decision variable evolve as the threshold is varied. Table 9.7 shows the numerical values for thresholds which are the midpoints between measured values.

The maximum of sensitivity + specificity is 1.5, and this occurs at the two thresholds $t = 44.00$ and $t = 46.65$. (In fact, the same value of the sum is obtained for thresholds around those values.) If only the sum is important, then we could choose either of those thresholds.

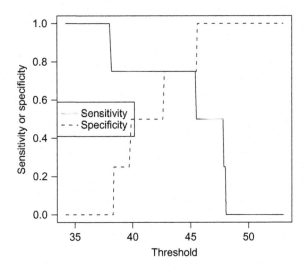

FIGURE 9.6   Sensitivity and specificity as a function of threshold.

TABLE 9.7 Sensitivity and Specificity for Various Threshold Values, for Decision Variable SPAD

Threshold t for SPAD	$-\infty$	38.15	39.00	41.15	44.00	45.45	46.65	47.90	∞
Sensitivity	1	0.75	0.75	0.75	0.75	0.50	0.50	0.25	0
Specificity	0	0.00	0.25	0.50	0.75	0.75	1.00	1.00	1

As the example shows, there is a trade-off between sensitivity and specificity. For a very low threshold, sensitivity is 1 but specificity is 0, and as the threshold increases, sensitivity decreases but specificity increases. Choosing a threshold involves choosing the optimal trade-off between the two.

3.3.3 Evaluating a Decision Variable, Independently of the Threshold

We can ask, more generally, how useful is a particular decision variable for discriminating between the situations $X = 0$ and $X = 1$. We now show that this question has an answer, independent of the choice of threshold.

We can divide the target population into two parts, those individuals for which $X = 0$ and those for which $X = 1$. The Y values just for those individuals with $X = 0$ is a new random variable, noted

$$Y0 = Y|(X = 0).$$

The variable with Y values for those individuals with $X = 1$ is noted

$$Y1 = Y|(X = 1).$$

Figure 9.7 shows three contrasting situations. In the top panel, the $Y0$ and $Y1$ populations are well separated. It is clear that we can find a threshold such that most of the $Y1$ values are above the threshold, and most of the $Y0$ values are below it. In the middle graph, there will necessarily be a fair amount of error. Whatever threshold we choose, there will be errors of classification; there will be a substantial part of $Y0$ above the threshold and/or a substantial part of $Y1$ below the threshold. In the bottom graph, no threshold value will largely separate $Y0$ and $Y1$. There will be almost as many errors as correct classifications.

We can summarize the differences between those situations in a single number: the probability that $Y1$ is greater than $Y0$, $P(Y1 > Y0)$, for randomly chosen individuals from $Y0$ and $Y1$. If this probability is large, as in the top plot, then the decision variable Y can be used to differentiate between $X = 0$ and $X = 1$ in most cases. If the probability is near 0.5, as in the bottom graph, then the decision variable Y is not very useful in classifying cases as belonging to $X = 0$ or $X = 1$. A probability of 0.5 means that the classification variable is completely useless for distinguishing $X = 0$ cases from $X = 1$ cases; it does not classify situations better than a random classification rule. A value below 0.5 means that the variable has some capacity to distinguish $X = 0$ cases from $X = 1$ cases, but classifies them incorrectly.

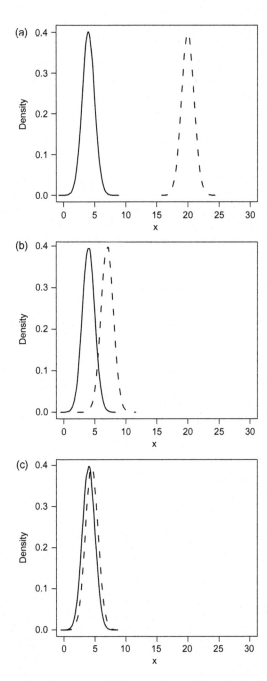

FIGURE 9.7   Probability distributions of Y|X = 0 (Y0) and Y|X = 1 (Y1) for three contrasting cases.

Example 3 cont.

For the small sample of Table 9.5, we can easily calculate the probability $P(Y1 > Y0)$. To do so, we take all possible combinations of $Y1$ and $Y0$ values (the number of possible combinations is noted C) and count the number of times $Y1 > Y0$ (noted $n_{Y1 > Y0}$). Then

$$P(Y1 > Y0) = n_{Y1 > Y0}/C.$$

Of the eight values in the data set, four have $X = 0$ and four have $X = 1$, so there are $4*4 = 16$ possible pairs. Of those, 11 are such that the SPAD value for the $X = 1$ member of the pair is larger than the SPAD value for the $X = 0$ member. Thus the probability that $Y1 > Y0$ here is $11/16 = 0.6875$. Using the AZODYN values, 10 pairs are such that the AZODYN value for the $X = 1$ member of the pair is larger than the AZODYN value for the $X = 0$ member. Thus the probability that $Y1 > Y0$ here is $10/16 = 0.625$. The conclusion is that, based on this sample, SPAD would be preferred as a decision variable because it has a larger value for $P(Y1 > Y0)$. Of course the sample size is quite small, so there might be substantial variability between possible samples.

3.3.4 Receiver Operating Characteristics Analysis

A receiver operating characteristics (ROC) analysis gives information about sensitivity and specificity for various thresholds, and also estimates $P(Y1 > Y0)$. Originally ROC analysis was used to depict the trade-off between true hits (true positive rates) and false alarms (false positive rates) in signal processing. That is the origin of the name ROC. Nowadays, ROC analysis is very widely used in medicine (Koepsell and Weiss, 2003) and other fields.

A ROC curve plots sensitivity on the y axis and $1 -$ specificity, from 1 to 0, on the x axis. The different points on the graph correspond to different threshold values. Points in the upper left hand corner of the graph correspond to large values of both sensitivity and specificity.

A ROC curve shows the possible values of sensitivity and specificity, but the most important feature of a ROC curve is the area under the curve (AUC). This is the sample estimate of $P(Y1 > Y0)$. Thus a perfect decision variable has AUC $= 1$, while a decision variable that has AUC $= 0.5$ is useless for discriminating between $X = 0$ and $X = 1$.

We do not show the general proof, but we can easily show that $P(Y1 > Y0)$ is equal to AUC in two limiting cases. Consider the case where $P(Y1 > Y0) = 1.0$ (like the top panel in Figure 9.7 but with perfect separation of the $Y0$ and $Y1$). Starting from low thresholds and increasing the threshold, the sensitivity remains 1 while the specificity increases from 0 to 1. Then, as the threshold continues to increase, the specificity remains at 1

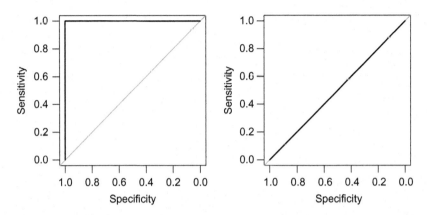

FIGURE 9.8   ROC diagrams for two limiting cases. Left: perfect discrimination. Right: No power of discrimination.

while the sensitivity decreases to 0. The ROC diagram is shown in Figure 9.8, left-hand panel. The area under the ROC curve is then 1.0, equal to $P(Y1 > Y0)$.

Consider next the case where $P(Y1 > Y0) = 0.5$ (similar to the bottom panel in Figure 9.7 but with the two curves exactly superposed). In this case, whatever the threshold, the true positive rate will exactly equal the false positive rate. That is, sensitivity $= 1 -$ specificity and so the ROC diagram will be a diagonal line and the area under the ROC curve will be 0.5 (Figure 9.8, right-hand panel).

The pROC package in R (Robin et al., 2011) does all the calculations of sensitivity, specificity, AUC, and confidence intervals for AUC (see Section 7.2). An alternative is the ROCR package (Sing et al., 2005).

Example 3 cont.

The ROC curve for the SPAD variable of this example is shown in Figure 9.9.

The AUC value is 0.6875 (95% confidence interval 0.21–1.0). For the AZODYN variable the AUC value is 0.625 (95% confidence interval 0.18–1.0).

Since a larger AUC value indicates better discrimination, we would prefer SPAD rather than AZODYN. Both values are above 0.5, the value for a variable with no discriminatory capacity at all. The sample size here is very small, so there is a very high degree of uncertainty attached to the estimated values.

When the full data set of 43 values from Barbottin et al. (2008) is used, the AUC values for the SPAD and AZODYN variables are, respectively, 0.76 (95% confidence interval 0.61–0.92) and 0.46 (95% confidence interval 0.28–0.64).

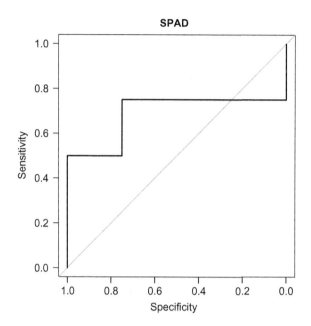

FIGURE 9.9   ROC diagram for SPAD as the decision variable.

3.4 Evaluating Probabilistic Predictions

In most cases, system model predictions in agronomy are deterministic. Given the same set of input variables, the model produces the same outputs. There are, however, situations where the results of a model are stochastic; for given values of the input variables, the model produces a probability distribution rather than a single prediction.

One case is where the model parameters are estimated using a Bayesian approach (see Chapter 7). The result in this case is a joint probability distribution for the parameters, which leads to a probability distribution for the predictions of the model.

Another case is where the input variables of the model are uncertain, and one has a probability distribution for the inputs (see Chapter 5). Once again, the result is a probability distribution for the predictions of the model.

A third situation is where one predicts using an ensemble of models. Then one can use the ensemble of predictions as a probability distribution. For example, one could predict that there is a 50% chance that yield will be within the range of yields obtained by removing the largest 25% and the smallest 25% of predictions.

Several different approaches to the evaluation of probability forecasts have been proposed. Here we consider the discrete ranked probability score (rps) (Toth et al., 2003), which is adapted to the probabilistic prediction of continuous variables. It is widely used in evaluating weather forecasting systems.

We divide the range of possible outcomes into K categories. For yield prediction, for example, the categories might be 0−1t/ha, ..., 9−10t/ha. For each situation, the observed result falls into one particular category. Thus it has probability 1 for that category and 0 probability elsewhere. The predictions, on the other hand, have probabilities which can be any-where from 0 to 1 for each category. For both predictions and observations, we can calculate the cumulative probability for each category, which is the probability that the observation or prediction is less than or equal to the upper value in the category.

The definition of the rps for a single situation is then as follows:

$$rps = \frac{1}{K-1} \sum_{k=1}^{K} \left(\hat{p}_k - o_k \right)^2 \tag{7}$$

where K is the number of categories and $\hat{p}_k$ and o_k are, respectively, the cumulative probabilities that the prediction and that the observation are less than or equal to the upper bound of category k. The larger the rps score, the worse the prediction. The factor $1/(K-1)$ is included in the formula so that the upper limit on rps is 1.

The smallest possible value of rps (best model) is 0, when the prediction is deterministic and falls into the same category as the observation. The worst possible case is where the prediction has a probability of 1 for the smallest category, $k = 1$, and the observation falls into the largest category, $k = K$. Then $\hat{p}_k = 1$ for $k = 1, \ldots, K$ (since $\hat{p}_k$ is a cumulative probability), $o_k = 0$ for $k = 1, \ldots, K-1$ and $o_K = 1$. Then

$$rps = \frac{1}{K-1} \left[\sum_{k=1}^{K-1} (1-0)^2 + 0 \right] = 1$$

Note that the farther a prediction is from the true value, the more it will contribute to the rps value. This is appropriate for the case where the categories have a natural order, like increasing yield.

We can transform the rps score into a skill score by comparing it with a naïve estimator. The ranked probability skill score is defined by

$$rpss = 1 - \frac{rps}{rps^{naive}}$$

where rps^{naive} is the rps value of the naïve predictor. One choice for the naïve estimator is the probabilistic predictor which uses the probability distribution of the observed values in the sample to predict for all situations. Another is a deterministic predictor that predicts the average of the observed values for all situations.

The R package verification (NCAR, 2012) can be used to calculate rps and rpss.

Example 4

The AgMIP project (Rosenzweig et al., 2013) aims at crop model intercomparison and improvement. An initial project was to run 27 different wheat models for the same situations. The prediction puts a probability of 1/27 on the value predicted by each model.

We illustrate rps with artificial data that have a structure similar to the AgMIP data, with yield predictions for wheat by 10 different models at four sites. The model predictions are presented in Table 9.8.

We divide the possible results into eight categories, with upper limits of 2,...,9 t/ha. The probability of having a prediction in each interval is the number of models whose prediction is in that interval divided by 10 (the total number of models). The resulting probabilities are presented in Table 9.9; the cumulative probabilities are shown in Table 9.10.

The observed values for the four sites are respectively 7.5, 6.1, 3.9, and 1.9 t/ha. The resulting cumulative probability table is given in Table 9.11.

TABLE 9.8 Simulated Yields (T/Ha) from an Ensemble of 10 Models for Four Situations

Site	M 1	M 2	M 3	M 4	M 5	M 6	M 7	M 8	M 9	M 10
1	8.4	6.8	7.3	7.6	8.1	7.0	7.5	8.7	6.6	6.9
2	5.7	5.4	5.8	6.1	5.8	4.9	6.6	5.8	6.4	4.9
3	3.7	3.5	3.9	3.8	3.7	3.6	4.0	5.6	4.3	2.5
4	3.6	2.8	2.7	2.5	4.0	3.0	2.2	1.9	1.9	2.6

TABLE 9.9 Probability for Each Yield Class Based on Ensemble of Models

Yield Class (yield range in t/ha)

Site	1 (1−2)	2 (2−3)	3 (3−4)	4 (4−5)	5 (5−6)	6 (6−7)	7 (7−8)	8 (8−9)
1	0.0	0.0	0.0	0.0	0.0	0.4	0.3	0.3
2	0.0	0.0	0.0	0.2	0.5	0.3	0.0	0.0
3	0.0	0.1	0.7	0.1	0.1	0.0	0.0	0.0
4	0.2	0.6	0.2	0.0	0.0	0.0	0.0	0.0

TABLE 9.10 Cumulative Probability for Each Yield Class, Based on Ensemble of Models

Site	1–2	2–3	3–4	4–5	5–6	6–7	7–8	8–9
1	0.0	0.0	0.0	0.0	0.0	0.4	0.7	1.0
2	0.0	0.0	0.0	0.2	0.7	1.0	1.0	1.0
3	0.0	0.1	0.8	0.9	1.0	1.0	1.0	1.0
4	0.2	0.8	1.0	1.0	1.0	1.0	1.0	1.0

TABLE 9.11 Cumulative Probability of Each Yield Class, for Measurements

Site	1–2	2–3	3–4	4–5	5–6	6–7	7–8	8–9
1	0.0	0.0	0.0	0.0	0.0	0.0	1.0	1.0
2	0.0	0.0	0.0	0.0	0.0	1.0	1.0	1.0
3	0.0	0.0	1.0	1.0	1.0	1.0	1.0	1.0
4	1.0	1.0	1.0	1.0	1.0	1.0	1.0	1.0

TABLE 9.12 Ranked Probability Scores

Site	RPS
1	0.25/7
2	0.53/7
3	0.06/7
4	0.68/7
Mean	(1/4)1.52/7 = 0.05428

The rps values for the four sites calculated using Equation (7), and their average, are shown in Table 9.12.

As a naïve estimator we use the predictor which has the same probability distribution as the observed values in the sample. Since the four observed values fall, respectively, into the categories 7–8, 6–7, 3–4, and 1–2, the naïve estimator has a probability of 0.25 for each of those categories. The resulting

TABLE 9.13 Cumulative Probability of Each Yield Class for Naïve Estimator

Site	1–2	2–3	3–4	4–5	5–6	6–7	7–8	8–9
1	0.25	0.25	0.5	0.5	0.5	0.75	1.0	1.0
2	0.25	0.25	0.5	0.5	0.5	0.75	1.0	1.0
3	0.25	0.25	0.5	0.5	0.5	0.75	1.0	1.0
4	0.25	0.25	0.5	0.5	0.5	0.75	1.0	1.0

cumulative probabilities are shown in Table 9.13. The resulting rps value for this naïve estimator is rps = 0.1875. Then the value of the skill score is rpss = 1−0.054/0.187 = 0.710. The probabilistic prediction in this artificial example is substantially better than the naïve predictor.

4. FROM THE SAMPLE TO THE POPULATION

In the preceding section we presented measures that summarize the agreement between the model and past measurements. In general, however, our real interest is not in how well the model reproduces data that has already been measured, but rather in how well it can predict new results.

Is model quality based on the sample an unbiased estimator of model quality for the population? (Unbiased means that the average over possible samples is equal to the population mean. See Chapter 2.) The answer is yes if two major requirements are met. The first is that the sample must be representative of the population in which we are interested (called the target population). The second is that the sample must not have been used for model development.

We consider each of these requirements below. These considerations apply to all of the measures of goodness of the previous section. In the next section we consider in more detail the specific measure of goodness-of-fit MSE.

We concentrate here on whether a measure of model quality calculated using a sample is an unbiased estimator of model quality for the target population. Remember, however, that unbiasedness isn't everything. The mean squared error of any estimator is the sum of bias^2 and variance of the estimator (Chapter 2). Unbiasedness is certainly desirable since it makes the first contribution 0, but there is still the second term. In particular, for small samples, the variance of the estimator can be large.

4.1 The Sample Must Be Representative of the Target Population

If the sample is not representative of the target population, then measures of goodness-of-fit for the sample cannot automatically be extrapolated to the population.

For example, the target population might be corn fields without significant weed or disease or pest problems in some region. If we included fields that do have substantial yield loss due to weeds, pests, or diseases, then that would be a different target population. Suppose we are interested in predicting yield. A model might predict well for the first target population, but not for the second. If we evaluate model quality using a sample from the first population, it will not in general give an unbiased estimate of model quality for the second population (and vice versa).

As another example, the target population might be sunflower fields in the European Union under current CO_2 levels. A different population would be future sunflower fields where CO_2 levels are 600ppm. An estimate of model quality based on a sample from the first population would not necessarily give an unbiased estimate of model quality for the second population.

In the case of variables such as LAI, soil moisture, or disease prevalence, which vary over time, there is the additional complication that model quality evaluated at one time might not be applicable to other times. One simple possibility is to avoid the problem by considering only responses with single values. For example, disease prevalence at each time could be replaced by integrated disease prevalence over the growing season of the crop. Soil moisture at each time could be replaced by some overall measure of water stress. Wallach et al. (1990), looked at N uptake integrated over time in a study of nitrogen uptake by the root systems of young peach trees.

Often in statistics one assumes that the sample is drawn at random from the target population. By construction then the sample is representative of the target population and sample quantities are unbiased estimators of the corresponding population quantities (see Chapter 2). For system models in agronomy, however, obtaining a random sample from the target population may not be possible. Often one has to make do with the historical data that is available, which may represent a different mix of conditions (weather, soils, management) than the target population. One possibility in that case is to define the target population to be consistent with the sample. That is, one describes the variability in the sample, and then defines the target population to encompass approximately that variability.

Example 5

Wang et al. (2003) evaluated the CROPGRO model on Missouri claypan soils for soybean. That is, the target population is soybean fields in Missouri for this specific type of soil (claypan). For this target population, the authors are interested in yield and also in other variables.

The authors note that average long-term rainfall for the 5 months of the soybean growing season is 507 mm. The sample covers three growing seasons, with accumulated rainfall of 428 mm, 628 mm, and 299 mm. The sample average is 452 mm, which seems reasonably close to the population average. One could claim that the sample here is representative of the weather conditions of the target population. However, it would be worthwhile looking more closely at the distribution of rainfall in the long-term historical record. In particular, is the 299 mm year a very rare occurrence or not? If so, then the sample gives heavy weight (1/3 of the sample) to a rare event in the population.

It is explicitly stated that herbicide and pesticide were applied as needed, so it seems that the target population has minimal levels of pest damage or weed competition. Other management practices are given for the experimental fields but it is not clear what the corresponding target population is. Perhaps the target population is meant to have standard management practices for this region.

A second problem situation is where one draws a sample from the target population, but the sample is relatively small and does not contain some important types of condition that exist in the target population. This is particularly a problem for weather, since one cannot be assured of sampling the desired range of weather conditions. For example, it may be that the sample does not include severe drought years though these are known to occur for the target population. Here again, it is probably more realistic to redefine the target population to be consistent with the sample. If the sample does not include drought years, one would specifically note that the model has not been tested for drought conditions.

A third problem situation is where the sample has the same range of conditions as the target population, but does not have the correct weighting of different types of conditions. This should be noted. Example 6 illustrates the problem. In some cases, one could reweight the sample values to give the correct target population weighting. For example, in the case of spatial data, one might want to weight each point by the area it represents (Willmott et al., 1985). Example 7 gives a different example.

Example 6

Rötter et al. (2012) compare nine models for barley in 'different climatic zones of Northern and Central Europe'. This clearly specifies the geographical area of interest, and indicates that the results are meant to be valid for a range of climates. The authors characterize the experimental sites by giving average rainfall and temperature and soil characteristics, which allows one to see the range of climates and soils that is represented in the sample.

Of the 44 growing seasons in the sample, 14 come from one Czech site and 13 from another Czech site, two sites with similar long-term average rainfall and temperature. This should be taken into account when extrapolating the performance of the model. The performance measures are heavily weighted toward the performance at these two sites.

Example 7

Chapman et al. (2000) consider the problem of selection of sorghum hybrids in north-eastern Australia. This is a variety selection problem rather than a modeling problem, but variety selection also requires the definition of a target population. They identify three general types of drought environment. Based on long-term historical weather data they calculate the fraction of years where each occurs. This they identify as their target population as concerns weather. Hybrid trials only cover a few years, and the drought environments encountered do not always reflect the long-term average prevalence of those environments. They show that one can reweight the results from the hybrid trials, in order to give each drought environment its long-term weight.

In conclusion, it is important both to characterize the sample (soil, weather, management) and to describe the target population. If the two differ appreciably, then it is not reasonable to assume that model predictive quality will be similar to model performance based on the sample.

4.2 The Sample Must Not Have Been Used for Model Development

If the sample has been used for model development, then model quality based on the sample will, in general, be more optimistic (smaller errors) than model quality for the full target population.

In fact, this is really the same problem as having a sample that is not representative of the target population. If the model is adjusted to the sample data, then the sample becomes special compared to the full population.

The most common way to use the sample data for model development is to use the data for parameter estimation. However, the requirement here also applies to other uses of the data. Suppose, for example, that one has several different possible equations for one of the processes in a system model, and one chooses the equation that gives the best fit to the sample data. The sample data have then been used for model development, and so sample-based values of model quality will tend to be over-optimistic compared to model quality for the target population.

How then can one obtain a realistic estimate of model error, in cases where one needs data for model calibration? This problem is of major interest in statistics. We discuss it in detail in the next section.

5. THE PREDICTIVE QUALITY OF A MODEL

5.1 A Criterion of Prediction Quality, MSEP

The standard criterion of prediction quality in statistics is the mean squared error of prediction or MSEP, defined as

$$MSEP = E\{(y-\hat{y})^2\} \tag{8}$$

where y is some system response, $\hat{y}$ is the simulated value, and the expectation is over the target population. Often one reports the root mean squared error of prediction (RMSEP) in order to deal with a quantity that has the same units as y. The definition is

$$RMSEP = \sqrt{MSEP}$$

MSEP is in fact the population version of MSE. MSE is a sample quantity — it concerns just the situations that have actually been measured, while MSEP is a population quantity — it involves an expectation over the full target population.

We discussed in the previous section the extrapolation of sample quantities to the target population. That discussion applies fully here. MSE will be a realistic estimator of MSEP if the sample is representative of the target population and if the sample isn't used for model development.

We illustrate below the difference between MSE and MSEP when the sample is used for parameter estimation. This example uses a simple artificial model and artificial data, but further on we will see examples based on real data.

Example 8

We suppose that we have a response variable y, an explanatory variable x, and that the relation between them is

$$y = \theta^{(0)} + \theta^{(1)}x + \theta^{(2)}x^2 + \theta^{(3)}x^3 + \theta^{(4)}x^4 + \varepsilon$$
$$\theta^{(0)} = -3.0, \quad \theta^{(1)} = 1.0, \quad \theta^{(2)} = 0.4, \quad \theta^{(3)} = 0.1, \quad \theta^{(4)} = 0.02$$
$$e \sim N(0, 3^2)$$

The fact that there is a random term ε means that knowing x is not sufficient to completely determine the value of y. The target population is defined by the distribution of ε and of x, which has a uniform distribution from 0 to 2: $x \sim U(0,2)$. A large sample from the resulting target population is shown in Figure 9.10a.

The best possible model is

$$E(y|x) = \theta^{(0)} + \theta^{(1)}x + \theta^{(2)}x^2 + \theta^{(3)}x^3 + \theta^{(4)}x^4$$

This curve is shown as a solid line in each panel of Figure 9.10.

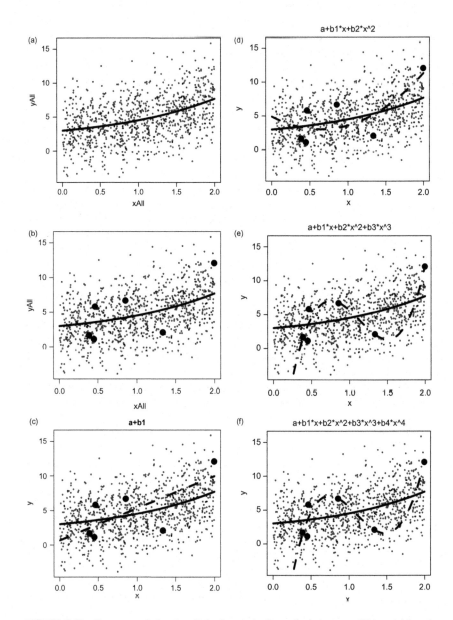

FIGURE 9.10 Target population (small dots), sample (large dots), best possible model based on x (solid line), and four models of increasing complexity (dashed lines).

TABLE 9.14 MSE and MSEP Values for Four Models of Increasing Complexity

Model	MSE	MSEP	MSEP/MSE
a + bx	7.311	11.05	1.51
a + bx + cx²	6.026	10.36	1.72
a + bx + cx² + dx³	1.590	46.16	29.04
a + bx + cx² + dx³ + ex⁴	1.587	53.41	33.66

We suppose that we have a random sample of size 6 from the target population. The sample is shown as large solid circles in Figure 9.10. We consider four different models of increasing complexity for y, as follows:

$$y^{(1)} = a + bx$$
$$y^{(2)} = a + bx + cx^2$$
$$y^{(3)} = a + bx + cx^2 + dx^3$$
$$y^{(4)} = a + bx + cx^2 + dx^3 + ex^4$$

where the coefficients a, b, c, d, and e are estimated by ordinary least squares using the sample data. The parameter values will be different for each equation. Figures 9.10c-f show the fit of each model.

Table 9.14 shows the MSE and MSEP values for the four models after parameter estimation using ordinary least squares. The apparent error MSE is calculated using the sample data. The prediction error, MSEP, is calculated using all 1000 points in Figure 9.10 (for this many data, it doesn't matter that the sample points are also used).

The major point is that MSE underestimates MSEP, and that the degree of underestimation increases as the number of parameters increases. For the cubic and quartic equations, the degree of underestimation is very large, even though the true model is quartic. The problem is quite clear from Figure 9.10. These cubic and quartic models are tracking the random errors in the sample points rather than the overall trend in the target population. As this example shows, when parameters are estimated from the data, MSE is not a good estimator of MSEP and the difference between the two can be very large.

Another point to note is the different behavior of MSE and MSEP as more flexibility is added to the model (more estimated parameters). MSE declines as extra parameters are added. This is always true. For a given data set, as more flexibility is added to a model, for example, by estimating additional parameters that had been treated as fixed, MSE must decrease or at worst remain unchanged. MSEP, on the other hand, here and in general, has a minimum value at some intermediate level of flexibility. Here MSEP is minimal for the quadratic equation. The number of parameters where MSEP is minimal will depend on the amount of data. In general, with more data, more parameters can be estimated before MSEP begins to increase with extra parameters.

5.2 Estimating MSEP When the Sample Is Used for Parameter Estimation

We have just shown that if the sample is used for parameter estimation (or other model development), then MSE is not a good estimator of MSEP. Below we present ways of estimating MSEP in this case.

5.2.1 Data Splitting

In this approach the data are split into two parts. One part (the training data) is used for model development; the other part (the evaluation data) is used to calculate MSE. This MSE value is then used as the estimate of MSEP. In this way, we have evaluation data that have not been used for model development.

However, splitting the sample into two exclusive parts is not by itself sufficient to ensure that MSE based on the evaluation sample is an unbiased estimator of MSEP. There are two additional conditions.

First, the evaluation data must be representative of the target population. Note that there are no requirements here for the training data. The only conditions are on the evaluation data.

Secondly, the evaluation data must not have been used for model development, and furthermore must be independent of the data used for model development. This independence requirement is new. More precisely, model residual error for the evaluation sample must be independent of model residual error for the situations used for model development. Independence is defined formally in Chapter 2. Typically, data from the same field or from the same year may not have independent residual errors. For an illustration see Example 11.

To see this intuitively, consider the case where data from the same field but different years are used as training data and as evaluation data. Suppose this field has some particularity. For example, it might have particularly shallow soil. Since data from this field enter into the training data, the model will be specifically adapted (at least in part) to very shallow soils. Then the model will fit the evaluation data from the same field better than it would fit a field chosen at random from the target population.

Example 9

The data used in Carberry et al. (1989) come from 13 different experiments at one location. There are three sowing dates, each with three irrigation treatments in 1984/1985 and two sowing dates each with two irrigation treatments in 1983/1984, so overall 13 treatments. Four of these treatments were used for model calibration (two from each growing season); the remainder were used for 'independent' (their term) model evaluation.

The training and evaluation data come from the same field, and so may not be independent. Furthermore, the fact that the same years of data appear in the training and evaluation data may mean that the errors in the two data sets are not fully independent.

If the required conditions are satisfied, data splitting gives an unbiased estimate of MSEP. However, this approach is not very satisfactory. First of all, it is not an efficient use of the data. Only part of the data is used for parameter estimation, which means that the parameter estimators will have larger variance than if all the data had been used. Also, only part of the data is used to estimate MSEP, so this estimator also has larger variance than if all the data had been used. Given that one is often working with quite small amounts of data, this inefficiency is a major drawback. A second problem with data splitting is that it has a large element of arbitrariness. The division of the data between training and evaluation data is subjective. A different choice could lead to quite different results.

Cross validation is a method of estimating MSEP that is based on data splitting, but it remedies to a large extent the above two drawbacks. This approach is presented below.

5.2.2 Cross Validation

The method of cross validation (Efron and Tibshirani, 1993) is a very general approach for estimating MSEP when the data are used for parameter estimation (or more generally for model development). It is a non-parametric approach that does not require any assumptions about the distributions of the random variables. Nor does it require that the model be correct in some sense. This method can be applied to any model. There are, however, some requirements that apply to the sample, as noted below.

We first describe the simplest form of cross validation, called leave-one-out (LOO) cross validation. We assume that each element of the sample is drawn independently at random from the target population. We will treat below the case where we do not have a simple random sample.

One begins by estimating the parameter values using all the data except one, say y_1. The result is the estimated parameter vector noted $\hat{\theta}_{-1}$. Since y_1 was not used to estimate $\hat{\theta}_{-1}$, the squared error $\left[y_1 - f(X_1; \hat{\theta}_{-1})\right]^2$ is an unbiased estimator of MSEP for this model. Then the procedure is repeated, this time basing parameter estimation on all the data except y_2 and calculating $\left[y_2 - f(X_2; \hat{\theta}_{-2})\right]^2$. The procedure is repeated for every one of the data points. Each adjusted parameter vector gives rise to an estimate of MSEP. The final estimator of MSEP is the average of the n unbiased estimators $\left[y_i - f(X_i; \hat{\theta}_{-i})\right]^2$ for $i = 1, \ldots, n$. This procedure can be written succinctly as an equation:

$$\hat{MSEP}_{LOO-CV} = \frac{1}{n} \sum_{i=1}^{N} \left[y_i - f(X_i; \hat{\theta}_{-i})\right]^2 \qquad (9)$$

where $\hat{MSEP}_{LOO-CV}$ is the LOO cross validation estimate of MSEP.

In this approach, all of the data are used to estimate MSEP. This is thus a more efficient use of the data than in data splitting. Furthermore, all the data are used in the same way; every data point is used once for evaluation and $n-1$ times as training data. The problem of deciding which data to use as training data and which data to use as evaluation data thus disappears.

At the end of the procedure, we have n different estimates of the parameter vector θ. Which should we use in future simulations? None, in fact. The best estimate of θ is the one based on all the data, and that is the value that should be used when the model is applied

Note that we haven't exactly estimated MSEP for this model. Rather, at each stage of the cross validation procedure, we estimate MSEP for a model which has the same form as the final model, and where the parameters are estimated in the same way, but where the data are slightly different than our full data set because one value has been removed. In general, $\hat{MSEP}_{LOO-CV}$ will tend to be larger than MSEP, because removing one data point will increase the variance of the parameter estimators. This may be important for small samples, but much less so for large samples.

Computation time in cross validation can be a problem. In LOO cross validation we must do the parameter estimation n times to estimate MSEP, plus once using all the data to get the best parameter estimates.

Parameter estimation for system models in agronomy is sometimes based on a trial and error procedure, with a subjective criterion (parameters are varied until the model gives a 'satisfactory' fit to the data). Such an approach has many disadvantages, but a further disadvantage is that cross validation becomes essentially impossible. Cross validation requires that parameter estimation be repeated several times. The implicit assumption is that the method of estimation does not vary; only the data vary. For this to be true, one needs a reproducible algorithm for parameter estimation (for example, ordinary least squares, see Chapter 6).

Example 8 cont

We return to the artificial example with linear regression models for predicting y. We apply LOO cross validation to the 2-parameter model of this example. Each line in Table 9.15 corresponds to leaving out one measurement, leaving five data points for parameter estimation. Column 2 shows the value of the explanatory value for the left-out point. Columns 3 and 4 show the parameter values obtained by OLS based on the five remaining data points. Column 5 shows the predicted value for the left-out point, using the parameter values for this row. Column 6 shows the measured value for the left-out point, and column 7 shows mean

TABLE 9.15 LOO Cross Validation Estimate of MSEP

i = data point left out	x_i	a_{-i}	b_{-i}	$f(x_i; \hat{\theta}_{-i})$ $= a_{-i} + b_{-i}x_i$	y_i	$\left[y_i - f(x_i; \hat{\theta}_{-i})\right]^2$
1	0.386	1.113	4.363	1.676	2.798	1.258
2	0.450	1.499	4.144	1.126	3.363	5.006
3	0.463	−0.842	5.553	5.815	1.729	16.699
4	0.854	0.186	4.723	6.688	4.221	6.087
5	1.333	0.528	5.966	2.086	8.481	40.894
6	1.994	3.246	0.334	12.076	3.911	66.665
Mean = $\hat{MSEP}_{LOO-CV}$						22.768

squared error (MSE) for the left-out point. The average of those MSEs is the LOO cross validation estimate of MSEP, which here is $\hat{MSEP}_{LOO-CV} = 22.768$.

From Table 9.14, the value of MSE based on the full six sample points is 7.311 and the value of MSEP is 11.05. Cross validation in this case gives an estimated value of MSEP larger than MSE, as expected, but it in fact over-predicts MSEP to a fairly substantial extent.

Example 10

Ennaïfar et al. (2007) evaluated 16 different models (dynamic and static models, but not system models) for their ability to predict take-all incidence in winter wheat as a function of cropping practices, soil, and climate. The data came from 25 site-year combinations, with 2 to 24 different cropping systems in each case, for a total of 204 site-year-cropping system combinations.

Model parameters were estimated using these data, and then for each model RMSE was calculated using all the data and RMSEP was estimated using LOO cross validation. The results for five of the models tested are shown in Table 9.16.

As expected, the estimated RMSEP values here are larger than the RMSE values. (This is also the case for the models not shown here.) Furthermore, the choice between models is quite different when based on RMSEP than when based on RMSE. The model with the lowest RMSE value among all 16 is model STAL1 −, but this is no longer the best model based on RMSEP. The model STAM1 − is one of the better models based on RMSE, but is the worst of all 16 based on RMSEP.

TABLE 9.16 RMSE and LOO Cross Validation Estimates of RMSEP for Various Models[1]

Model	Number of Parameters	RMSE	RMSEP
DynL0 −	29	21.46	38.12
DynL0 +	21	22.43	34.28
DynL1 +	23	13.42	16.34
STAL1 −	20	11.22	19.38
STAM1 −	20	17.77	65.93

[1](Ennaïfar et al., 2007).

A generalization of LOO cross validation is K-fold cross validation, where the data are split into K roughly equal-sized parts. There can be several data points in each part. Parameter estimation is done K times. At each stage of the cross validation procedure, the data in one part are set aside, the parameters are estimated using the remaining data, and the part set aside is used as the evaluation data.

If the data set is large, then LOO cross validation may not be practical because it requires estimating the parameters n times. In this case one can do K-fold cross validation with K < n. For very large data sets, one can choose K = 2. The data are split into two approximately equal parts, and parameter estimation is done twice, each time using as evaluation data the half of the data that wasn't used for parameter estimation.

Another important use of K-fold cross validation is the case where the residual errors for the sample are not independent. If one can split the data into K parts that are independent, then one can do K-fold cross validation.

The independence structure of agronomic data is often complicated, and requires a variant of standard K-fold cross validation. As an example, consider the case where the data come from five sites, six years, and three treatments for each site-year combination (Figure 9.11). Residual errors for the same site or same year may not be independent. We split the data into 5*6 = 30 parts, each with one site-year combination. At each stage one site-year furnishes the evaluation data. The training data used for parameter estimation would be the data that doesn't refer to that site or year. Note that in this case, there are data at each step that are neither in the training data nor in evaluation data. Wallach et al. (2001) used this approach.

The following example illustrates the importance of assuring independence.

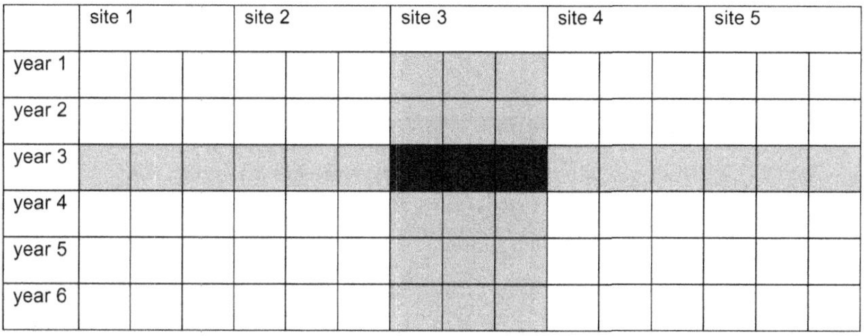

	site 1			site 2			site 3			site 4			site 5		
year 1															
year 2															
year 3															
year 4															
year 5															
year 6															

FIGURE 9.11   Cross validation for data with multiple measurements for each site and year. At each step of the cross validation approach, the data from one site-year are used for evaluation. The white area is the training data, the black area is the evaluation data. The gray area is not used in this stage.

Example 11

Landschoot et al. (2012) used cross validation to evaluate forecasting systems for fusarium head blight of wheat. Specifically, they forecast the concentration of the mycotoxin deoxynivalenol (DON) at harvest. The forecasting systems were static models, developed using multiple linear regression, ridge regression, and three other procedures. The data come from nine growing seasons and eight different sites for a total of 78 site-year combinations (not every site was represented every year). In each site-year, at least 10 different wheat varieties were grown.

They compared different cross validation approaches to estimating MSEP. We consider here two of these approaches. The first approach is LOO cross validation. The second approach is that described in Figure 9.11, where the training data and the evaluation data have neither the same field nor the same year in common. They consider that this second approach gives a realistic estimation of MSEP, and that the difference between the two approaches indicates to what extent LOO cross validation underestimates MSEP.

The results are shown in Table 9.17. The MSEP estimate is smaller with LOO cross validation than when all the data from the same site or year as the evaluation data are left out of the training data. The difference is huge in the case of multiple linear regression. The authors suggest that this is because the model here is very largely over parameterized. LOO cross validation does not correctly penalize for the excessive number of parameters. All the other approaches include built-in limitations on the number of parameters, and thus have smaller differences between the two cross validation methods, although the differences can still be substantial. The conclusion is that LOO cross validation underestimates MSEP if the residuals are not independent. (The example also emphasizes the dangers of over-parameterizing a model, and shows that the problem may not be recognized if an inappropriate method of estimating MSEP is used.)

TABLE 9.17 Comparison of Two Different Cross Validation Schemes for Various Models[1]

	Forecasting System					
	Multiple linear regression	Ridge regression	Regression tree	Boosting	SVR linear	SVR RBF
LOO cross validation	0.039	0.039	0.046	0.041	0.046	0.036
Leave out site and year	27883943	0.073	0.100	0.073	4.230	0.100

[1](Landschoot et al., 2012)

5.2.3 Bootstrap Estimation

The bootstrap, like cross validation, is a data re-sampling approach. That is, the same data are used several times to provide an estimator of the quantity of interest, here MSEP. Like cross validation, it is a very general approach which does not require distributional assumptions nor assumptions about the true model.

The basic idea is to treat the original data set, with n data points, as though it were the full target distribution. Then we draw samples of size n from this population. The samples are drawn with replacement (the same data point can be drawn multiple times), so in general many different samples are possible. For example, if $n = 8$, three bootstrap samples could be $(y_6, y_7, y_3, y_7, y_6, y_7, y_6, y_8)$, $(y_3, y_1, y_7, y_8, y_5, y_4, y_7, y_2)$, and $(y_1, y_7, y_8, y_7, y_5, y_2, y_4, y_7)$. We estimate sample quantities (for example, MSE) from each bootstrap sample, and population quantities (for example, MSEP) from the original sample.

For prediction error, it has been suggested that it is better not to estimate MSEP directly, but rather to estimate the difference

$$op = MSEP - MSE \qquad (10)$$

(Efron and Tibshirani, 1994; Efron, 1983). The notation 'op' comes from 'optimism' and was chosen to emphasize the fact that when the parameters are estimated from the data, MSE is, in general, smaller (more optimistic about the quality of the model) than MSEP.

For each bootstrap sample b, we estimate the model parameters, noted $\hat{\theta}^b$, and then calculate MSE for that sample as:

$$MSE^b = \frac{1}{n} \sum_{i=1}^{n} [y_i^b - f(X_i^b; \hat{\theta}^b)]^2$$

Here y_i^b is the ith data point in bootstrap sample b, and $f(X_i^b; \hat{\theta}^b)$ is the corresponding simulated value. We calculate $MSEP^b$, the bootstrap MSEP value for the model based on sample b, by evaluating that model using the initial sample:

$$MSEP^b = \frac{1}{n} \sum_{i=1}^{n} [y_i - f(X_i; \hat{\theta}^b)]^2$$

where y_i is the ith data point in the initial sample. Then we calculate op^b, the bootstrap estimate of op for bootstrap sample b, as

$$op^b = MSEP^b - MSE^b$$

The final estimator of op is the average of op^b over the B bootstrap samples:

$$\hat{op} = \frac{1}{B} \sum_{b=1}^{B} op^b$$

Finally, the bootstrap estimator of MSEP is

$$\hat{MSEP}_{bootstrap} = MSE + \hat{op}$$

where MSE is mean squared error based on the original sample.

It has been suggested that a sufficient number of bootstrap samples is often in the range $25-200$. One can monitor the final bootstrap estimate as a function of the number of bootstrap samples. Once the estimate no longer changes appreciably with additional samples, the bootstrap sample size is sufficient.

Since the parameter vector must be estimated for each sample, the overall calculation time can be quite long. Furthermore, there can be numerical difficulties in adjusting the parameters for certain bootstrap samples, especially if the original sample is quite small. On the positive side, in a simulation study Efron (1983) found that this bootstrap method gave better predictions of MSEP than did cross validation. However, as the example below shows, the bootstrap may not be feasible with relatively small sample sizes.

Example 8 cont

Table 9.18 shows five bootstrap samples for this example, together with the coefficients of the linear model and the values of MSE^b, $MSEP^b$, and $op^b = MSEP^b - MSE^b$ for each bootstrap sample.

The results are very variable between bootstrap samples, because of the small number of data points. In fact the bootstrap estimator here does not stabilize even with 10,000 samples, because of occasional very large values of op^b. This is simply too small a sample (six data points) for the non-parametric bootstrap.

TABLE 9.18 A Few Bootstrap Samples for Estimating op

Indices of Data Points In Bootstrap Sample	a^b	b^b	MSE^b	$MSEP^b$	op^b
2,3,5,3,1,2	3.472	−0.900	4.151	22.921	18.770
1,1,6,2,1,4	−0.834	6.714	0.761	8.909	8.149
5,6,6,3,4,4	2.490	4.069	6.904	9.106	2.201
1,3,4,4,6,1	0.965	5.838	1.789	9.713	7.924
6,2,6,4,2,5	−1.708	6.421	5.519	8.946	3.428

Like cross validation, the simple bootstrap must be modified if the data are not independent. The bootstrap sample should retain the dependence structure of the original sample. One common approach in the case of dependent data is to sample blocks of data rather than individual data points (Davison and Hinkley, 1997). Suppose for example that the data are multiple measurements from each of N randomly chosen individuals. Then bootstrap sampling would choose N individuals with replacement, but once an individual is chosen one uses all the data for that individual. In this case, the individual is the block.

5.2.4 Parametric Estimation of MSEP

One can obtain an analytical expression for MSEP in the context of linear regression, if one assumes that the assumptions underlying OLS are satisfied. The point here is not really to use this expression to estimate MSEP for system models; the models of course are not linear, and the assumptions are very probably not satisfied. Nevertheless, the analytical expression for MSEP can give some insight into MSEP and the difference between MSE and MSEP.

We assume that

$$y = x\beta + \varepsilon$$
$$\varepsilon \sim N(0, \sigma^2)$$

where x is the vector of explanatory variables and σ^2 is the model residual variance.

Let X be the matrix of explanatory values for the sample used for parameter estimation, and x the explanatory vector for a new situation for which we want to predict. Our predictor is $\hat{y} = x\hat{\beta}$, where $\hat{\beta}$ is the OLS estimator of the parameter vector. Since the assumptions underlying OLS are satisfied we have

$$E\left(\hat{\beta}\right) = \beta$$
$$\text{var}\left(\hat{\beta}\right) = \left(X^T X\right)^{-1} \sigma^2 \tag{11}$$

The mean squared error of prediction for this specific x is obtained by taking an expectation over y and over $\hat{\beta}$:

$$MSEP(x) = E\left[(y - \hat{y})^2\right] = E\left[\left(x^T \beta + \varepsilon - x^T \hat{\beta}\right)^2\right]$$
$$= \sigma^2 + x^T \text{var}(\hat{\beta}) x \tag{12}$$

MSEP is then obtained by taking the expectation over the x values of the target population:

$$MSEP = \sigma^2 + E\left\{\left[x^T \text{var}(\hat{\beta}) x\right]\right\} \tag{13}$$

To estimate MSEP we would use the sample estimate of σ^2. The expression for MSEP has two terms. The first (σ^2) is the residual variance and represents the fact that the explanatory variables don't explain all the variability in y. The second term represents the contribution of errors in estimating the parameter values. We shall see in Section 5.4 how this generalizes to system models.

MSEP summarizes prediction error in a single number; it is average prediction error over the target population. Equation (12) gives more detail, since it shows how prediction error depends on the explanatory variables. We work with MSEP because we can estimate it even in complex cases, but we should be aware that prediction error can be very variable for different x values.

The linear model case can also provide information about the difference between MSE and MSEP. Suppose that we are working with fixed values of the explanatory variables. The target population then has the same values for the explanatory variables as the sample. In this case one can show that

$$E(MSE) = (1 - p/n)$$
$$MSEP = \sigma^2(1 + p/n)$$
$$MSEP - E(MSE) = \sigma^2 2p/n$$

where the expectation is over samples, p is the number of parameters in the linear model, and n the number of data points in the sample. Here we see that the difference between MSEP and the expectation of MSE depends on the ratio of the number of parameters to the number of data points. For very large ratios, the difference is negligible and both E(MSE) and MSEP are approximately equal to σ^2. When $n = 20p$, so that there are 20 data points for every parameter, then MSE underestimates MSEP by $0.1\sigma^2$.

The above expression for MSEP is only exact in the case of a linear model, but it can also provide an approximation to MSEP for non-linear models. In this case the model is replaced by a linear approximation, which involves replacing X by the derivative of the non-linear function with respect to the parameters (see Chapter 6). In principle then we could use Equation (13) even for complex system models. However as mentioned it is probably unrealistic to assume that the assumptions underlying OLS are satisfied.

5.3 MSEP Is Not Just Due to Model Error

5.3.1 Measurement Error and MSEP

MSEP measures the mean squared error of predicting measurements. However, this error is not solely the result of errors in the predictions. The measurements themselves are not perfect, but only approximations to the true value of a variable. Part of MSEP is the result of measurement error.

In fact, we can easily evaluate the contribution of measurement error to MSEP. We assume that

$$y^{measure} = y^{true} + \eta$$
$$E(\eta) = 0 \quad var(\eta) = \sigma_\eta^2$$

The assumption $E(\eta) = 0$ implies that the measured values are centered around the true value, which is often reasonable. The variance of measurement error can be estimated if one has repetitions. Remember that if we use the mean of repetitions as the measured value, σ_η^2 represents the variance of that mean. This is $\sigma_\eta^2 = \sigma_{reps}^2/n_{reps}$ where σ_{reps}^2 is the variance between repetitions and n_{reps} is the number of repetitions. We assume here that the variance of measurement error is constant, the same for all measurements. Finally, we assume that measurement error and model error are independent. This is usually reasonable. Then

$$MSEP = E(y^{measured} - \hat{y})^2$$
$$= E(y^{measured} - y^{true} + y^{true} - \hat{y})^2$$
$$= \sigma_\eta^2 + E(y^{true} - \hat{y})^2 \qquad (14)$$
$$= \sigma_\eta^2 + MSEP^{y\,true}$$

Thus MSEP is the mean squared error of prediction of the true y value ($MSEP^{y\,true}$) plus the variance of measurement error. If the variance of measurement error is not constant, then σ_η^2 in the above equation would be replaced by the expectation of the variance over the target population. Note that the expectation above is both over the target population and over the possible measured values for each value of y^{true}. That is, this is MSEP averaged over possible measurement errors.

Measurement error thus simply adds on to MSEP. Put another way, the variance of measurement error is a lower bound on MSEP, since MSEP is that variance plus $MSEP^{y\,true}$ (which cannot be negative).

5.3.2 Error in Input Variables and MSEP

In Chapter 5 we treated the case where some of the explanatory variables X of the model are not perfectly well known. For example, this might be the case for soil water at field capacity, initial soil mineral nitrogen, daily weather, etc. Here we look at how uncertainty in X contributes to MSEP.

Let $\hat{X}$ represent the estimated value of the explanatory vector. We assume that we know the distribution of $\hat{X}|X$, noted $g(\hat{X}|X)$. For example, if the uncertainty is due to measurement error, we might assume that the measured value has a normal distribution with some variance σ^2, and is centered around the true value. That is, $g(\hat{X}|X) \sim N(X, \sigma^2)$. The model predictions using the estimated explanatory variables are noted $f(\hat{X}; \theta)$. The model parameters θ are assumed fixed. That is, we have in some way estimated the model parameters, and we are now interested in MSEP for the model with those fixed parameter values.

We can now show how the uncertainty in X contributes to MSEP. This derivation is analogous to that of Equation (14).

$$
\begin{aligned}
MSEP &= E\left\{ \left[y - f(\hat{X}; \theta) \right]^2 \right\} \\
&= E\left\{ \left[y - Ef(\hat{X}; \theta) + Ef(\hat{X}; \theta) - f(\hat{X}; \theta) \right]^2 \right\} \quad (15) \\
&= E\left\{ \left[y - Ef(\hat{X}; \theta) \right]^2 \right\} + E\left\{ \left[Ef(\hat{X}; \theta) - f(\hat{X}; \theta) \right]^2 \right\}
\end{aligned}
$$

The expectation is now both over the target population and, for each X of the target population, over $\hat{X}|X$.

MSEP is a sum of two terms. The second depends only on the model, not on the measured values of the response variable y. It is the variance of $f(\hat{X}; \theta)$ for fixed X, averaged over the target population. This could be calculated in a sensitivity analysis. The first term in the final line of Equation (15) depends on how well the model, averaged over $\hat{X}$, predicts the observed value y.

An important point to notice here is that sensitivity analysis only gives part of MSEP. It allows one to quantify the effect of errors in the explanatory variables (the second contribution to MSEP), but does not reflect model error (the first contribution).

5.4 The Relation Between Model Complexity and MSEP

A recurring question in system modeling concerns the choice of model complexity. Models are always simplifications of reality, but there is in general a lot of debate for each model, as to how extreme the simplification should be.

For example, crop models can describe roots just by depth or in more detail by a root distribution, the soil can be assumed laterally homogenous or variable in all three spatial dimensions, LAI can be described by a single state variable or one can have separate state variables for each leaf, etc. What level of detail is best?

We cannot provide an answer to this problem, but we can help understand what determines the optimal level of detail. That is the purpose of this section. Our analysis is based on a decomposition of MSEP (Bunke and Droge, 1984; Wallach and Goffinet, 1987).

We write MSEP as the sum of three terms, and then analyze how each is affected by the level of model detail.

The decomposition of MSEP is

$$
MSEP = E_X\left\{ E_y\left[y|X - E_y(y|X) \right]^2 \right\} + E_X\left\{ \left[E_y(y|X) - E_{\hat{\theta}}\left(f(X;\hat{\theta})|X \right) \right]^2 \right\}
$$
$$
+ E_X\left\{ E_{\hat{\theta}}\left[E_{\hat{\theta}}\left(f(X;\hat{\theta})|X \right) - f(X;\hat{\theta})|X \right]^2 \right\}
$$

(16)

The random variables here are y, X, and $\hat{\theta}$. The above expression is obtained quite simply by adding and subtracting $E_y(y|X)$ and $E_{\hat{\theta}}\left(f(X;\hat{\theta})|X \right)$ in the original expression for MSEP.

The first term in Equation (16), called the population variance, is

$$
\Lambda = E_X\{E_y[y - E_y(y|X)]^2|X\} = E_X[\text{var}(y|X)]
$$

(17)

This term depends on how much y varies for fixed values of the explanatory variables in the model. When X is fixed y still varies within the target population, because not all the variables that affect y are included in the model. That variability is then averaged over X. Note that Λ does not involve $f(X;\hat{\theta})$. That is, the exact equations of the model are irrelevant here. It is only the choice of the explanatory variables that is important. If the explanatory variables in the model do not explain much of the variability in y then the remaining variability in y for fixed X is large and Λ is large. Consider, for example, a model which does not include initial soil mineral nitrogen. If y (for example yield) for the target population is strongly affected by initial soil nitrogen, then Λ will be large.

We see that the choice of explanatory variables is a major decision as far as prediction accuracy is concerned. That choice determines one of the contributions to MSEP and thus sets a minimum value for it. Adding more explanatory variables to a model necessarily reduces the population variance, or at worst leaves it unchanged (if the new explanatory variables are totally useless for predicting y).

The second term in Equation (15), called the squared bias, is

$$\Delta = E_X \left\{ \left[E_y(y|X) - E_{\hat{\theta}}\left(f(X; \hat{\theta})|X \right) \right]^2 \right\}$$

This term does depend on the form of the model. It is zero if

$$E_y(y|X) = E_{\hat{\theta}}\left(f(X; \hat{\theta})|X \right),$$

which is the definition of an unbiased model. The best model, once the explanatory variables are chosen, is the model that for each X value equals $E_y(y|X)$.

The last term in Equation 15 is the model variance:

$$\Gamma = E_X \left\{ E_{\hat{\theta}} \left[E_{\hat{\theta}}\left(f(X; \hat{\theta})|X \right) - f(X; \hat{\theta})|X \right]^2 \right\}$$

This term measures how variable the model predictions are, because of the variability in the estimated model parameters.

We can add complexity to a model in two ways: by adding more explanatory variables (for example, weed population to account for weed competition), or by adding more detailed equations (for example, individual leaves rather than an average leaf). If we add more explanatory variables, the population variance term decreases or at worst is unchanged, but we then need to add additional equations and parameters to the model. This will tend to increase the squared bias and the model variance.

If we change just the complexity of the equations without adding explanatory variables, then the population variance remains unchanged. We may decrease the squared bias term; after all, the reason for using more complex equations is to get a better approximation to E(y|X). However, the additional complexity will require additional parameters, and that will tend to increase model variance.

The overall result will depend on how much population variance or squared bias is reduced, compared to how much error is added due to parameter estimation. Eventually, as more complexity is added, the gain from reducing population variance or squared bias will become smaller than the cost of estimating extra parameters, and MSEP will begin to increase. The point at which this happens will depend on many factors and in particular on p/n, the ratio of the number of estimated parameters to the number of data points. With more data one can estimate more parameters.

Suppose that one has a preliminary model and wants to decide whether or not to add additional explanatory variables. There is a better chance that the additional explanatory variables will reduce MSEP if:

1. they play an important role in determining the variability in y in the target population, so that adding them to the model reduces population variance by a substantial amount.

2. the associated equations and parameters can be well estimated from the available data, so that the additional detail does not cause a substantial increase in squared bias and model variance. This depends both on the data and on the role of the parameters in the model.

An analogous argument can be made for adding more detailed equations. That will be worthwhile if the additional equations play an important role in reducing squared bias, and if the additional parameters can be well estimated.

Example 8 cont

Here we consider MSEP for the fixed sample of this example. The decomposition of MSEP in this case is

$$MSEP = E_X\left\{ E_y\left[y|X - E_y(y|X)\right]^2 \right\} + E_X\left\{ \left[E_y(y|X) - f(X;\hat{\theta})|X\right]^2 \right\} \quad (18)$$

There are now only two terms in the decomposition, the population variance and the squared error between the model and $E_y(y|X)$. We will show how MSEP varies as the complexity of the model is increased.

Table 9.19 shows population variance and the squared error from Equation (18) for the four models of this example. All the models in the example have the same explanatory variable and thus the same value for the population variance. The value is $\Lambda = 9$. (This is just the variance of the random term ε.) Thus MSEP cannot be less than 9 in this example. The remainder of MSEP is squared error.

We see that squared error decreases at first, in going from the linear to the quadratic model. However, as more parameters are added, squared error

TABLE 9.19 MSEP and Its Two Components, for Four Models of Increasing Complexity

Model	MSEP	Λ	Squared Error
$a + bx$	11.05	9	2.05
$a + bx + cx^2$	10.36	9	1.36
$a + bx + cx^2 + dx^3$	46.16	9	37.16
$a + bx + cx^2 + dx^3 + ex^4$	53.41	9	44.41

increases very substantially. This is because the increase in squared error due to parameter error more than offsets the decrease due to being closer to the correct dependence on x. The quartic model has the correct dependence on x. However, this is not the best model because of errors in the parameters.

As this example shows, parameter error limits the complexity that is advisable for a model, if the goal is to reduce prediction error.

6. SUMMARY

It is always important to first analyze the data, and then to evaluate the fit of the model to the data. There is no compelling reason to choose a small set of goodness-of-fit measures in the initial analysis. The different measures provide different perspectives on model error, and are easily calculated. Therefore, it is good practice to look multiple measures of goodness-of-fit.

There are some specific cases that require specific criteria of goodness. One is the use of the model to sort situations into two categories. Another is the case where one makes probabilistic predictions.

In general one wants to estimate how well the model will predict new data. This requires as preliminary steps the definition of the target population – the range of conditions at which the model is aimed – and the definition of a criterion of model quality. This will often be MSEP.

If the sample is representative of the target population and the data have not been used for model development, then MSE is a reasonable estimator of MSEP. If this is not the case, then MSE in general underestimates MSEP and is not a reasonable estimator of MSEP.

To obtain a more reliable estimate of MSEP one can split the two into separate parts, one for training and one for evaluation. A more efficient use of the data is to estimate MSEP using cross validation. The bootstrap is another possible approach to the estimation of MSEP. In each case the overall sample must be representative of the target population. Cross validation further requires that at each stage the residual errors of the evaluation sample be independent of the residual errors in the training sample.

MSEP is not just the result of model errors. Measurement error and errors in the explanatory variables also contribute to MSEP. It is worthwhile estimating those contributions.

MSEP can be expressed as the sum of three contributions: population variance, squared bias and the model variance. As model complexity increases with more explanatory variables) the first term systematically decreases, while the second and third may initially decrease, but then increase. Additional model complexity is warranted if the additional explanatory variables and/or additional equations are important and if the additional parameters can be well estimated.

7. R FUNCTIONS

7.1 Goodness-of-Fit Measures

These measures require only basic R instructions. Nevertheless, it can be convenient to have a function that does these calculations. The `goodness. of.fit` function in the ZeBook R package calculates all the measures presented in this chapter except those based on a threshold of accuracy. These are calculated using the `threshold.measures` function of the ZeBook package. Other software for calculating evaluation measures also exists, see for example Fila et al. (2003).

The usage of the `goodness.of.fit` function is:

```
goodness.of.fit (Yobs,Ypred,draw.plot = FALSE)
```

The first two arguments are, respectively, the vectors of observed and predicted values. They must be of the same length. An NA in either vector will cause that case to be ignored. If draw.plot = TRUE, then the function will draw plots of observed versus predicted values and of residuals versus predicted values. It will also produce a bar plot of the components bias^2, SDSD, and LCS of MSE. The value of the function is a data frame containing all of the goodness-of-fit measures.

The usage of the `threshold.measures` function is:

```
threshold.measures (Yobs,Ypred,p,d,units = ' ' ' ')
```

The first two arguments are, respectively, the vectors of observed and predicted values. They must be of the same length. An NA in either vector will cause that case to be ignored. The p and d arguments define the thresholds. The function calculates the absolute error such that a percentage p of cases has less than that absolute error. The function also calculates the percentage of cases with absolute error less than d. The value of the function is a data frame containing p, d, and the corresponding absolute errors. The units argument takes a character string giving units of the observed values. This is used in printing out the results. It has as its default value the null string.

7.2 ROC

The pROC R package (Robin et al., 2011) contains several functions for drawing a ROC curve and analyzing the results. An alternative package is ROCR (Sing et al., 2005).

The central function of the pROC package is entitled roc. A simple use of the function is as follows:

```
roc(response = X,predictor = Y,ci = TRUE,plot = TRUE)
```

The `response` argument in roc takes a vector with the X values (0 or 1) of the sample, and the `predictor` argument takes a vector with the values of the decision variable. The ci = TRUE argument says to print confidence intervals for the AUC, and the plot = TRUE argument says to create a ROC graph.

The value of the function includes the components thresholds (the vector of thresholds that are used to calculate the ROC curve), sensitivities (the corresponding sensitivities), and specificities (the corresponding specificities).

The program to obtain the results for Example 3 with the results is as follows:

```
library(pROC)
# Put X values in X vector, Y values in Y vector
Y <-c(38.0,38.3,39.7,42.6,45.4,45.5,47.8,48.0)
GPC <-c(14.6,10.6,11.3,10.3,11.9,11.4,12.4,14.1)
X <-GPC = 11.5
print(rbind(Y,GPC,X))
     [,1] [,2] [,3] [,4] [,5] [,6] [,7] [,8]
Y    38.0 38.3 39.7 42.6 45.4 45.5 47.8 48.0
GPC  14.6 10.6 11.3 10.3 11.9 11.4 12.4 14.1
X    1.0  0.0  0.0  0.0  1.0  0.0  1.0  1.0
  roc(response = X,predictor = Y,ci = TRUE,plot = TRUE)
Call:
roc.default(response = X, predictor = Y, ci = TRUE, plot =
TRUE)
Data: Y in 4 controls (X FALSE) < 4 cases (X TRUE).
Area under the curve: 0.6875
95% CI: 0.2078-1 (DeLong)
myroc <-roc(response = X,predictor = Y,ci = TRUE,plot = TRUE)
myroc$specificities
[1] 0.00 0.00 0.25 0.50 0.75 0.75 1.00 1.00 1.00
    print(rbind(myroc$thresholds, myroc$sensitivities,
    myroc$specificities))
     [,1]  [,2]  [,3]  [,4]  [,5]  [,6]  [,7]  [,8]  [,9]
[1,] -Inf 38.15 39.00 41.15 44.00 45.45 46.65 47.90  Inf
[2,]    1  0.75  0.75  0.75  0.75  0.50  0.50  0.25    0
[3,]    0  0.00  0.25  0.50  0.75  0.75  1.00  1.00    1
```

7.3 Ranked Probability Scores

The verification R package (NCAR, 2012) contains the function rps, which calculates ranked probability scores. The form of the function is:

```
rps(obs, pred,baseline).
```

The user decides on how to divide the response values into K classes, numbered $1,\ldots,K$. The argument obs is a vector of length n (n is the number of observations) which has the class number (not the observed value) for each observation. The pred argument is a matrix of size nxK. Each line corresponds to one observed value. The columns represent the classes. Element i,j contains the predicted probability that the observation i will be in class j.

The argument `baseline` is a vector of length n with the probabilities for each class of the naïve predictor. If `baseline = NULL` is used, then the climatological naïve predictor is used. This has a probability of m_i/n in each class, where m_i is the number of observations that fall into class i. One can also define other naïve estimators. In that case, one would set `baseline` to a vector with the probability of each class according to the naïve estimator. For example, if there are four classes and the naïve estimator puts probability 0.5 on classes 2 and 3, we would use `baseline = c(0,0.5,0.5,0)`. The result is the rps score, the rps skill score, and the rps score for the naïve predictor.

The use of this function is illustrated below for the data of Example 4.

```
library(verification)
# yield classes are 1–2,2–3,3–4,4–5,5–6,6–7,7–8,8–9 t/ha.
   K is number of classes
K <- 8
# measured values are c(7.5,6.1,3.9,1.9). obs gives class
   of each observation
obs <- c(7,6,3,1)
# probModel is matrix with probability of prediction in
each yield class
# Columns are yield classes, rows correspond to observations
nModels <- 10
probModel <- matrix(c(0,  0,  0,  0,  0,  4,  3,  3,
+ 0,  0,  0,  2,  5,  3,  0,  0,
+ 0,  1,  7,  1,  1,  0,  0,  0,
+ 2,  6,  2,  0,  0,  0,  0,  0),
+ ncol = K,byrow = T)/nModels
   rps(obs,probModel)
$rps
[1]  0.05428571
$rpss
[1]  0.7104762
$rps.clim
[1]  0.1875
```

EXERCISES

The data sets referred to in the exercises are available in the ZeBook R package.

Most of the exercises can be run with subsets of data. Thus these exercises can have many possible variants. In addition, it may be of interest to do the same exercise with subsets of different sizes, to see how the results vary with sample size.

1. Goodness-of-fit. Use the `Wheat_GPC` data.frame of the ZeBook package. The first column contains observed grain protein content (GPC, g protein/ g seed weight*100), and the third column contains GPC simulated using the AZODYN model.
   a. Calculate all the goodness-of-fit measures using the `goodness.of.fit` function of the ZeBook package.
   b. Is the model biased?
   c. Which is larger, RMSE or MAE? Why?
   d. What is the value of EF? What does this tell you about the quality of the model?
   e. What is the major contribution to MSE in the decomposition into bias^2, SDSD, and LCS?
   f. What is the relation between the bias^2 term and the graph of residuals? What is the relation between the SDSD term and the graph of observed versus simulated values?
2. Goodness-of-fit based on a threshold of model accuracy. Use the same data as in problem 1.
   a. Calculate the goodness-of-fit measures using the `threshold.measures` function of the ZeBook package. Choose reasonable values for p and d.
   b. Explain the results in words.
3. Ranked probability score. Suppose that we have yield data and simulated values by four models for three situations. We want to use the model results as probabilistic predictions, and to use rps as our measure of the quality of those probabilistic predictions.
   The observed values for the three situations are 3, 7, 2 t/ha. The corresponding model predictions are 2, 5, 3 t/ha (model 1), 3, 3, 4 t/ha (model 2), 3, 6, 1 t/ha (model 3), and 4, 7, 2 t/ha (model 4). We will divide yield into seven categories, namely 0.5−1.5 t/ha, 1.5−2.5 t/ha, ..., 6.5−7.5 t/ha.
   a. Use the `rps` function of the verification R package to calculate the ranked probability score for these probabilistic predictions.
   b. Comment on the skill score value.
   c. Calculate a new skill score using a different naïve estimator: the estimator that uses the average of the observed values as the deterministic estimator for all situations. Compare the results with the default naïve estimator.
4. Cross validation. In the Wheat_GPC data frame the AZODYN model is biased compared to observed GPC. This suggests creating a new model, AZNEW = AZODYN + θ. We consider all the parameters in AZODYN as fixed, so this new model has just one adjustable parameter, θ. In this exercise we adjust this parameter to remove the bias, and then estimate MSEP for this new model using LOO cross validation.
   a. Calculate the OLS estimate of θ. This is given by

$$\hat{\theta} = (1/n) \sum_{i=1}^{n} (y_i - \hat{y}_i^{AZODYN})$$

where the y_i are the measured values and the $\hat{y}_i^{AZODYN}$ the corresponding values simulated with AZODYN.

   b. Calculate MSE for AZODYN and for AZNEW. Which is smaller? Comment.

   c. Estimate MSEP for AZNEW using LOO cross validation.

   d. Which model is preferable, AZODYN or AZNEW? Justify.

   e. How do you expect MSEP for AZNEW to vary as sample size varies? How do you expect MSEP for AZODYN to vary as sample size varies?

5. Bootstrap. For the same AZNEW model as in the previous exercise, estimate MSEP using the bootstrap. Use the `sample` function with `replace = TRUE` to create the bootstrap samples. Do 100 bootstrap samples.

   a. Calculate MSE for AZODYN and for AZNEW. Which is smaller? Comment.

   b. Estimate op and MSEP for AZNEW using bootstrap.

   c. Which model is preferable, AZODYN or AZNEW? Justify.

   d. Do several studies with different values of B (the number of bootstrap samples). What value of B should you choose? Justify.

6. ROC. Phomopsis stem canker is a worldwide fungal disease of sunflower, which causes stem girdling lesions and a consequent reduction in yield. One wants to decide if the number of girdling lesions at harvest in the absence of early treatment will exceed 15% (Debaeke and Estragnat, 2009). If the answer is yes (population X1), then one should treat the field with a fungicide; if the answer is no (population X0), then a treatment is not needed. Debaeke and Estragnat (2009) propose, as an early indicator of the final number of lesions, the fraction of intercepted photosynthetically active radiation (IPAR) at the early E2 growth stage. Larger values of IPAR correspond to greater vegetative development, which leads to higher humidity in the canopy. This, in turn, favors disease development. We want to quantify the usefulness of early IPAR for determining if the number of girdling lesions at harvest in the absence of early treatment will exceed 15%.

   The data frame Sunflower_Phomopsis consists of a sample of fields with values for IPAR (Sunflower_Phomopsis $IPAR) and for the percent of girdling lesions at harvest (Sunflower_Phomopsis $lesions).

   a. Plot percent girdling lesions versus IPAR.

   b. Do the roc analysis. What is the AUC for this decision variable? What is the confidence interval for AUC? Is this a good decision variable? Justify.

   c. If sensitivity and specificity are of equal importance, what is the best decision threshold? What is the sum of sensitivity and specificity at this threshold?

   d. Suppose that the cost of failing to spray when one should (false negative) is twice the cost of spraying unnecessarily (false positive). What is the optimum threshold?

REFERENCES

Barbottin, A., Makowski, D., Le Bail, M., Jeuffroy, M.-H., Bouchard, C., Barrier, C., 2008. Comparison of models and indicators for categorizing soft wheat fields according to their grain protein contents. Eur. J. Agron. 29, 159–183.

Bunke, O., Droge, B., 1984. Estimators of the mean squared error of prediction in linear regression. Technometrics 26, 145–155.

Carberry, P.S., Muchow, R.C., McCown, R.L., 1989. Testing the CERES-Maize simulation model in a semi-arid tropical environment. Field. Crop. Res. 20, 297–315.

Chapman, S.C., Hammer, G.L., Butler, D.G., Cooper, M., 2000. Genotype by environment interactions affecting grain sorghum. III. Temporal sequences and spatial patterns in the target population of environments. Aust. J. Agric. Res. 51, 223–234.

Davison, A.C., Hinkley, D.V., 1997. Bootstrap Methods and Their Application. Cambridge University Press, Cambridge, p. 594.

Debaeke, P., Estragnat, A., 2009. Crop canopy indicators for the early prediction of Phomopsis stem canker (Diaporthe helianthi) in sunflower. Crop Protect. 28 (9), 792–801.

Efron, B., 1983. Estimating the error rate of a prediction rule: improvement on cross validation. J. Am. Stat. Soc. 78, 316–331.

Efron, B., Tibshirani, R.J., 1993. An Introduction to the Bootstrap (Chapman & Hall/CRC Monographs on Statistics & Applied Probability), first ed. Chapman and Hall/CRC, Boca Raton.

Ennaïfar, S., Makowski, D., Meynard, J.-M., Lucas, P., 2007. Evaluation of models to predict take-all incidence in winter wheat as a function of cropping practices, soil, and climate. Eur. J. Plant Pathol. 118, 127–143.

Fila, G., Bellocci, G., Acutis, M., Donatelli, M., 2003. IRENE: a software to evaluate model performance. Eur. J. Agron. 18, 369–372.

Gauch, H.G., Hwang, J.T.G., Fick, G.W., 2004. Model evaluation by comparison of model-based predictions and measured values. Agron. J. 95, 1442–1446.

Gent, M.P.N., 1994. Photosynthate reserves during grain filling in winter wheat. Agron. J. 86, 159–167.

Haefner, J.W., 2005. Modeling Biological Systems: Principles and Applications. Springer, New York, p. 475.

Jeuffroy, M.-H., Recous, S., 1999. Azodyn: a simple model simulating the date of nitrogen deficiency for decision support in wheat fertilization. Eur. J. Agron. 10, 129–144.

Kobayashi, K., Salam, M.U., 2000. Comparing simulated and measured values using mean squared deviation and its components. Agron. J. 92, 345–352.

Koepsell, T.D., Weiss, N.S., 2003. Epidemiologic Methods: Studying the Occurrence of Illness. Am. J. Epidemiol., vol. 159. Oxford.

Landschoot, S., Waegeman, W., Audenaert, K., Vandepitte, J., Haesaert, G., De Baets, B., 2012. Toward a reliable evaluation of forecasting systems for plant diseases: a case study using Fusarium head blight of wheat. Plant Dis. 96, 889–896.

Loehle, C., 1987. Errors of construction, evaluation, and inference: a classification of sources of error in ecological models. Ecol. Modell. 36, 297–314.

Makowski, D., Tichit, M., Guichard, L., Van Keulen, H., Beaudoin, N., 2009. Measuring the accuracy of agro-environmental indicators. J. Environ. Manage. 90 (Suppl 2), S139–S146.

Mayer, D.G., Butler, D.G., 1993. Statistical validation. Ecol. Modell. 68, 21–32.

Mitchell, P.L., 1997. Misuse of regression for empirical validation of models. Agr. Syst. 54, 313–326.

Mohanty, M., Probert, M.E., Reddy, K.S., Dalal, R.C., Mishra, A.K., Rao, A.S., et al. (2012). Simulating soybean-wheat cropping system. APSIM model parameterization and validation. Agric. Ecosyst. Environ. 152, 68–78.

NCAR. (2012). verification: Forecast verification utilities. cran.r-project.org/web/packages/verification/verification.pdf.

Pang, X.P., 1997. Yield and nitrogen uptake prediction by CERES-maize model under semiarid conditions. Soil Sci. Soc. Am. J. 61, 254–256.

Robin, X., Turck, N., Hainard, A., Tiberti, N., Lisacek, F., Sanchez, J.-C., et al., 2011. pROC: an open-source package for R and S + to analyze and compare ROC curves. BMC Bioinformatics 12, 77.

Rosenzweig, C., Jones, J.W., Hatfield, J.L., Ruane, A.C., Boote, K.J., Thorburn, P., et al., 2013. The Agricultural Model Intercomparison and Improvement Project (AgMIP): Protocols and pilot studies. Agri. Forest Meteorol. 170, 166–172.

Rötter, R.P., Palosuo, T., Kersebaum, K.C., Angulo, C., Bindi, M., Ewert, F., et al., 2012. Simulation of spring barley yield in different climatic zones of Northern and Central Europe: A comparison of nine crop models. Field Crop Res. 133, 23–36.

Rykiel, E.J., 1996. Testing ecological models: the meaning of validation. Ecol. Modell. 90, 229–244.

Sinclair, T.R., Seligman, N., 2000. Criteria for publishing papers on crop modeling. Field Crop Res. 68, 165–172.

Sing, T., Sander, O., Beerenwinkel, N., Lengauer, T., 2005. ROCR: visualizing classifier performance. R. Bioinformatics (Oxford, England) 21, 3940–3941.

Swartzman, G.L., Kalzuny, S.P., 1987. Simulation model evaluation. In: Swartzman, G.L., Kalzuny, S.P. (Eds.), Ecological Simulation Primer. Macmillan, New York, pp. 209–251.

Tedeschi, L.O., 2006. Assessment of the adequacy of mathematical models. Agr. Syst. 89, 225–247.

Toth, Z., Talagrand, O., Candille, G., Zhu, Y., 2003. Probability and ensemble forecasts. In: Jolliffe, I.T., Stephenson, D.B. (Eds.), Forecast Verification: A Practitioner's Guide in Atmospheric Science. John Wiley & Sons, Chichester.

Wallach, D., Goffinet, B., 1987. Mean squared error of prediction in models for studying ecological and agronomic systems. Biometrics 43, 561–573.

Wallach, D., Goffinet, B., Bergez, J.E., Debaeke, P., Leenhardt, D., Aubertot, J.-N., 2001. Parameter estimation for crop models: a new approach and application to a corn model. Agron. J. 93, 757–766.

Wallach, D., Loisel, P., Goffinet, B., Habib, R., 1990. Modeling the time dependence of nitrogen uptake in young trees. Agron. J. 82, 1135–1140.

Wang, F., Fraisse, C.W., Kitchen, N.R., Sudduth, K.A. (2003). Site-specific evaluation of the CROPGRO-soybean model on Missouri claypan soils. Agr. Syst., 76, 985–1005.

Willmott, C.J., 1981. On the validation of models. Physical Geography 2, 184–194.

Willmott, C.J., Ackleson, S.G., Davis, R.E., Feddema, J.J., Klink, K.M., Legates, D.R., et al., 1985. Statistics for the evaluation and comparison of models. J. Geophys. Res. 90, 8995–9005.

Willmott, C.J., Matsuura, K., 2005. Advantages of the mean absolute error (MAE) over the root mean square error (RMSE) in assessing average model performance. Clim. Res. 30, 79–82.

Zweig, M.H., Campbell, G., 1993. Receiver-operating characteristic (ROC) plots: a fundamental evaluation tool in clinical medicine. Clin. Chem. 39, 561–577.

## Chapter 10

Putting It All Together in a Case Study

1. INTRODUCTION

In the previous chapters we discussed various methods for working with dynamic system models. The purpose of this chapter is to show how these different methods can be used together in a modeling project. To do so, we use artificial (invented) data and a very simplified crop growth model. The emphasis is on the use of the different methods, not on the agronomic aspects of the problem. All the steps can be easily executed using the demonstration R scripts (demos) in the R package ZeBook.

2. DESCRIPTION OF THE CASE STUDY

2.1 Objective

The assumed objective in this artificial example is to use the simple maize model (see Chapter 1 and Appendix 1) to estimate potential final biomass production of maize, and its year to year variability, at representative locations in Europe. Potential final biomass refers to situations where only temperature and radiation affect growth. The results will be output as maps.

The target population here (the range of conditions where we want to apply the model) is that part of Europe where maize can be grown, and only fields grown under potential conditions. We further restrict the target population to fields sown on day 100 (April 9 or 10).

As a supplementary application, we present a data assimilation application where in-season leaf area index (LAI) measurements are used to estimate LAI later in the season.

2.2 Data

The artificial data are biomass ($g.m^{-2}$) and LAI measurements for a random sample of $n = 15$ site-year combinations in Europe. All fields are sown on day = 100, and managed to eliminate water and nutrient stresses and losses due to pests, diseases, and weeds. The site locations are shown in Figure 10.1.

Working with Dynamic Crop Models. DOI: http://dx.doi.org/10.1016/B978-0-12-397008-4.00010-1

FIGURE 10.1   Map of the sites in the sample. In black, sites with only B at day = 240. In gray, sites with dynamics of B and LAI.

For 10 site-years, only final biomass values (at day = 240) are available. For each of the remaining five site-years, there are LAI and biomass measurements at six different dates including final date (for day = 140, 160, 180, 200, 220, 240). The available data are presented below.

```
maize.data_EuropeEU
    idsite year  day LAIobs   Bobs      sy
1      18 2006  240    NA  3578.1  18-2006
2      64 2004  240    NA  2992.9  64-2004
3      27 2010  240    NA  2944.0  27-2010
4      23 2003  240    NA  3301.0  23-2003
5      75 2007  240    NA  3230.2  75-2007
6      81 2009  240    NA  3100.6  81-2009
7      84 2004  240    NA  2006.8  84-2004
8      38 2004  240    NA  3040.6  38-2004
9      81 2002  240    NA  2954.1  81-2002
10      2 2004  240    NA  2477.7   2-2004
11     45 2006  140   0.4   100.2  45-2006
12     45 2006  160   4.7   771.7  45-2006
13     45 2006  180   6.4  1700.8  45-2006
14     45 2006  200   7.0  2755.1  45-2006
```

```
15   45 2006 220   5.7 3308.7 45-2006
16   45 2006 240   5.8 3301.1 45-2006
17   79 2001 140   0.0   26.2 79-2001
18   79 2001 160   0.6  138.9 79-2001
19   79 2001 180   4.2  584.6 79-2001
20   79 2001 200   7.2 1208.7 79-2001
21   79 2001 220   6.3 1893.4 79-2001
22   79 2001 240   7.3 2385.5 79-2001
23   92 2010 140   0.0   24.2 92-2010
24   92 2010 160   0.4  152.6 92-2010
25   92 2010 180   3.0  672.7 92-2010
26   92 2010 200   6.2 1628.8 92-2010
27   92 2010 220   7.2 2504.3 92-2010
28   92 2010 240   6.5 2904.5 92-2010
29   56 2002 140   0.1   68.7 56-2002
30   56 2002 160   3.1  535.6 56-2002
31   56 2002 180   6.4 1467.8 56-2002
32   56 2002 200   6.1 2318.7 56-2002
33   56 2002 220   6.5 3204.3 56-2002
34   56 2002 240   6.9 3995.7 56-2002
35   72 2002 140   0.0   28.6 72-2002
36   72 2002 160   1.4  205.8 72-2002
37   72 2002 180   7.1  831.9 72-2002
38   72 2002 200   6.7 1455.3 72-2002
39   72 2002 220   6.4 2120.3 72-2002
40   72 2002 240   6.5 2642.9 72-2002
```

These site-years were chosen by drawing sites at random from grid locations in maize growing areas of Europe and years at random from the years 2001−2010. In Chapter 2, we discuss the difficulty in practice of obtaining a random sample from the target population in agronomic studies.

The climate data were obtained from the NASA POWER Climatology resource for agro-climatology (http://power.larc.nasa.gov) and saved to a data frame included in the ZeBook package (weather_EuropeEU). The weather data and observed data both include the identifiers idsite and year, which allows us to connect the correct weather data to the observed data.

2.3 The Steps in the Study

The following steps were carried out:

- Step 1. Examine the data. Compare the variability of weather in the sample to the variability in the target population.
- Step 2. Compare simulated results with available data. Based on Chapter 4 (simulation) and Chapter 9 (evaluation).
- Step 3. Evaluate the uncertainty in simulated final biomass due to parameter uncertainty. Based on Chapter 5 (uncertainty analysis).

- Step 4. Rank the contributions of the different parameters to overall final biomass uncertainty. Based on Chapter 5 (sensitivity analysis).
- Step 5. Calibrate the model using classical statistical methods. Based on Chapter 6 (parameter estimation).
- Step 6. Estimate the mean squared error of prediction (MSEP) of the calibrated model. Based on Chapter 9 (evaluation).
- Step 7. Use the calibrated model to attain the objectives of the study. In particular, produce the maps of maize production and inter-annual variability.

We also carried out a variant of the above approach, using a Bayesian approach to estimate the model parameters. Here steps 5 and 7 were replaced by:

- Step 5′. Calibrate the model using a Bayesian approach. Based on Chapter 7 (Bayesian parameter estimation).
- Step 7′. Use the calibrated model to attain the objectives of the study.

Finally, to illustrate data assimilation for these same data, we used in-season LAI measurements and the model to estimate later LAI.

- Data assimilation to predict LAI. Based on Chapter 8 (data assimilation).

2.4 Going Through the Steps in the Study

2.4.1 Step 1. Examining the Data

It is important to examine the sample data. One objective is to examine how much variability between individuals exists in the sample.

A histogram of final maize biomass values is shown in Figure 10.2. Summary statistics are shown

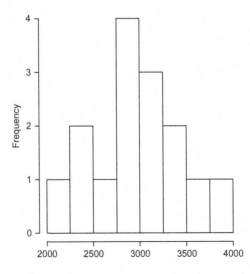

FIGURE 10.2   Histogram of biomass at day = 240 for the sample.

```
summary(maize.data_EuropeEU[maize.data_EuropeEU$day = = 240, "Bobs"])
  Min. 1st Qu.  Median  Mean 3rd Qu.  Max.
  2007  2774    2993    2990  3266    3996
```

The final biomass values range from 2t/ha to 4t/ha. That is substantial variation, and we would want the model to explain a good part of that.

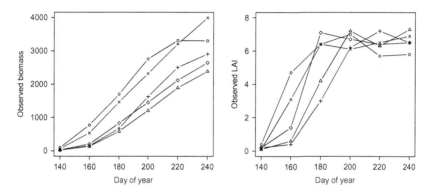

FIGURE 10.3 Dynamics of biomass (left) and LAI (right) for the site-years of the sample with in-season data.

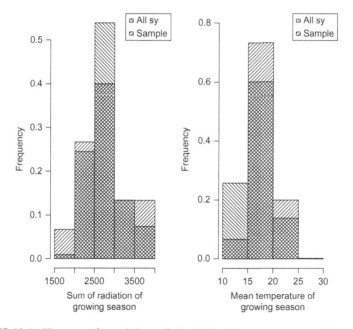

FIGURE 10.4 Histogram of cumulative radiation (left) and mean temperature (right) during the growing season for the sample of 15 site-years and for the all site-years available in the weather database.

The dynamics of biomass and LAI for the five sites with in-season measurements are shown in Figure 10.3.

It is also of interest to examine the range in weather conditions in the sample, and compare with the full range in the target population (represented by all the site-years in the ZeBook package). Average temperatures and total radiation values for the sample and for the target population are shown in Figure 10.4. There is no egregious discrepancy between the two.

2.4.2 Step 2. Comparing Simulated Results with Available Data

Based on Chapter 4 (simulation) and Chapter 9 (evaluation). The model was run for each site-year for which there are data using default parameter values.

```
maize.define.param()["nominal",]
    Tbase      RUE        K      alpha     LAImax      TTM        TTL
7.00e+00 1.85e+00 7.00e-01 2.43e-03 7.00e+00 1.20e+03 7.00e+02
```

The following table shows final simulated LAI and final biomass values, together with the observed values.

```
simobs240
```

	sy	day	TT	LAI	B	idsite	year	LAIobs	Bobs
141	18-2006	240	2036.50	6.969051	2894.291	18	2006	NA	3578.1
292	64-2004	240	1230.85	6.970798	2425.892	64	2004	NA	2992.9
443	27-2010	240	2024.20	6.977846	2663.207	27	2010	NA	2944.0
594	23-2003	240	1736.15	6.979192	2920.571	23	2003	NA	3301.0
745	75-2007	240	1311.55	6.967723	2555.934	75	2007	NA	3230.2
896	81-2009	240	1401.75	6.976406	2663.286	81	2009	NA	3100.6
1047	84-2004	240	951.90	6.971458	1825.426	84	2004	NA	2006.8
1198	38-2004	240	1600.10	6.975111	2562.649	38	2004	NA	3040.6
1349	81-2002	240	1562.10	6.968158	2389.159	81	2002	NA	2954.1
1500	2-2004	240	2221.15	6.981230	2549.210	2	2004	NA	2477.7
1651	45-2006	240	1493.25	6.976809	2531.336	45	2006	5.8	3301.1
1802	79-2001	240	1280.10	6.979118	2345.359	79	2001	7.3	2385.5
1953	92-2010	240	1385.40	6.974352	2669.931	92	2010	6.5	2904.5
2104	56-2002	240	1201.75	6.972435	3017.428	56	2002	6.9	3995.7
2255	72-2002	240	1089.50	6.975231	2227.381	72	2002	6.5	2642.9

Graphical comparisons of observed and simulated biomass values for each site-year are shown in Figure 10.5. Figure 10.6 shows observed versus simulated values for final biomass (at day = 240). Figure 10.7 shows the same information, but presented as a graph of residuals versus predicted values. It is clear that the model is biased; almost all of the residuals are positive (simulated values too low). However, there is no obvious trend in the

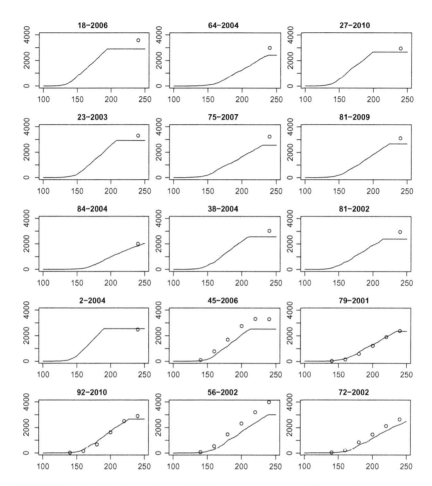

FIGURE 10.5   Graphical comparisons of all observed and simulated biomass values.

residuals with simulated biomass. Thus, there is no obvious indication of how the model equations should be changed. Therefore, we will rely on calibration to improve the model. The figures also show that it is reasonable to assume that the variance of model residual error is constant, since there is no obvious trend in the spread of the residual errors (Figure 10.7).

Several measures of goodness-of-fit proposed in Chapter 9 were calculated using the goodness.of.fit function in the ZeBook package (Table 10.1).

It is useful to look at all the measures, but we only comment on those that seem the most informative in this case. The evaluation criteria indicate poor agreement between simulated and observed values. Efficiency is negative

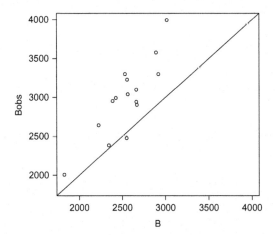

FIGURE 10.6   Observed versus simulated values for biomass at day = 240.

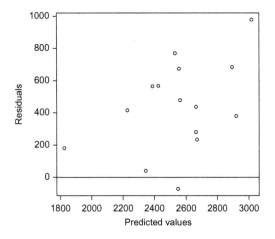

FIGURE 10.7   Graph of residuals (observed − predicted final biomass) versus predicted final biomass.

(EF = −2.278), indicating that the model is a worse predictor than just using the average of observed values for all predictions. RMSE (517.9) is fairly large, and is about 20% of the average final biomass (relative.RMSE = 0.20).

The decomposition of MSE into bias.squared, SDSD, and LCS shows that all three terms make a substantial contribution. We note in particular that squared bias makes the largest contribution. Since calibration is often quite effective in reducing bias, this again suggests that calibration may well lead to significant improvement in the model.

TABLE.10 1 Measures of Goodness-of-Fit for Final Biomass

N	mean.Yobs	mean.Ypred	bias	sd.Yobs	sd.Ypred
15	2549.404	2990.38	−440.9761	296.0649	489.0983
MSE	RMSE	relative.RMSE	MAE	relative.RMAE	relative.MAE.prime
268199.4	517.8797	0.2031376	450.5107	0.1767122	0.1735928
EF	Willmott.index	bias.squared	SDSD	LCS	bias.squared.again
−2.278288	0.6536824	194459.9	34777.78	38961.73	194459.9
NU	LC	MSE.systematic	MSE.unsystematic		
51854.64	21884.87	208473.5	59725.86		

2.4.3 Step 3. Evaluating the Uncertainty in Simulated Final Biomass Due to Parameter Uncertainty

In this step we evaluate the total uncertainty in the calculated final biomass values that results from uncertainty in the parameters. We assume that the explanatory variables are known without error, so there is no contribution to uncertainty from that source. If the total uncertainty due to parameter uncertainty is large, this means that changes in the parameter values can lead to large changes in the simulated responses. This implies that calibration can potentially lead to substantial improvement in the model. If, on the other hand, the total uncertainty due to parameter uncertainty is small, then calibration will not substantially change the model responses and so cannot substantially improve the model agreement with the data.

The uncertainty in the parameter values, based on the range of values found in the literature, is shown below.

```
maize.define.param()
        Tbase  RUE   K   alpha  LAImax  TTM  TTL
nominal    7  1.85  0.7  0.00243     7  1200  700
binf       6  1.50  0.6  0.00200     6  1100  600
bsup       8  2.50  0.8  0.00300     8  1400  850
```

Each parameter is assigned a uniform distribution between a lower bound (binf) and an upper bound (bsup). We assume independence between parameters.

We generated N = 10,000 parameter vectors, by sampling randomly from the distribution of each parameter independently. For each parameter vector the model was run, thus generating 10,000 final biomass values. The distribution of these values represents the uncertainty in final biomass.

One way of summarizing these results is in a histogram for each site-year. Another way is to report a confidence interval, for example, the 90% confidence interval.

A histogram of the uncertainty in final biomass for the site-year 18–2006 is shown in Figure 10.8. The 90% confidence interval is [2403; 4368], as shown below. (The vector simUnc240 contains the 10,000 simulated final biomass values for this site-year.)

```
mean(simUnc240)
[1] 3337.684
quantile(simUnc240, prob = c(0.05,0.95))
    5%      95%
2403.598 4368.985
```

The following table shows mean values and 90% confidence intervals for all sites.

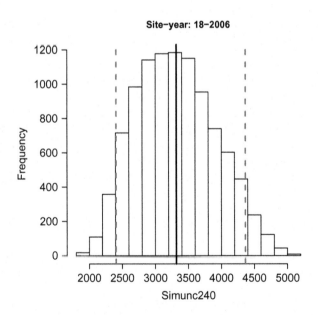

FIGURE 10.8 Histogram representing the distribution of final biomass for the site-year = 18–2006 (latitude = 42, longitude = 15).

```
t(sumary.uncertainty.all.sy)
           sy         mean       q05        q95
  [1,]  "18-2006"  3316.727  2406.158  4362.871
  [2,]  "64-2004"  2595.813  1949.189  3312.707
  [3,]  "27-2010"  3003.995  2215.941  3897.914
  [4,]  "23-2003"  3336.257  2425.18   4359.068
  [5,]  "75-2007"  2825.158  2097.185  3641.794
  [6,]  "81-2009"  3016.573  2200.413  3945.112
  [7,]  "84-2004"  1960.783  1418.818  2564.956
  [8,]  "38-2004"  2967.428  2156.538  3898.654
  [9,]  "81-2002"  2678.143  1956.992  3492.672
 [10,]   "2-2004"  2963.32   2129.8    3922.963
 [11,]  "45-2006"  2878.469  2104.452  3760.653
 [12,]  "79-2001"  2567.253  1908.557  3303.291
 [13,]  "92-2010"  2957.672  2185.612  3831.036
 [14,]  "56-2002"  3197.651  2410.281  4062.553
 [15,]  "72-2002"  2398.387  1783.663  3085.358
```

The uncertainty in all cases is very large and could easily explain the discrepancy between observed and simulated values.

2.4.4 Step 4. Ranking the Contributions of the Different Parameters to Overall Final Biomass Uncertainty

We have shown that the uncertainty in the parameters leads to a large uncertainty in simulated final biomass. We now want to know which parameters are principally responsible for the uncertainty in the results. The maize model is a model of a system with interactions, and as for all such models, it is difficult to determine from examination of the model which parameters have the largest effect on a particular response, here final biomass. The difficulty is increased because this depends not only on the way each parameter enters into the model, but also on the possible range of values of each parameter.

Sensitivity analysis allows us to rank the parameters in terms of their effect on final biomass. We will use two common methods of carrying out sensitivity analysis in this step.

An important choice here involves the situations for which we estimate sensitivity coefficients. Since we are interested in the prediction of final biomass for all site-years, we evaluate the contribution of each parameter to final biomass averaged over the site-years of the data set (using the functions maize.simule_multisy240, maize.multisy240 and maize.simule240).

2.4.4.1 Full Factorial Design and Analysis of Variance

Because the maize model has only seven parameters, it is possible and recommended to use analysis of variance (ANOVA) with a full factorial design for the sensitivity analysis. If each parameter can have two values (minimum and maximum), the total number of parameter vectors is $2^7 = 128$. Since there are 15 site-years in the database, the model must be run 15 times for each parameter vector. Thus $1920 = 15 * 128$ simulations are required for two levels per parameter. This already requires substantial computing time, so we use a design with two levels per parameter. To give three levels (minimum, maximum, and nominal) to each parameter would require $32805 = 15 * 3^7$ model runs.

The ANOVA function that does the calculation is:

$$aov(simX240 \sim Tbase^*RUE^*K^*alpha^*LAImax^*TTM^*TTL)$$

This will provide the full decomposition of variance, including contributions of all the factors and all their interactions.

```
# 2 levels for 7 parameter: 128*sites-years simulations (up to 20min)
subset(TabIndices, Indices > 0.1)
                Df      Sum Sq      Mean Sq       Indices
RUE              1  60670657.3   60670657.3    71.0688127
TTM              1   9838627.3    9838627.3    11.5248391
alpha            1   7397764.6    7397764.6     8.6656445
LAImax           1   4963965.5    4963965.5     5.8147241
RUE:TTM          1    614914.2     614914.2     0.7203024
K                1    540312.4     540312.4     0.6329148
RUE:alpha        1    462360.3     462360.3     0.5416028
RUE:LAImax       1    310247.8     310247.8     0.3634203
Tbase:TTM        1    205452.1     205452.1     0.2406639
alpha:LAImax     1    113393.1     113393.1     0.1328272
```

The table shows that RUE is the parameter with the greatest effect on final biomass with a contribution of 71% (Indices = 71.07), followed by TTM (11.52%), alpha (8.67%), and LAImax (5.81%). Tbase, K, and TTL have little importance. The interactions between parameters also contribute little. The largest contributor among interactions is the interaction between RUE and TTM, with 0.72%.

2.4.4.2 Morris Elementary Effects Screening Method

The previous analysis gives all the information needed, but as an additional illustration, we do a second sensitivity analysis using the Morris method. The method is quite easy to implement using the Morris function in the sensitivity package in R. When a model has many parameters, the Morris method is often used to rank the parameters according to their effect on an output.

The parameters of the Morris function that we use are `levels = 6`, `grid.jump = 3` (Morris's recommendation is to set `grid.jump = levels/2`) and number of repetitions `r = 20`. Note that r is chosen as a compromise between space-filling and acceptable computation time. It is worthwhile in practice to run this function with more than one value of r, to verify the stability of the result (we also tested greater values for r, and found results very similar to $r = 20$). With 20 paths and 7 parameters, the number of parameter vectors tested is r * (nfac + 1) = 160. The number of model runs for the 15 site-years is then 15 * 160 = 2400. This required about 20 minutes of computation time.

```
morris(model = maize.simule240, factors = row.names(t(param))[1:nfac],
r = 20, design = list(type = "oat", levels = 6, grid.jump = 3), scale = T,
binf = as.vector(param["binf",]), bsup = as.vector(param["bsup",]),
weather = weather, sdate = 100, ldate = 250)
# sort the table by mu.star
r = 20
table.morris[order(table.morris$mu.star,decreasing = TRUE),]
                 mu        mu.star        sigma
RUE      0.849271142  0.849271142  0.113242454
TTM      0.368360369  0.368360369  0.079775599
alpha    0.267637585  0.267637585  0.061188162
LAImax   0.217877558  0.217877558  0.048361135
K        0.066942138  0.066942138  0.015919060
Tbase    0.036181319  0.051238870  0.045057345
TTL      0.003151231  0.003151231  0.007380606
```

The column *mu.star* gives the sensitivity ranking of each parameter. Morris is not based on a decomposition of variance, so the results give a different kind of information than the analysis of variance. Nevertheless, in this case, the ranking is very similar for Morris and for the analysis of variance. The most important parameter by a substantial margin is RUE, followed in order by TTM, alpha, and LAImax.

2.4.5 Step 5. The Model Using Classical Statistical Methods

The sensitivity analysis suggests that we should add parameters to the calibration procedure in the following order: RUE, TTM, alpha, and LAImax. We start with the most influential parameter, and work our way down.

We begin first with parameter estimation using only the final biomass data. In a second step, we will use all the data, including the in-season data.

2.4.5.1 Using Ordinary Least Squares to Fit the Model to the Final Biomass Data

Here we use ordinary least squares (OLS) to estimate the parameter values. We first estimate just the single parameter RUE, then the two parameters RUE and TTM, then RUE, TTM, and alpha.

The optimization was run with multiple starting values (at least two), to increase confidence that a global minimum has been found.

1) Gradient Method (nls)

OLS parameter estimation is easily implemented using the R function nls. Part of the script using nls for the estimation of 1 parameter (RUE) is shown below:

```
# The estimated parameter here param is RUE
param_opti = c("RUE")
OLS1p   <-try(nls((Obs~maize.ValCalNLS(param,param_opti,   param_
default,data_n_B240,list_sy = list_n_sy)), start = list(param = param_
init), trace = T,control = nls_cont))
summary(OLS1p)
Formula: Obs ~ maize.ValCalNLS(param, param_opti, param_ default,
data_n_B240,
   list_sy = list_n_sy)
Parameters:
      Estimate Std.  Error t value Pr(>|t|)
param    2.17567     0.04893     44.46   <2e-16 ***
- - -
Signif. codes: 0 '***' 0.001 '**' 0.01 '*' 0.05 '.' 0.1 ' ' 1
Residual standard error: 262.7 on 14 degrees of freedom
Number of iterations to convergence: 1
Achieved convergence tolerance: 4.024e-08
sum(summary(OLS1p)$residuals^2)
[1] 966180.2
```

The estimated value is RUE = 2.18.

We then ran nls to estimate the two parameters RUE and TTM. In this case the algorithm didn't converge, and gave the error message 'singular gradient matrix at initial parameter estimates'. This indicates that there is a problem in calculating model derivatives with respect to the parameters. This is not too surprising, since the effect of TTM is discontinuous; biomass accumulation stops when total thermal time exceeds TTM.

We therefore switched to a different function, optim, to estimate the two parameters RUE and TTM.

2) Simplex Method (optim)

Here we use the optim function and specify method = "Nelder-Mead". This algorithm (also called the simplex algorithm) does not require derivatives. Part of the script using optim for the estimation of 1 parameter (RUE) is shown below:

```
# definition of the function for OLS with optim
maize.LS.B240.optim<-function(param,param_opti,param_default,data_sy,list_sy){
    param_all = param_default
    param_all[param_opti] <- param
    sim = maize.multisy(param_all,list_sy, 100, 250)
    data240 = data_sy[data_sy$day = = 240,]
```

```
    simobs240 = merge(sim,data240,by = c("sy","day"))
    return(sum((simobs240$Bobs-simobs240$B)^2,na.rm = TRUE))
    }
# estimation of RUE
param_opti = c("RUE")
param_init <-param_default[param_opti]*0.85
optim_OLS1p<-optim(param_init, maize.LS.B240.optim, method = " Nelder-Mead",control =
list(trace = 0,maxit = 1000), param_opti = param_opti, param_default = param_default,
data = data_n_B240,list_sy = list_n_sy )
optim_OLS1p
$par
    RUE
2.175675
$value
[1] 966180.2
$counts
function gradient
      28         NA
$convergence
[1] 0
$message
NULL
# compute AIC
param_all = param_default
param_all[param_opti] <- optim_OLS1p$par
simobs240 = merge(maize.multisy(param_all,list_n_sy, 100, 250),data_n_B240,by = c
("sy","day"))
AICf(simobs240$Bobs,simobs240$B,length(param_opti) + 1)
AICcomplete    AICshort
   212.6640   170.0958
goodness.of.fit(simobs240$Bobs,simobs240$B)
   N  mean.Yobs  mean.Ypred      bias  sd.Yobs  sd.Ypred      MSE      RMSE
1  15   2990.38   2998.027  -7.646608  489.0983  348.1844  64412.02  253.7952
   relative.RMSE      MAE  relative.RMAE  relative.MAE.prime        EF
1    0.08487056  208.4179    0.06969613          0.07154138  0.7115054
   Willmott.index  bias.squared     SDSD      LCS  bias.squared.again        NU
1     0.8954264      58.47062  18532.96  45820.58            58.47062  4627.683
      LC  MSE.systematic  MSE.unsystematic
1  59725.86      34143.68          30268.34
```

Use of `optim` allowed us to estimate RUE and TTM, and also the three parameters RUE, TTM, and alpha. In the case of three parameters, despite the use of two different starting points, the algorithm converged to a local minimum. (We know this because the RMSE value was larger than for two parameters.) We resolved this problem by starting the three parameter calibration with the values for RUE and TTM obtained from the two parameter calibration.

The resulting parameter values for one, two, or three estimated parameters are shown in Table 10.2.

3) **Choose the best model**

It is very important to avoid estimating too many parameters, because that can lead to poor predictive quality. Here we use the criterion AIC (Akaike Information Criterion) to decide how many parameters to calibrate. The number of parameters, used in the penalization term of AIC, is the number of calibrated parameters plus one for residual variance. We calculate AIC for each of the calibrated models (with one, two, or three calibrated parameters) and choose the model with the smallest AIC value.

AIC was calculated using the function `AICf` of the `ZeBook` package. We also calculated RMSE values for each model, using the function `goodness.of.fit` of the `ZeBook` package. The results are shown in Table 10.2.

The smallest AIC value is that for the model where only RUE is estimated. This then is our choice of model.

Table 10.2 shows that RMSE is 253.8 after estimation of RUE, compared to the value of 517.9 with the initial value of RUE. It seems that estimation of this parameter has substantially improved the goodness-of-fit of the model. The decomposition of MSE shows that most of the reduction is due to a reduction in the squared bias term (from 194460 before calibration to 58 after estimation of RUE).

We did not constrain RUE to lie within its uncertainty range, in order to see where the true optimum value is. In fact, the estimated value of 2.176 is within the uncertainty range of the parameter. However, the new value of RUE should not be considered an improved estimate of the true value of RUE. It should rather just be considered an adjustment factor that compensates for many errors in the model.

TABLE 10.2 Results of Calibration with OLS Considering Biomass at Day 240

Number of Calibrated Parameters	Number of Iterations	Parameter Values	RMSE B240	AIC B240
1	28	RUE = 2.176	253.8	212.7
2	85	RUE = 2.232 TTM = 1165	245.7	213.7
3	314	RUE = 2.812 TTM = 1165 alpha = 0.001601	229.3	213.6

2.4.5.2 Using All the Data, Including the In-Season Data

Here we redo the parameter estimation, using all the data including the in-season data for LAI and biomass. This is tempting; improving model fit to the dynamics of biomass and LAI might improve prediction of final biomass. However, we also know that using data on other variables than the target variable is not guaranteed to improve prediction of the target variable.

We cannot use OLS here, since: 1) the same variable is measured at several dates and may have different variances at different dates; and 2) the measurements correspond to different output variables whose residual variances will not be equal.

For the first problem, it is often found that residual variance is approximately proportional to the size of the response. In that case, taking the log of both observed and simulated values will lead to approximately equal residual variances. In our case we did the calculations both with and without a log transformation. Examination of the residuals in both cases was not conclusive, so we keep both analyses. To address the second problem, we use the concentrated likelihood method (Chapter 6).

The concentrated likelihood method was used to estimate just RUE, then RUE and TTM, and finally RUE, TTM, and alpha. Part of the script using optim and concentrated likelihood for one parameter (RUE) is shown below. The results are presented in Table 10.3.

```
# log of concentrated likelihood for LAI and Biomass
# with log transformation of variable.
maize.logConclikelihood.optim<-function(param,param_opti,param_
default,data_sy,list_sy, transf = function(x){x}){
    param_all = param_default
    param_all[param_opti] <- param
    sim = maize.multisy(param_all,list_sy, 100, 250)
    simobs = merge(sim,data_sy,by = c("sy","day"))
    n_Bobs = length(na.omit(simobs$Bobs))
    n_LAIobs = length(na.omit(simobs$LAIobs))
    MSE_B = sum((transf(simobs$Bobs)-transf(simobs$B))^2,na.
    rm = TRUE)/n_Bobs
    MSE_LAI = sum((transf(simobs$LAIobs)-transf(simobs$LAI))^2,
    na.rm = TRUE)/n_LAIobs
    Conclikelihood = log(MSE_B^(n_Bobs/2)) + log( MSE_LAI^(n_LAIobs/2))
    return(Conclikelihood) }
# estimation of RUE
optim_ConclLp <-optim(param_init, maize.logConclikelihood.optim,
method = "Nelder-Mead",control = list(trace = 1,  maxit = 1000),param_
opti = param_opti,param_default = param_default,data = maize.
data_EuropeEU,list_sy = list_n_sy, transf = log )
```

```
optim_OLS1p
$par
    RUE
2.175662
$value
[1] 1.167486e + 36
$counts
function gradient
        30        NA
$convergence
[1] 0
```

We can calculate RMSE and AIC values for all biomass or LAI values, or just for final biomass, and with or without the logarithmic transformation. In every case, we see the same behavior (Tables 10.2 and 10.3). RMSE decreases with increasing number of parameters. This is always the case; adding flexibility to the model by estimating additional parameters always reduces RMSE. AIC, on the other hand, is minimal for just a single estimated parameter for all cases, i.e., with only final biomass or with all data, with or without a logarithmic transformation. Thus the best model (smallest AIC value) is that with a single estimated parameter, RUE. However, the estimated value of RUE here is not exactly the same as when only final biomass data are used for parameter estimation. We thus need to choose between two different models; the model with RUE = 2.176 that resulted from using OLS and just the final biomass data, and the model with RUE = 2.184 that results from using concentrated likelihood and all data. (We choose to use the results that involve a log transformation.)

AIC is only applicable for comparing models based on the same data, so it is not applicable for choosing between our two candidates. We need a different procedure here.

2.4.6 Step 6. Estimating the MSEP of the Calibrated Model

To choose between our two candidate models, we use cross validation to estimate mean squared error of prediction (MSEP) for each model. We will choose the model with the lower estimated value of MSEP. We use leave-one-site-year-out cross validation (Chapter 9), because residual errors are independent between site-years (by construction, since each site-year was drawn independently).

Part of the code used to implement cross validation in the case of estimation of only one parameter is shown below. The script does the parameter estimation 15 times, leaving out a different site-year each time.

TABLE 10.3 Results of Calibration with Concentrated Likelihood Considering All the Data (Including the In-Season Data)

Number of Calibrated Parameters	Parameter Values	Number of Iterations	RMSE B240	RMSE B	RMSE LAI	AIC B240	AIC B	AIC LAI
1 (no transformation)	RUE = 2.189	25	254.5	252.9	1.019	212.7	560.2	90.27
2 (no transformation)	RUE = 2.217 TTM = 1179	79	244.5	250.0	1.019	213.5	561.2	92.26
3 (no transformation)	RUE = 2.215 TTM = 1179 Alpha = 0.002434	115	244.4	250.0	1.019	215.5	563.2	94.26
1 (log transformation)	RUE = 2.184	26	254.1	253.0	1.019	-26.87	13.04	41.94
2 (log transformation)	RUE = 2.198 TTM = 1178	71	245.1	250.7	1.019	-26.03	14.95	43.94
3 (log transformation)	RUE = 2.228 TTM = 1178 Alpha = 0.002395	136	243.8	249.9	1.021	-23.87	17.09	45.63

```
param_opti = c("RUE")
best_param = param_default
best_param[param_opti] = 2.175675
param_init <- best_param[param_opti]
#for each site-year
MSEPcvB240 <- data.frame()
for (sy in list_n_sy)
{
    # show site-year left out for computation of componcnent of MSEP
    print(paste("site year left out ",paste(sy, collapse = " - ")) )
    list_n_sy_w_sy <- list_n_sy[list_n_sy%in%sy]
    # exclude this site-year for calibration
    list_n_sy_wo_sy <- list_n_sy[!list_n_sy%in%sy]
    #estimate parameter on sites-years without sy
    print(paste("estimate parameter on sites-years :", paste(list_n_sy_wo_sy,
collapse = " ; ")))
    system.time(result_opti <- optim(param_init, maize.LS.B240.optim, method =
"Nelder-Mead",control = list(trace = 1, maxit = 1000),param_opti = param_opti,param_
default = param_default,data = data_n_B240,list_sy = list_n_sy_wo_sy ))
    # show results of estimation, to see if there are problems with convergence
    print(result_opti)
    best_param_sy <- param_default
    best_param_sy[param_opti] <- result_opti$par
    #simulation for year y and with param optim on other years
simobs = merge(maize.multisy(best_param_sy,list_n_sy,100,250),maize.data_EuropeEU,
by = c("sy","day"))
    simobs240 = subset(simobs,day = = 240)
    # compute componnent of MSEPcv
    MSEPcvB240_i = goodness.of.fit(simobs240$Bobs,simobs240$B)["MSE"]
    MSEPcvB240 <- rbind(MSEPcvB240, MSEPcvB240_i)
}
#show mean of MSE values
mean(MSEPcvB240)
```

The results are shown in Table 10.4. Estimated RMSEP (square root of estimated MSEP) for final biomass is lower for the model based just on final

TABLE 10.4 Results of Cross Validation for Ols Based Only on Final Biomass Data and for Concentrated Likelihood Based on All the Data

Method for Parameter Estimation	B240 RMSE	B RMSE	LAI RMSE	B240 RMSEP	B RMSEP	LAI RMSEP
OLS (RUE = 2.175675)	253.8	253.3	1.019	254.5	253.7	1.019
Concentrated likelihood (RUE = 2.183994)	254.1	253.0	1.019	266.9	260.0	1.019

biomass (RMSEP = 254.5) than for the model based on all data (RMSEP = 266.9). The final conclusion then is to use the model where only a single parameter is estimated, namely RUE, and where that estimation is based only on final biomass.

We note that calibration has substantially improved predictive accuracy. For the uncalibrated model RMSE is an unbiased estimator of RMSEP. The value is RMSE = 517.9. Estimation of RUE has reduced this to RMSEP = 254.5.

2.4.7 Step 7. Using the Calibrated Model to Attain the Objectives of the Study

Finally, we run the model using the full weather data file, which includes sites across Europe and the years 2001−2010.

For each site we calculate final biomass averaged over years (B240.mean), the inter-annual standard deviation of final biomass (B240.sd), and the coefficient of variation (B240.coefvar = B240.sd/B240.mean). A sample of results is shown in the table below.

```
head(by_idsite)
   idsite GPSlatitude GPSlongitude B240.mean B240.sd B240.coefvar
1      1          38           -7  3063.381 162.7063   0.05311333
2      2          38           -5  3080.135 158.9017   0.05158920
3      4          38           -1  3187.348 120.9839   0.03795755
4      6          40           -7  3387.824 214.0689   0.06318771
5      7          40           -5  3311.792 196.9963   0.05948328
6     16          42            3  3520.959 141.2140   0.04010668
```

Maps of average final biomass and of the coefficient of variation at each site are shown in Figure 10.9.

2.4.8 Step 5'. Calibrating the Model Using a Bayesian Approach

An alternative to the classical statistical approach for model calibration is a Bayesian approach, which we apply here. In this case the parameters are treated as random variables, whose distribution represents our knowledge or rather lack of knowledge about the parameters. Our knowledge about the parameters before taking into account the information in the experimental data is summarized in a prior distribution for the parameters. This is updated with the experimental data, to give a posterior distribution.

Here we use all the data, and estimate the posterior distribution of all the parameters; this includes the seven model parameters plus the variances of residual error (one for biomass and one for LAI) (Chapter 7). The algorithm that we use is a Metropolis-Hastings within Gibbs algorithm.

There is an underlying assumption that the residual variances for all LAI values, on the one hand, for all biomass values, on the other hand, are equal.

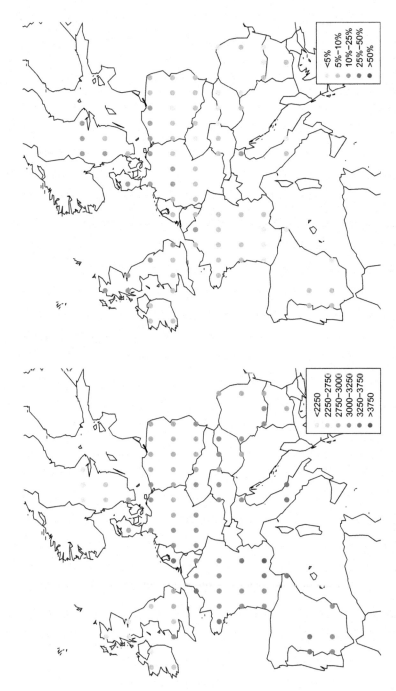

FIGURE 10.9 Map of mean final biomass (left) and of coefficient of variation of final biomass (right). The mean is over the years 2001–2010. The coefficient of variation is the inter-annual standard deviation at the site, divided by the mean at that site.

Legend (left map):
- <2250
- 2250–2750
- 2750–3000
- 3000–3250
- 3250–3750
- >3750

Legend (right map):
- <5%
- 5%–10%
- 10%–25%
- 25%–50%
- >50%

As discussed in Step 5, this often leads to doing a logarithmic transformation for data with a wide range of values. We do not use a variable transformation here because we saw that residuals are quite constant in Step 5 without any transformation, but in general one should. Consider carefully the hypothesis and this possibility of a logarithmic transformation when working with variables that increase over time such as LAI and biomass.

2.4.8.1 Prior Distribution

We have already shown reasonable ranges for the parameters, based on the literature, in Step 3. These are repeated below, where binf and bsup are, respectively, the lower and upper bound for each parameter. In the absence of information to the contrary, we assign a uniform distribution to each parameter and assume that the parameters are independent.

```
maize.define.param()
          Tbase   RUE    K    alpha  LAImax   TTM   TTL
nominal     7   1.85   0.7  0.00243     7    1200   700
binf        6   1.50   0.6  0.00200     6    1100   600
bsup        8   2.50   0.8  0.00300     8    1400   850
```

2.4.8.2 Posterior Distribution

The algorithm was run twice, with two different starting points. In each case, 20,000 iterations were carried out, giving rise to two chains of 20,000 parameter vectors. The two chains were run in parallel. The calculations took ~14 hours.

```
gelman.diag(MCMC.L, confidence= 0.95, transform= FALSE, autoburnin = TRUE)
Potential scale reduction factors:
                Point est.   Upper C.I.
Tbase              1.16         1.57
RUE                1.01         1.01
K                  1.02         1.11
alpha              1.22         1.74
LAImax             1.04         1.18
TTM                1.13         1.46
TTL                1.00         1.00
Multivariate     psrf
1.18
```

A test of whether the chains have converged to the posterior distribution can be based on the upper limit of the criterion of Gelman et al. (2004), as calculated by the coda R package software. The criterion should be below 1.1. Values for each parameter, as well as the multivariate value, are shown above. Some of the values are slightly above 1.1, but we also examined graphs of each parameter versus iteration number and those seemed to indicate that it was reasonable to assume that convergence had occurred.

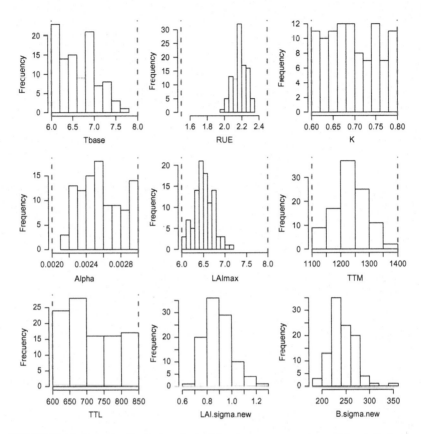

FIGURE 10.10   Posterior distribution obtained for the seven parameters and for residual standard error (sigma) for LAI and B.

We eliminated the first half of each chain (to eliminate the effect of the starting values), and then combined the remaining vectors in each chain to give a single chain of 20,000 vectors. We then thinned the chains, keeping only one vector out of 200, to reduce auto-correlation. Finally, this left 101 vectors of parameters.

Figure 10.10 shows the posterior distribution of each of the seven parameters of the model and of the residual variances of LAI and B.

The value of RMSE, calculated as the square root of MSE averaged over the posterior distribution, is shown below for B and LAI. The values are quite similar to those obtained using the classical approach (Table 10.2).

```
mean(MCMC.R1.R2.garde_ssautocor[,"B.MSE"])^0.5
[1] 259.2675
mean(MCMC.R1.R2.garde_ssautocor[,"LAI.MSE"])^0.5
[1] 0.9467981
```

In this case, we only estimate the quality of adjustment by calculating an MSE value. Estimating the quality of prediction using a cross-validation procedure is possible in theory, as in Step 6, but would require a great amount of computation time (several days).

2.4.9 Step 7'. Using the Calibrated Model to Attain the Objectives of the Study

Given the posterior distribution, we can calculate a distribution for any quantity that depends on the parameters of the model or the residual variances. We simply calculate the quantity in question for each parameter vector, to obtain the distribution of that quantity.

Here we do the calculation for final biomass, for each of the site-years in our data set. The code below shows the calculation for the site-year $18-2006$. We first calculate final biomass using the model parameters of the posterior distribution. Then we add to those final biomass values residual errors drawn at random from the distribution of residual error. For each model parameter vector, we use the residual variance for the same position in the chain.

```
# uncertainty from parameters
simUnc240 = maize.simule240(param.mat,  weather =  maize.weather(working.year = 2006,
working.site = 18,weather_all = weather_EuropeEU), sdate = 100, ldate = 250)
# add residual error drawn at random from normal with posterior residual std dev
B.epsilon.pred <- matrix(rnorm(length(id_param)*100,mean = 0,  sd = Xparam[,"B.sigma.
new"]),length(id_param),100)
B240.pred = B.epsilon.pred + as.vector(simUnc240)
B240.pred[B240.pred < 0] = 0
mean(B240.pred)
[1] 3414.214
quantile(B240.pred, prob = c(0.05,0.95))
      5%        95%
2984.334 3842.052
```

Uncertainty for a given site-year can be represented as a histogram (Figure 10.11). For all site-years, we can summarize the information by giving the mean and chosen quantiles in a table (Table 10.5). In our example the width of the (q95-q05) credible intervals are quite similar among site-years (mean = 830.2 and sd = 12.4).

We have presented credible intervals for the site-years for which we have observed data, but we can easily predict credible intervals for any other site-years if we have the weather data.

Finally, we proceed to a verification of the credible intervals. If our statistical model is realistic, then we should find that in about $\alpha\%$ of cases, the observation lies within the calculated $\alpha\%$ credible interval. Results are shown in Table 10.6 for final biomass. The agreement between

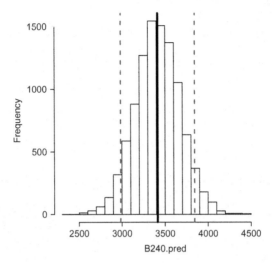

FIGURE 10.11 Histogram of biomass at day 240 with mean value (thick line) and [5%; 95%] credible interval (dotted lines).

TABLE 10.5 Summary of Uncertainty Analysis for All the Site-Years

Site-year	mean	q05	q95	q95−q05
18-2006	3417	3001	3834	832
64-2004	2832	2421	3247	826
27-2010	3143	2734	3560	825
23-2003	3445	3019	3871	852
75-2007	2970	2553	3390	837
81-2009	3120	2707	3535	828
84-2004	2189	1776	2605	829
38-2004	3033	2631	3438	807
81-2002	2787	2371	3205	834
2-2004	3040	2617	3468	852
45-2006	2998	2592	3399	807
79-2001	2740	2327	3155	829
92-2010	3110	2697	3525	829
56-2002	3540	3125	3960	835
72-2002	2666	2255	3088	833

TABLE 10.6 Results of Verification of Credible Intervals with Data

Credible Intervals	[5%–95%]	[10%–90%]	[25%–75%]
Number of observations in the interval (out of 15)	13	12	6
Percentage of observations in the interval	86.7%	80%	40%

the nominal and the observed percentage of observations in each interval is acceptable.

2.5 Data Assimilation to Predict LAI

As a supplementary illustration, we use the in-season LAI measurements to predict subsequent LAI values. The approach uses a dynamic linear model as described in the soil water example of Chapter 8. The example here is exactly analogous, with LAI in place of soil water content.

The measurements of LAI, noted Y_t, are related linearly to the LAI output of the original non-linear maize model O_t (observation equation) by:

$$Y_t = \alpha_{0t} + \alpha_{1t}O_t + e_t.$$

The intercept and slope of the regression equation are treated as the two state variables,

$$Z_t = (\alpha_{0t}, \alpha_{1t}),$$

assumed to vary over time according to a stochastic process defined by the system equation:

$$Z_t = Z_{t-1} + \eta_{t-1}$$

For more details on the form of the model and statistical assumptions, refer to the water balance example in Chapter 8.

The code for one site-year (site = 92 and year = 2010) is shown below. We make the hypothesis that error of observation is known and fix its variance at dV = 0.75.

```
library(ZeBook)
library(dlm)
# choose a year and a site
year = 2010
idsite = 92
```

```
# build a table with ALL simulation and observation
simobs = merge(maize.data_EuropeEU[(maize.data_EuropeEU$idsite
= = idsite)&(maize.data_EuropeEU$year = =year),],maize.model2
(maize.define.param()["nominal",],maize.weather(working.year =
year,working.site − idsite,weather_all − weather_EuropeEU), 100,
250), by = c("day"), all.y = TRUE)[,c("day","LAIobs","Bobs","TT",
"LAI","B")]
# function of DLM
x < −matrix(simobs$LAI,ncol = 1)
buildFun < −function(theta) {
    modMAIZE < −dlmModReg(x,dV = 0.75,dW = c(exp(theta[1]),exp
(theta[2])))
    return(modMAIZE)
    }
Y < −simobs$LAIobs
fit < −dlmMLE(Y,parm = c(0,0),build = buildFun)
fit
fitted.modMAIZE < −buildFun(fit$par)
modFilter < −dlmFilter(Y,mod = fitted.modMAIZE)
modSmooth < −dlmSmooth(Y,mod = fitted.modMAIZE)
SmoothedLAI < −modSmooth$s[,1][−1] + modSmooth$s[,2][−1]*x
SmoothedLAI[SmoothedLAI < 0] < −0
plot(1:length(x),Y,ylim = c(0,10),pch = 19,   xlab = "Time   (days)",
ylab = "LAI")
lines(1:length(x),x,lty = 1, lwd = 1)
lines(1:length(x),modFilter$m[-1,1] + modFilter$m[−1,2]*x,
lty = 2, lwd = 1)
lines(1:length(x),SmoothedLAI, lty = 3, lwd = 1)
```

The LAI values estimated by the Kalman filter are closer to the measurements of LAI than the initial simulations of the maize model (Figure 10.12).

3. HOW DIFFICULT AND TIME-CONSUMING IS EACH STEP?

Table 10.7 gives our evaluation of the complexity of the different methods
implemented in this case study, and the time we spent on each. It's not a
general case, but this report can provide guidance to the reader.

4. R CODE USED IN THIS CHAPTER

All the R code used in this chapter is included in the ZeBook R package as
demonstration R scripts (demos) (Table 10.8). Instructions on how to use a
demo can be found in Appendix 2. The names of the demos for all steps are
given below. Note that some of these calculations are long.

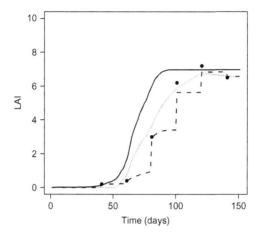

FIGURE 10.12 Simulation (solid line), observations (points), correction with Kalman filter (dotted line), and the smoother (thick dotted line) using a dynamic linear model.

TABLE 10.7 Time and Complexity Evaluation of the Steps in This Case Study

Steps and Methods	Time	Complexity
Step 1. Explore data and examine the variability in the sample	+	+
Step 2. Compare simulated results with available data	+	+
Step 3. Evaluate the uncertainty due to parameter uncertainty	++	+
Step 4. Rank the parameters using sensitivity analysis	+	++
Step 5. Calibrate the model using classical statistical methods	+++	++
Step 6. Estimate the mean squared error of prediction of the calibrated model	++	++
Step 7. Use the calibrated model to answer the initial question	+	+
Step 5'. Calibrate the model using a Bayesian approach, using MCMC	++++	++++
Step 7'. Use the calibrated model to answer the initial question. Include the uncertainty represented in the posterior distribution	++	++
Supplementary application. Data assimilation using in-season LAI measurements and the model to estimate later LAI	+	++

TABLE 10.8 List of Demonstration R Scripts Used to Obtain Results of This Package

Steps	Name of Demo in ZeBook	Description
Step 1	ch10maize1data.r	Exploration of data
Step 2	ch10maize2evaluation.r	Evaluation
Step 3	ch10maize3uncertainty.r	Uncertainty analysis (up to 20 min)
Step 4	ch10maize4sensitivity.r	Sensitivity analysis (up to 40 min)
Step 5	ch10maize5parameterOLS_gradient.r ch10maize5bparameterOLS_simplex.r ch10maize5cparameterConcL.r ch10maize5cparameterConcL_log.r	Parameter calibration with OLS and gradient method (up to 10 min) Parameter calibration with OLS simplex method (up to 120 min) Parameter calibration with concentrated likelihood – simplex method (up to 120 min) Parameter calibration with concentrated likelihood with logarithmic transformation of variable – simplex method (up to 120 min)
Step 6	ch10maize6aMSEP_OLS.r ch10maize6bMSEP_ConcL.r	Estimation of MSEP by cross validation – OLS (up to 120 min) Estimation of MSEP by cross validation – concentrated likelihood (up to 120 min)
Step 7	ch10maize7scenario.r	Use of final model for multi-year multi-site simulations (few minutes)
Step 5'	ch10maizeB5aBayesEstimRUE.r ch10maizeB5bBayesEstimAllparam.r	Calibrate the model using a Bayesian approach, using MCMC – only RUE (up to few hours) Calibrate the model using a Bayesian approach, using MCMC – all parameters (up to a few days)
Step 7'	ch10maizeB7BayesUncertainty.r	Use the posterior distribution for multi-site multi-year simulations. You must run ch10maizeB5bBayesEstimAllparam before this demo in the same session or adapt both scripts (up to a few hours)
Supplementary application	ch10maizeS8DataAssimilation.r	Data assimilation using DLM (few minutes)

The Models Included in the ZeBook R Package: Description, R Code, and Examples of Results

1 INTRODUCTION

This appendix describes the models used in this book, for which the R code is available in the ZeBook R package (Table A1.1). The models come from the domains of agronomy, plant protection, ecology, and environment (linked to agronomy). All are dynamic, with the exception of the SeedWeight and the Magarey model. The more complex models referred to in this book are not included in the R package. For more details about those models, see the references given in the chapters.

For each model we present the equations, the available information on the parameters, the R code for the associated simulators, and typical examples of simulation.

2 SEEDWEIGHT MODEL

2.1 Model Description

The SeedWeight model is a logistic model of grain weight over time in wheat. The model was proposed by Darroch and Baker (1990) in a study of grain filling in three spring wheat genotypes. The model is

$$seed.weight = \frac{W}{1 + e^{B - C*DD}}$$

This model has a single input variable, degree days after anthesis, noted DD, and three parameters, noted W, B, and C. Parameters are estimated from observations.

TABLE A1.1 Models Included in the ZeBook R Package Used in This Book

Short Name	Description	Field	R Function in the ZeBook Package
SeedWeight	Grain filling for wheat (from Darroch and Baker, 1990)	Agronomy	seedweight.model
Magarey	Generic model of infection for foliar diseases caused by fungi (from Magarey et al., 2005)	Plant protection	magarey.model
CarbonSoil	Model of soil carbon content, annual time step (adapted from Hénin-Dupuis model, Jones et al., 2004)	Agronomy environment	carbonsoil.model
WaterBalance	Budget-based soil water balance model for grass (from Woli, 2010)	Agronomy	watbal.model
Maize	Model of crop growth for maize cultivated in potential conditions (based on widely used concepts for crop modeling)	Agronomy	maize.model
Verhulst	Homogenous population with limited food supply	Ecology	verhulst.model
PopulationAge	Population dynamics model with age classes	Ecology	population.age.model population.age.matrix.model population.age.model.ode
PredatorPrey	Predator-prey population dynamics model	Ecology	predator.prey.model
Weed	Model of wheat yield under conditions of a weed infestation of blackgrass (from Munier-Jolain et al., 2002)	Agronomy Plant protection	weed.model
Epirice	Disease model for rice (from Savary et al., 2012)	Plant protection	epirice.model
Cotton	Model of dynamic for numbers of cotton fruiting points (Wallach, 1980). See Chapter 6, Exercises section, for a full description	Agronomy	cotton.model

2.2 R Code

The R function for evaluating the model is

```
seedweight.model <-function(DD, W, B, C){ return(W/(1 + exp
(B-C*DD))))}exp(B-C*DD))) }
```

2.3 Description of Data

The data used for fitting the model are contained in the `seedweight.data` object of the ZeBook package.

```
head(seedweight.data)
   year  DD seedweight
1 1986 189         10
2 1986 220         15
3 1986 280         21
4 1986 320         25
5 1986 360         31
```

3 MAGAREY MODEL

3.1 Model Description

The Magarey model is a generic model of infection for foliar diseases caused by fungi (from Magarey, Sutton, and Thayer 2005). This model has been used to compute the wetness duration requirement as a function of temperature for many species and was included in a disease forecast system (Magarey et al., 2005, 2007).

This model has a single input, the daily mean temperature (T, °C). It simulates a single output variable, the wetness duration (W, hour) required to achieve a critical disease intensity (5% disease severity or 20% disease incidence).

$$W = \frac{W_{min}}{f(T)} \text{ if } T_{min} \leq T \leq T_{max} \tag{1}$$

$$= W_{max} \text{ otherwise}$$

$$f(T) = \left(\frac{T_{max} - T}{T_{max} - T_{opt}}\right) \left(\frac{T - T_{min}}{T_{opt} - T_{min}}\right)^{(T_{opt} - T_{min})/(T_{max} - T_{opt})} \tag{2}$$

The model has five parameters whose values were estimated from experimental data and expert knowledge for different foliar pathogens (e.g., Magarey et al., 2005; EFSA 2008b). However, for some species, these parameters are uncertain due to the limited number of available data (Magarey et al., 2005). Ranges of parameter values for two fungi are shown in Table A1.2 and Table A1.3.

TABLE A1.2 Minimum (binf) and Maximum (bsup) Values for the Parameters of the Magarey Model for Pycnidiospores of the Fungus Guignardia Citricarpa Kiely by EFSA (2008)

Name	binf	bsup	Unit	Description
Tmin	10	15	°C	Minimal temperature for infection
Topt	25	30	°C	Optimal temperature for infection
Tmax	32	35	°C	Maximal temperature for infection
Wmin	12	14	hour	Minimal wetness duration for infection
Wmax	35	48	hour	Maximal wetness duration for infection

TABLE A1.3 Nominal (nominal), Minimum (binf), and Maximum (bsup) Values for the Parameters of the Magarey Model for Kaki Fungus

Name	nominal	binf	bsup	Unit	Description
Tmin	7	2	13	°C	Minimal temperature for infection
Topt	18	14	26	°C	Optimal temperature for infection
Tmax	30	27	35	°C	Maximal temperature for infection
Wmin	10	5	17	hour	Minimal wetness duration for infection
Wmax	42	18	90	hour	Maximal wetness duration for infection

3.2 R Code

The R function for evaluating the model is

```
# the Magarey model function
magarey.model <-function(T, Tmin, Topt, Tmax, Wmin, Wmax){
  fT <-((Tmax-T)/(Tmax-Topt))*(((T-Tmin)/(Topt-Tmin))^
((Topt-Tmin)/(Tmax-Topt)))
  W <-Wmin/fT
  W[W>Wmax | T<Tmin | T>Tmax] <-Wmax
  return (W) }
```

The input T can be a single temperature or a vector of temperatures (it can be a time series of daily temperatures from weather observation). In the latter case the result is a vector of wetness duration requirements, but each calculation is independent of the other.

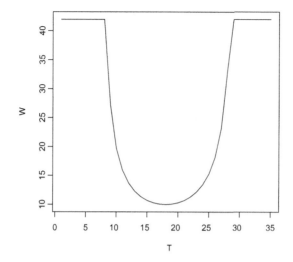

FIGURE A1.1 Required wetness duration (W) versus daily average temperature (T) according to the Magarey model.

3.3 Example of Model Output

Figure A1.1 shows wetness duration requirement for temperatures from 1 to 35°C for the Kaki fungus, calculated using the magarey.model function and the nominal parameter values.

```
plot(1:35, magarey.model(1:35,7, 18, 30, 10, 42), type = "l", xlab = "T",
ylab = "W")
```

4 SOIL CARBON MODEL

4.1 Model Description

An important aspect of the study of the contribution of agricultural emissions and storage of CO_2 is the determination of carbon stored in the soil. This can be estimated by measurements or by using mathematical simulation models (e.g., Andriulo et al., 1999). Another possibility, proposed by Jones and Graham (2006), is to combine model simulations with measurements.

The model here is a very simple dynamic model of soil carbon content in the top 20 cm of soil in a field, with a time step of one year. The equations that describe the dynamics of this system are adapted from the Hénin-Dupuis model described in Jones, Graham, Wallach, Bostick, and Koo (2004). The soil carbon content is represented by a single state variable: the mass of carbon per unit land area in the top 20 cm of soil in a given year (Z, kg.ha^{-1}). It is assumed that soil C is known in some year, taken as the initial year.

The yearly change in soil C is the difference between input from crop biomass and loss. The model equations are then

$$Z(year + 1) = Z(year) + dZ(year) \qquad (1)$$

$$dZ(year) = -R \cdot Z(year) + b \cdot U(year) \qquad (2)$$

These can be combined into a single equation

$$
\begin{aligned}
Z(year + 1) &= Z(year) - R \cdot Z(year) + b \cdot U(year) \\
&= G \cdot Z(year) + b \cdot U(year)
\end{aligned}
\qquad (3)
$$

where $G = 1 - R$ and where U(year) is the amount of C in crop biomass production in the given year ($kg[C]ha^{-1}$).

This model uses only one input (explanatory) variable U(year). For some examples, we will consider U(year) as a constant.

The model has two parameters: R, the fraction of soil carbon content lost per year $(-)$, and b, the fraction of yearly crop biomass production left in the soil $(-)$.

4.2 R Code

For the use of this model with data assimilation, it is convenient to write it as two separate functions (Table A1.4). The 'update function' (carbonsoil.update in the ZeBook R package) calculates Z(year + 1) given Z(year). The 'integration function' (carbonsoil.model in the ZeBook R package) calculates Z(year) using recursively the previous function for multiple years, starting from some given initial value.

5 WATERBALANCE MODEL

5.1 Model Description

This model is a budget-based soil water balance approach used to compute a generic agricultural drought index based on the soil-plant-atmosphere continuum (ARID, Woli, 2010).

The model simulates the amount of water present in the root zone of a homogenous grass canopy growing on a well-drained and homogenous soil. The model assumes a root depth of 400 mm with a homogenous distribution of roots. The input is water from rain or irrigation. Water is lost through runoff, transpiration by the plant, and deep drainage (Figure A1.2). Soil evaporation is assumed negligible (because of total soil coverage by vegetation) and it is assumed that there is no capillary rise of water.

The model equations are shown in Table A1.5. The model has a single state variable: the amount of water at the beginning of the day (WAT, mm).

TABLE A1.4 The Soil Carbon Model Functions

```
# the Soil Carbon model function - Update Z
# Arguments: Zy = Z(year), 2 parameters, Uy = U(year)
carbonsoil.update < -function(Zy,R,b,Uy)
    {
    # Calculate the rates of state variables (dZ)
    dZ = -R*Zy + b*Uy
    # Update the state variables Z
    Zy1 = Zy + dZ
    # Return Zy1 = Z(year + 1)
    return(Zy1)
    }
```

```
# the Soil Carbon model function - integration loop
# Arguments: 2 parameters, U table, Z1: initial soil C, duration
carbonsoil.model < -function(R,b,U,Z1,duration)
{
    # U : if U is a constant (not a dataframe), convert into
    dataframe.
    if(!is.data.frame(U)){U = data.frame("year" = 1:duration,
"U" = rep(U,duration))}
    # Initialize variables
    # 1 state variable, as 1 vectors initialized to NA
    # Z : mass of carbon per area (top 20 cm of soil) (kg.ha-1)
    Z = rep(NA,duration)
    # Initialize state variable when starting simulation on year 0
    Z[1] = Z1
    # Integration loop
    for (year in 1:(duration-1))
    {
    # using the update function.
    Z[year + 1] = carbonsoil.update(Z[year],R,b,U$U[year])
    }
    # End loop
    return(data.frame(year = 1:duration,Z = Z[1:duration]))
}
```

The first function is the update function that calculates Z(year + 1) depending on Z(year). The second is the integration loop. This function has as arguments the two parameters (R and b), a table or a constant of amount of C in crop biomass each year (U), the initial content of carbon in the soil (Z1), and the last date of the simulation (ldate). The state variable is returned to the main program as a table (data.frame) with two columns (day, Z).

The explanatory variables are daily rainfall (RAIN(day)), daily crop reference evapotranspiration (ETr(day)), soil water content at wilting point (WP, $cm^3.cm^{-3}$), soil water content at field capacity (FC, $cm^3.cm^{-3}$), and initial water content (WAT0, $cm^3.cm^{-3}$).

The model has five parameters, which are described in Table A1.6.

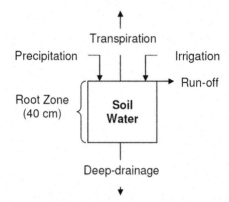

FIGURE A1.2   The processes taken into account in the WaterBalance model (from Woli, 2010).

TABLE A1.5 Equations of the WaterBalance Model

Change in Water Before Drainage (Precipitation-Runoff)

(1) if (RAIN > IA){RO = (RAIN-0.2*S)^2/(RAIN + 0.8*S)}else{RO = 0}

Calculating the amount of deep drainage

(2) if (WAT(day) + RAIN-RO > WATfc){DR = DC*(WAT(day) + RAIN-RO-WATfc)}

else{DR − 0}Calculating the amount of water lost by transpiration (after drainage)

(3) TR = min(MUF*(WAT(day) + RAIN-RO-DR-WATwp), ETr)

(4) dWAT = RAIN-RO-DR-TR

(5) WAT(day + 1) = WAT(day) + dWAT

TABLE A1.6 Nominal (nominal), Minimum (binf), and Maximum (bsup) Values for Parameters of the Water Balance Model

Name	nominal	binf	bsup	Unit	Description
WHC	0.13	0.05	0.18	$cm3.cm^{-3}$	Water holding capacity of the soil
MUF	0.096	0.06	0.11	$mm3.mm^{-3}$	Water uptake coefficient
DC	0.55	0.25	0.75	$mm3.mm^{-3}$	Drainage coefficient
z	400	300	600	mm	Root zone depth
CN	65	15	90	−	Run-off curve number

5.2 R Code

It is convenient to write the model as two separate functions (Table A1.7). The update function (`watbal.update` in the ZeBook R package) calculates Z (year + 1) given Z(year). The integration function (`watbal.model` in the ZeBook R package) calculates Z(year) for multiple years, starting from some given initial value.

5.3 Example of Model Output

Figure A1.3 shows weather input and the simulated dynamics of soil water content.

6 MAIZE CROP MODEL

6.1 Model Description

This model is a dynamic model of crop growth for maize cultivated under potential conditions. It is based on key concepts included in most crop models for describing potential growth, that is, growth determined by temperature and solar radiation. The model does not take into account effects of soil water, nutrients, pests, or diseases. See Chapter 1 for a detailed description of the model and underlying formalisms.

The model has three state variables: leaf area per unit ground area (leaf area index, LAI), total biomass (B, $g.m^{-2}$), and cumulative thermal time since plant emergence (TT). These state variables are dynamic variables which depend on days after emergence: TT(day), B(day), and LAI(day). The model has a time step of one day (dt = 1).

The model equations are presented in (Table A1.8). The model has a total of seven parameters, described in Table A1.9.

The explanatory variables are the daily solar radiation (I (day), $MJ.m^{-2}.day^{-1}$), daily minimum (TMIN(day)) and maximum temperature (TMAX(day), °C), and starting day (sdate). A date of end of simulation (ldate) is needed too.

6.2 Simulation

The simulation has the following steps:

1) Initialize all state variables, parameters, and other variables necessary to perform the calculations.
2) Repeat the following two steps at each time step (each day, in our case), up to the end of the simulation:
   - Calculate the rate of change of each state variable. See Equations (4), (5), and (6) in Table 8.

TABLE A1.7 The WaterBalance Model Functions[1]

```
# the water balance model function — Update
watbal.update = function(WAT0,RAIN, ETr,param,FC,WP){
   WHC = (param["WHC"])
   MUF = (param["MUF"])
   DC = (param["DC"])
   z = (param["z"])
   CN = (param["CN"])
   # Maximum abstraction for run off
   # (potential maximum soil moisture retention after runoff begins)
   S = 25400/CN-254
   # Initial Abstraction for run off (the amount of water before runoff)
   IA = 0.2*S
   # WATfc : Maximum Water content at field capacity (mm)
   WATfc = FC*z
   # WATwp : Water content at wilting Point (mm)
   WATwp = WP*z
   # Change in Water Before Drainage (Precipitation — Runoff)
   if (RAIN > IA){RO = (RAIN-0.2*S)^2/(RAIN + 0.8*S)}else{RO = 0}
   # Calculating the amount of deep drainage
   if (WAT0 + RAIN-RO > WATfc){DR = DC*(WAT0 + RAIN-RO-WATfc)}else
      {DR = 0}
   # Calculating the amount of water lost by transpiration (after
    drainage)
   TR = min(MUF*(WAT0 + RAIN-RO-DR-WATwp), ETr)
   dWAT = RAIN - RO -DR -TR
   WAT1 = WAT0 + dWAT
   return(WAT1)
}


# the water balance model function — Integration loop
watbal.model = function(param, weather, WP, FC, WAT0 = NA) {
   z = (param["z"])
   # input variable describing the soil
   # WP : Water content at wilting Point (cm^3.cm^-3)
   # FC : Water content at field capacity (cm^3.cm^-3)
   # WAT0 : Initial Water content (cm^3.cm^-3)
   if (is.na(WAT0)) {WAT0 = z*FC}
   # Initialize variable
   # WAT : Water at the beginning of the day (mm) : State variable
   WAT = rep(NA, nrow(weather))
   # initialisation use Amount of water at the beginning
   WAT[1] = WAT0
   # integration loops
   for (day in 1:(nrow(weather)-1)){
     WAT[day + 1] = watbal.update(WAT[day],weather$RAIN[day],weather$
         ETr[day],
         param,FC,WP)[1]
   }
   # Volumetric Soil Water content (fraction : mm.mm-1)
   WATp = WAT/z
   return(data.frame(day = weather$day, RAIN = weather$RAIN,
ETr = weather$ETr, WAT = WAT, WATp = WATp))
}
```

[1]See Chapter 4 for a variant, with another order of calculation.

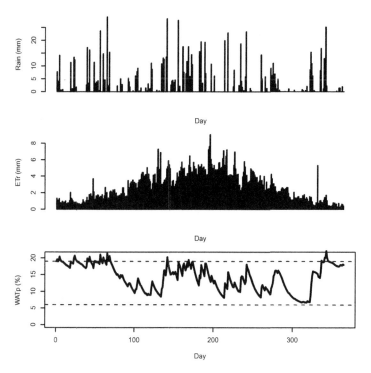

FIGURE A1.3 Inputs to the WaterBalance model (daily rainfall RAIN and daily evaporation ETr), and simulated values of soil water (WAT).

TABLE A1.8 Equations of the Maize Crop Model

(1) $TT(day + 1) = TT(day) + dTT(day)$

(2) $B(day + 1) = B(day) + dB(day)$

(3) $LAI(day + 1) = LAI(day) + dLAI(day)$

with

(4) $dTT(day) = max\left[\frac{TMIN(day) + TMAX(day)}{2} - Tbase\,;\,0\right]$

(5) $\begin{aligned}dB(day) &= RUE \cdot (1 - e^{-K \cdot LAI(day)}) \cdot I\,(day) & TT(day) \leq TTM\\ &= 0 & TT(day) > TTM\end{aligned}$

(6) $\begin{aligned}dLAI(day) &= alpha \cdot dTT(day) \cdot LAI(day) \cdot max[LAImax - LAI(day)\,;\,0] & TT(day) \leq TTL\\ &= 0 & TT(day) > TTL\end{aligned}$

    — Update each state variable by adding the rate of change multiplied by the time step to the value of the state variable from the previous time step. This is called Euler integration (Equations (1), (2), and (3) in Table A1.8).

3) Store the simulation results in a table for later analysis.

TABLE A1.9 Nominal (nominal), Minimum (binf), and Maximum (bsup) Values for Parameters of the Maize Crop Model

Name	nominal	binf	bsup	Unit	Description
Tbase	7	6	8	°C	Baseline temperature for growth
TTM	1200	1100	1400	°C. day	Temperature sum for crop maturity
TTL	700	600	850	°C. day	Temperature sum at the end of leaf area increase
K	0.7	0.6	0.8	–	Extinction coefficient (relation between LAI and intercepted radiation)
RUE	1.85	1.50	2.50	g. MJ^{-1}	Radiation use efficiency
alpha	0.00243	0.00200	0.00300	$(°C. day)^{-1}$	The relative rate of LAI increase for small values of LAI
LAImax	7	6	8	–	Maximum LAI

6.3 R Code

This model has been programmed in R as a function (maize.model), to facilitate its use in the various examples (Table A1.10). A similar structure (first calculate changes, then update state variables) could be used for more complex models. The arguments of the function are the seven parameters (Tbase, RUE, K, alpha, LAImax, TTM, TTL), the daily weather data in a table (weather), the starting day (sdate), and the last simulation day (ldate). The function returns a table (data.frame) whose columns are the daily values of day, TT, LAI, and B.

The ZeBook R package also has weather data sets for multiple grid points and years for western France (weather_FranceWest) or Europe (weather_EuropeEU) taken from the NASA POWER data, that can be used as weather input for the Maize model (see Table A1.11 for an illustration). The function maize.weather reads weather data (day, I, Tmax, and Tmin) for one year and one site from the indicated data set. Examples of graphs of daily temperatures (Tmax and Tmin) and of solar radiation (I) are shown in Figure A1.4.

The data set maize.data_EuropeEU has artificial (invented by simulation) observation data for several sites and years. These data can be used in exercises for model evaluation or calibration.

To actually do a simulation, one would need a main program that calls the model function. Table A1.12 presents the outline of a program (see demo

TABLE A1.10 The Maize Model Function

```
# the Maize model function.
# Arguments : 7 parameters, weather table, sowing date and last date
maize.model < -function(Tbase,RUE,K,alpha,LAImax,TTM,TTL,
weather,sdate,ldate)
  {
  # Initialize variables
  #3 states variables, as 3 vectors initialized to NA
  # TT : temperature sum (°C.d)
  TT < -rep(NA,ldate)
  # B : Biomass (g/m2)
  B < -rep(NA,ldate)
  # LAI : Leaf Area Index (m2 leaf/m2 soil)
  LAI < -rep(NA,ldate)
  # Initialize state variable when sowing on day sdate
  TT[sdate] <- 0
  B[sdate] <- 1
  LAI[sdate] <- 0.01
  # Integration loop
  for (day in sdate:(ldate-1))
    {
    # Calculate the rates of state variables (dTT, dB, dLAI)
    dTT <- max((weather$Tmin[day]+weather$Tmax[day])/2-Tbase, 0)
    if (TT[day]< = TTM) {dB <- RUE*(1-exp(-K*LAI[day]))*weather$I[day]}
    else {dB <- 0}
    if (TT[day]< =TTL){dLAI <- alpha*dTT*LAI[day]*max(LAImax-LAI[day],0)}
    else {dLAI <-0 }
    # Update all of the state variables
    TT[day+1] <- TT[day] + dTT
    B[day+1] <- B[day] + dB
    LAI[day+1] <- LAI[day] + dLAI
    }
  # End loop
return(data.frame(day=sdate:ldate,TT=TT[sdate:ldate], LAI=LAI
[sdate:ldate],B=B[sdate:ldate]))
  }
```

(ch04Simulation.maize) in ZeBook package). The main program is listed last as a reminder that all functions need to be defined before the main program can use them.

6.4 Example of Model Output

Table A1.13 shows a partial listing of the output of simulation of maize. model (for year 2010 and site 30 from weather_EuropeEU). The output starts

TABLE A1.11 Partial Listing of the Weather Data for One Site in Europe (latitude = 38, longitude = − 7) and Year 2010[1]

```
help(weather_EuropeEU)
weather=maize.weather(working.year=2010,working.site=30,
weather_all=weather_EuropeEU)
head(weather)
      idsite GPSlatitude GPSlongitude year day   I Tmax Tmin
61720   30         46            -1 2010  1 5.2  7.3  1.5
61721   30         46            -1 2010  2 6.0  4.6 -0.8
61722   30         46            -1 2010  3 2.3  2.4  0.5
61723   30         46            -1 2010  4 5.3  1.1 -3.5
61724   30         46            -1 2010  5 5.1  1.4 -4.7
61725   30         46            -1 2010  6 4.2  5.0 -1.9
```

[1]Data obtained from the NASA POWER agroclimatology site (http://power.larc.nasa.gov).

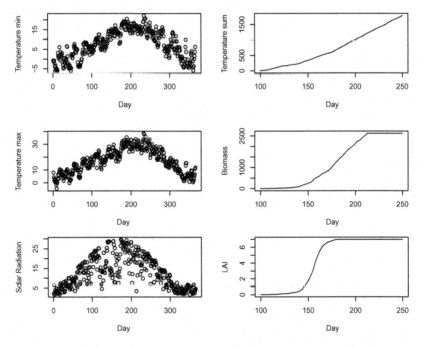

FIGURE A1.4 Inputs to the maize model on left-hand side, top to bottom: daily minimum temperature (°C), daily maximum temperature (°C), and daily solar radiation (MJ.m^{-2}). Right-hand side, top to bottom, Simulated values of temperature sum (°C.day), aboveground biomass (g.m^{-2}), and LAI (-).

TABLE A1.12 Structure of a Main Program to Use the Maize.Model in the Simulation Exercises

Create useful functions and data set

maize.model: to simulate the maize model

maize.weather: read weather data (day, I, Tmax, and Tmin) for one specific year and site from a weather data set included in the ZeBook package

Main program

Defines and sets the values of all model parameters

Sets starting day (sdate) and last simulation day (ldate)

Runs the maize model, storing results in the output table

Writes the output table in a file in the working directory

Displays graphs of output variables (TT, LAI, and B)

Displays graphs of input variables (Tmin, Tmax, and I)

TABLE A1.13 Partial Listing of the Output[1]

```
sim = maize.model(Tbase = 7, RUE = 1.85, K = 0.7, alpha = 0.00243,
LAImax = 7, TTM = 1200, TTL = 700, weather, sdate = 100, ldate = 250)
sim
      day      TT         LAI          B
1     100    0.00  0.01000000    1.000000
2     101    5.70  0.01096818    1.280034
3     102    8.85  0.01155496    1.570097
...
80    179  669.85  6.94770413  1363.660196
81    180  686.00  6.96196308  1411.572232
82    181  702.60  6.97264505  1458.386460
...
114   213 1186.85  6.97264505  2830.213894
115   214 1200.45  6.97264505  2861.975959
116   215 1214.55  6.97264505  2861.975959
...
151   250 1694.55  6.97264505  2861.975959
```

[1]Note that beyond day 115, LAI and B remain constant. This is because crop maturity has been reached (TT ≥ 1200).

on day 100, taken as the day that the crop emerged. The initial values of the state variables (LAI = 0.01, B = 1.0) are appropriate for emergence day. TT is initialized at 0.

Graphical presentation of input and output can be obtained by running the following instructions (Figure A1.4).

```
par(mfcol = c(3,2), mar = c(4,4,0.5,0.5))
# Produce graphical output of the input variables
plot(1:365,weather$Tmin, xlab = "day", ylab = "Temperature min")
plot(1:365,weather$Tmax, xlab = "day", ylab = "Temperature max")
plot(1:365,weather$I, xlab = "day", ylab = "Solar Radiation")
# Produce graphical output of the state variables
plot(sim$day, sim$TT, xlab = "day", ylab = "Temperature sum",type = "l")
plot(sim$day, sim$B, xlab = "day", ylab = "Biomass",type = "l")
plot(sim$day, sim$LAI, xlab = "day", ylab = "LAI",type = "l")
```

7 VERHULST MODEL

7.1 Model Description

This model is a classical dynamic model of a homogenous population with some limitation to population growth. It is a logistic growth model as proposed by Verhulst (1838). The model can be expressed as an ordinary differential equation for the size of population, Y (individuals.m^{-2}).

$$\frac{dY}{dt} = a \cdot y \cdot \left(1 - \frac{Y}{k}\right)$$

The two parameters are growth rate, a, and carrying capacity, k, (the maximum number of individuals that the environment can support per unit area). Both parameters are positive numbers. An analytical solution (logistic function) exists, but we use here a numerical solution with a time step of one day, written as:

$$Y(t = 1) = Y_0$$

$$dY = a \cdot Y \cdot \left(1 - \frac{Y}{k}\right)$$

$$Y(t + 1) = Y(t) + dY$$

7.2 R Code

We write the model as two separate functions (Table A1.14). The update function (verhulst.update in the ZeBook R package) calculates Y(day + 1)

TABLE A1.14 Listing of the Verhulst Model Function Verhulst.Model

```
# the Verhulst model function-Update function
verhulst.update<-function(Y,a,k)
{
    # Calculate the rates of state variables (dZ)
    dY = a*Y*(1-Y/k)
  # Update the state variables Z
    Y1 = Y + dY
  # Return Y1 = Y(t + 1)
    return(Y1)
}

# the Verhulst model function – integration loop
verhulst.model <-function(a,k,Y0,duration)
{
    # Initialize variables
    # 1 state variable, as 1 vectors initialized to NA
    # Y : Size of the population
    Y = rep(NA,duration)
    # Initialize state variable when starting simulation on year 0
    Y[1] = Y0
    # Integration loop
    for (day in 1:(duration−1))
    {
    # using the update function.
    Y[day + 1] = verhulst.update(Y[day],a,k)
    }
    # End loop
  return(data.frame(day = 1:duration,Y = Y[1:duration]))
}
```

given Y(day). The integration function (verhulst.model in the ZeBook R package) calculates Y(day) for multiple days, starting from some given initial value.

7.3 Example of Model Output

Figure A1.5 shows the simulated population dynamics over 100 days for the parameter values a = 0.08 day^{-1} and k = 100 individuals.m^{-2}.

8 POPULATION AGE MODEL

8.1 Model Description

This model is a population dynamics model for a population with seven age classes: E: egg stage, L1: larval1 stage, L2: larval2 stage, L3: larval3 stage, L4:

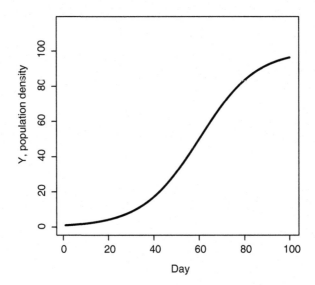

FIGURE A1.5 Dynamics of population density (individuals.m^{-2}) simulated with the Verhulst model.

larval4 stage, P: pupal stage, A: adult stage. The model has seven state variables corresponding to the density of population (number per ha) for each class.

The model has 15 parameters which determine the rates of transition between classes, mortality, and reproduction. See Chapter 1 for a more complete description.

8.2 R Code

The simulation process is similar as for previous dynamic models and can be written as before (see `population.age.model` in ZeBook package). But there are now more states variables (seven), so we propose to write it as a matrix computation (Table A1.15): it is possible for this model and quite simple. Note that it is really more efficient and reduces computer time significantly.

8.3 Example of model output

Figure A1.6 shows the simulated population dynamics for parameter values rb = 3.5, mE = 0.017, rE = 0.172, m1 = 0.060, r12 = 0.217, m2 = 0.032, r23 = 0.313, m3 = 0.022, r34 = 0.222, m4 = 0.020, r4P = 0.135, mP = 0.020, rPA = 0.099, mA = 0.027, iA = 0 for 40 days and an integration time step of dt = 0.01.

TABLE A1.15 Listing of the Model of Dynamic of Population with Age
Classes (population.age.matrix.model)

```
# Population Dynamics Model with Age Classes for an insect –
matrix form
population.age.matrix.model = function(rb = 3.5, mE = 0.017,
rE = 0.172, m1 = 0.060, r12 = 0.217, m2 = 0.032, r23 = 0.313,
m3 = 0.022, r34 = 0.222, m4 = 0.020, r4P = 0.135, mP = 0.020,
rPA = 0.099, mA = 0.027, iA = 0, duration = 100, dt = 1){
   # V : matrix of state variable (one per column), initialized to NA
   V = matrix(NA,ncol = 7,nrow = duration/dt + 1,dimnames = list
   (NULL,c("E","L1","L2","L3", "L4","P","A")))
   # initiation of state variable
   V[1,] <- c(5,0,0,0,0,0,0)
   # defining matrix transition
   M = matrix(c(-rE-mE, 0,0,0,0,0,rb,
   rE, -r12-m1,0,0,0,0,0,
   0,r12,-r23-m2,0,0,0,0,
   0,0,r23,-r34-m3,0,0,0,
   0,0,0,r34,-r4P-m4,0,0,
   0,0,0,0,r4P,-rPA-mP,0,
   0,0,0,0,0,rPA,-mA),ncol = 7,nrow = 7, byrow = TRUE)
   # input/output rate matrix
   IN = matrix(c(0, 0,0,0,0,0,iA),ncol = 1,nrow = 7,
   byrow = FALSE)
   # Simulation loop
   for (k in 1:(duration/dt)){
   # Calculate rates of change of state variables
   dV = (M %*% V[k,] + IN)*dt
   # Update state variables
   V[k+1,] <- V[k,] + dV
   }
   # End simulation loop
   return(round(as.data.frame(cbind(time = (1:(duration/
dt + 1))*dt-dt, V)),10))
}
```

9 PREDATOR-PREY MODEL

9.1 Model Description

This is the Lotka-Volterra model of predator-prey population dynamics
(with a logistic function for the prey population), applied to ladybeetles
(the predator) and aphids (the prey). There are only two state variables:
H is the aphid population density (number per ha) and A is the ladybeetle
population density (number per ha). See Chapter 1 for a more complete
description.

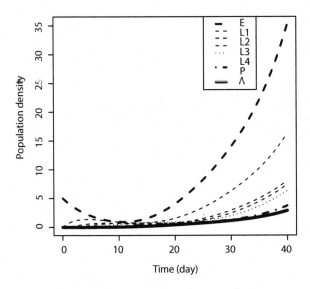

FIGURE A1.6 Dynamics of population density (individuals.m^{-2}) for a population with age structure, simulated with the PopulationAge model. E: egg stage, L1: larval stage 1, L2: larval stage 2, L3: larval stage 3, L4: larval stage 4, P: pupal stage, A: adult stage.

9.2 R Code

The model is expressed as a system of two ordinary differential equations. Here we use the ode function of the deSolve package for the integration (Table A1.16), which includes various different methods of numerical integration as discussed in Chapter 4.

9.3 Example Model Output

A graphical representation of model output is shown in Figure A1.7. It was obtained by running the following lines of code as a main program:

```
library(deSolve)
library(ZeBook)
sim <- predator.prey.model(grH = 1,kH = 10,mrH = 0.2,eff = 0.5,
mrA = 0.2, H0 = 1, A0 = 2, duration = 200, dt = 1, method = "euler")
plot(sim$time,sim$H,type = "l",xlab = "time (day)",ylab = "population
density",lty = 1)
lines(sim$time, sim$A, lty = 2)
legend("topright", c("H, prey", "A, predator"), lty = 1:2)
```

TABLE A1.16 Predator-Prey Model with a Logistic Function for the Prey Population Dynamics (predator.prey.model)

```
# Predator—Prey Lotka-Volterra model (with logistic prey)
predator.prey.model = function(grH = 1, kH = 10, mrH = 0.2,
eff = 0.5, mrA = 0.2, H0 = 1, A0 = 2, duration = 200, dt = 1,
method = "euler"){
# 2 states variables
# H : prey, Aphids, homogenous population (density) (number per ha)
# A : predators, Ladybeetles, homogenous population (density)
(number per ha)
# definiting the model as an ordinary differential equation
system
  predator.prey.ode <- function(Time, State, Pars) {
  with(as.list(c(State, Pars)), {
  dH <- grH*H*(1−H/kH) − mrH*H*A
  dA <- mrH*H*A*eff − mrA*A
  return(list(c(dH, dA)))
  })
  }
  sim = ode(y = c(H = H0, A = A0), times = seq(0, duration, by = dt),
func = predator.prey.ode, parms = c(mrH, grH, mrA, eff, kH),
method = rkMethod(method))
  return(as.data.frame(sim))
  }
```

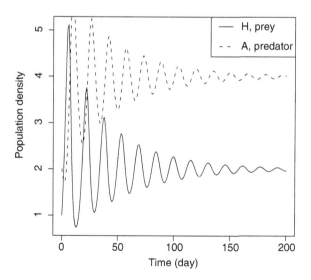

FIGURE A1.7 Dynamics of prey and predator population densities (individuals.m^{-2}) simulated with the PredatorPrey model.

10 WEED MODEL

10.1 Model Description

This model is a dynamic model of a weed infestation of blackgrass (*A. myo-suroides*) in wheat and the consequent loss of wheat yield. It is based on the weed model of Munier-Jolain, Chauvel, and Gasquez (2002) and includes standard population dynamics elements, namely survival, reproduction, and flows between classes.

The system is represented by four state variables: weed density at emergence (d, plants.m^{-2}), seed production (S, seeds.m^{-2}), seed bank in the upper soil after tillage (surface seed bank SSBa, seeds.m^{-2}), and seed bank in the lower soil after tillage (depth seed bank DSBa, seeds.m^{-2}).

Weed density at maturity (dm, plants.m^{-2}) and wheat yield (Yield, t. ha^{-1}) are two auxiliary variables.

The model has eight equations (Table A1.17). The time step is one year, with an intermediate calculation after tillage.

The model has 16 parameters (Table A1.18). Four parameters determine the effect of tillage on transfers between surface and depth seed banks (beta.0, beta.1, chsi.0, and chsi.1). Five parameters determine the production of seeds (delta.new, delta. old, Smax0, Smax1, and v). Four parameters determine seed losses (mh, mc, phi, and mu). Three parameters relate to wheat yield and loss due to weeds (Ymax, rmax, and gamma).

The explanatory variables are the management decisions each year, which determine tillage, herbicide application, and the type of wheat crop. In each case there are two possibilities each year (Table A1.19):

- if there is deep tillage, Soil = 1; if superficial work, Soil = 0.
- if herbicide is applied, Herb = 1; else Herb = 0.
- if the crop is winter wheat, Crop = 1; else Crop = 0.

10.2 R Code

We write the model as two separate functions. The update function (weed. update in the ZeBook R package) calculates the state variable in year + 1 based on the values in year. The integration function (weed.model in the ZeBook R package) calculates the state variables for multiple years, starting from given initial values.

10.3 Example of Model Output

Table A1.20 shows output in the case where herbicide is used every year.

Annual wheat yield is shown in Figure A1.8 for the two management sequences of Table A1.19.

TABLE A1.17 Description and Equations of the Weed Model

Before tillage

The surface seed bank depends on the stock of the previous year, natural mortality, germination, and seeds produced. The depth seed bank depends on the stock of the previous year and natural mortality.

(1) $SSBb(year + 1) = [1\text{-}mu] \cdot [SSBa(year)\text{-}d(year)] + v \cdot [1\text{-}phi] \cdot S(year)$

(2) $DSBb(year + 1) = [1\text{-}mu] \cdot DSBa(year)$

After tillage

Transfers between surface and depth zones due to tillage are calculated as:

(3) $SSBa(year + 1) = (1\text{-}beta) \cdot SSBb(year + 1) + chsi \cdot DSBb(year + 1)$

(4) $DSBa(year + 1) = (1\text{-}chsi) \cdot DSBb(year + 1) + beta \cdot SSBb(year + 1)$

with
$$beta = beta.1 \cdot Soil(year) + beta.0 \cdot (1\text{-}Soil(year))$$
$$chsi = chsi.1 \cdot Soil(year) + chsi.0 \cdot (1\text{-}Soil(year))$$

Density of young plants at emergence is calculated from seeds produced in the previous year and from seed stock in the surface zone after tillage.

(5) $d(year + 1) = delta_{new} \cdot v \cdot [1\text{-}phi] \cdot [1\text{-}beta] \cdot S(year) + delta_{old} \cdot$
$[SSBa(year + 1)\text{-}S(year) \cdot v \cdot [1\text{-}phi] \cdot [1\text{-}beta]]$

Density of weed plants at maturity (dm) depends on seedling mortality and on herbicide treatment.

(6) $dm(year + 1) = [1\text{-}mh \cdot herb(year)] \cdot [1\text{-}mc] \cdot d(year + 1)$

with
$$herb(year) = 1 \quad \text{if herbicide treatment}$$
$$= 0 \quad \text{if no herbicide treatment}$$

Seed production depends on weed plants at maturity.

(7) $S(year + 1) = Smax \cdot \dfrac{dm(year + 1)}{1 + alpha \cdot D(year + 1)}$

$Smax = Smax.1 \cdot Crop(year) + Smax.0 \cdot (1\text{-}Crop(year))$

with
$$alpha = \frac{Smax}{160000}$$

Wheat yield is calculated as yield in the absence of weeds (Y_{max}) reduced by a factor that depends on weed density at maturity.

(8) $Y(year + 1) = Y_{max} \cdot \left[1 - \dfrac{r_{max} \cdot dm(year + 1)}{1 + gamma \cdot D(year + 1)}\right]$

TABLE A1.18 Parameters of the Weed Model: Nominal Value (nominal), Lower Limit (binf), and Upper Limit (bsup)

Name	nominal	binf	bsup	Unit	Description	Equations
Mu	0.84	0.756	0.924	-	Rate of annual seedbank decline	1,2
V	0.6	0.540	0.66	-	Proportion of viable seeds	1
Phi	0.55	0.495	0.605	-	Fresh seed loss	1, 5
beta.1	0.95	0.855	1.0	-	Proportion of SSB buried (if soil = 1)	3, 4, 5
beta.0	0.2	0.180	0.22	-	Proportion of SSB buried (if soil = 0)	3, 4, 5
chsi.1	0.3	0.270	0.33	-	Proportion of DSB carried up (if soil = 1)	3, 4
chsi.0	0.05	0.045	0.055	-	Proportion of DSB carried up (if soil = 0)	3, 4
delta.new	0.15	0.135	0.165	-	Seedling recruitment from SSB (fresh seeds)	5
delta.old	0.3	0.270	0.33	-	Seedling recruitment from SSB (old seeds)	5
Mh	0.98	0.882	1.0	-	Mortality of emerged weed due to herbicide (herbicide efficacy)	6
Mc	0	0.0	0.1	-	Seedling mortality due to cold	6
Smax.1	445	400.5	489.5	-	Seeds per plant without density stress (autumn sown crops)	7
Smax.0	296	266.4	325.6	-	Seeds per plant without density stress (spring sown crops)	7
Ymax	8	7.2	8.8	ton. ha^{-1}	Potential production of wheat without weed	8
Rmax	0.002	0.0018	0.0022	-	Crop yield losses due to weed	8
Gamma	0.005	0.0045	0.0055	-	Reduction of production of wheat due to weed	8

TABLE A1.19 Examples of Management Variables

	With Herbicide Every Year				Without Herbicide in Year 3		
Year	Soil	Crop	Herb	Year	Soil	Crop	Herb
1	1	1	1	1	1	1	1
2	0	1	1	2	0	1	1
3	1	1	1	3	1	1	0
4	0	1	1	4	0	1	1
5	1	1	1	5	1	1	1
6	0	1	1	6	0	1	1

TABLE A1.20 Output of the Weed.Model Function, Including All State Variables, for Management with Herbicide Every Year

Year	D	s	SSBa	DSBa	Yield
0	400	68000	3350	280	7.88
1	148.81	1313.55	955.04	17821.76	7.95
2	116.53	1030.44	530.03	2820.84	7.96
3	43.69	387.98	152.62	643.14	7.98
4	18.29	162.67	102.89	122.19	7.99
5	2.29	20.39	8.74	68.27	7.99
6	1.07	9.54	5.78	11.68	7.99

11 EPIRICE MODEL

11.1 Model Description

EPIRICE is a model used to simulate potential epidemics in rice (Savary, Nelson, Willocquet, Pangga, and Aunario, 2012). It is a Susceptible-Exposed-Infectious-Removed (SEIR) model, combined with a simple model of plant growth that describes establishment, growth, and senescence. Five diseases with different disease sites (from leaf fraction to entire plant) but common, universal, epidemiological attributes are considered. The five state variables (with same unit, Nsites: number of sites) are healthy sites (H),

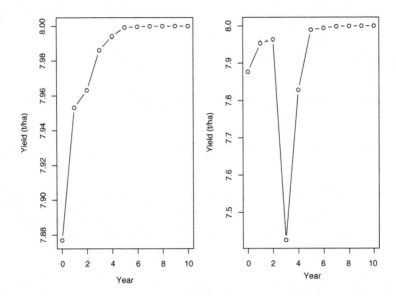

FIGURE A1.8 Simulated annual wheat yields with herbicide application every year (left) or without herbicide in year 3 (right), using the Weed model.

senesced sites (S), latent sites (L), infectious sites (II), and post-infectious (i.e., removed) sites (P). The model has a time step of one day (dt = 1). The system modeled is 1 m^2 of a rice crop stand.

The equations are listed in Table A1.21 .

The version here has a few simplifications compared to the model published by Savary et al. (2012). This concerns the original cohort modeling for RTransfer and RRemoval (Equations 9 and 10) and the choice of a single date of crop establishment of rice crops, which was fixed for the whole region of interest.

To illustrate, we run the model for just one rice disease, namely leaf blast, which causes lesions on the leaf blades, which produce propagules that spread the disease. The values of the 12 parameters are based on values found in the literature (Table A1.22).

The input variables are:

— Minimum daily temperature (TMIN(day), °C), maximum daily temperature (TMAX(day), °C), daily rainfall (RAIN(day), mm), and daily relative humidity (RH2M(day), %)
— Day of the year for crop establishment (sdate) and for the end of simulation (ldate).

The ZeBook R package has weather data for South Asia that can be used with this model (weather_SouthAsia).

TABLE A1.21 Equations of the EPIRICE Model for Plant Disease

(1) $H(day + 1) = H(day) + RGrow - RInf - RSenescedn$
(2) $S(day + 1) = S(day) + RSenesced$
(3) $L(day + 1) = L(day) + RInf - RTransfer$
(4) $II(day + 1) = II(day) + RTransfer - RRemoval$
(5) $P(day + 1) = P(day) + RRemoval$

The rate equations are as follows (dace = day-sdate is days after crop establishment).

(6) $RG = RRG \cdot H(day) \cdot \left(1 - \frac{TS(day)}{Sx}\right)$

(7) $Rc = RcOpt \cdot fRcA(dace) \times fRcT\left(\frac{TMIN(day) + TMAX(day)}{2}\right) \times$

$fRcW(RH2M(day), RAIN(day))$

with $fRcW(RH2M(day), RAIN(day)) = 1$ if $RH2M(day) \geq 90$ or $RAIN(day) \geq 5$
$= 0$ if not

(8) $RInf = Rc \cdot II(day) \cdot \left(\frac{H(day)}{TS(day)}\right)^a + Starter$ with Starter $= 1$ if dace $=$ EODate
$= 0$ if dace $\neq$ EODate

(9) $RTransfer = \frac{L(day)}{p}$

(10) $RRemoval = \frac{I(day)}{i}$

(11) $RSenesced = RRS \cdot H(day) + RRemoval \cdot SenescType$

Additional auxiliary variables of interest are site with disease (TOTDIS), total number of sites (TS) and severity (sev):

(12) $TOTDIS(day) = L(day) + II(day) + P(day)$

(13) $TS(day) = H(day) + TOTDIS(day)$

(14) $sev(day) = \frac{TOTDIS(day) - P(day)}{TS(day) - P(day)}$

11.2 R Code

The function for simulating with this model is `epirice.model` in the ZeBook R package.

11.3 Example of Model Output

Table A1.23 shows partial output for a dummy weather data set with a constant daily weather (that is optimal for infection): TMIN = 10, TMAX = 30,

TABLE A1.22 Parameters of EPIRICE Model for Leaf Blast

Name	Value	Description	Unit
A	1	Disease spatial aggregation coefficient	-
I	20	Duration of infectious period	day
P	5	Duration of latent period	day
RcOpt	1.14	Potential basic infection rate corrected for removals	NSites NSites^{-1} day^{-1}
fRcA	[0,1]	Modifier for Rc for crop age; See Figure A1.9	-
fRcT	[0,1]	Modifier for Rc for mean temperature; See Figure A1.9	-
fRcW	0 or 1	Modifier for Rc for wetness (rain and relative humidity) fRcW (RH2M,RAIN) = 1 if RH2M ≥ 90 or RAIN ≥ 5; fRcW = 0 otherwise	-
RRG	0.1	Relative rate of growth	NSites NSites^{-1} day^{-1}
RRS	0.01	Relative rate of senescence	NSites NSites^{-1} day^{-1}
SenescType	1	Rate of senescence induced by disease	NSites day^{-1}
Sx	30000	Maximum number of sites (site size = 45 mm leaf)	NSites
EODate	15	Epidemic onset date	day

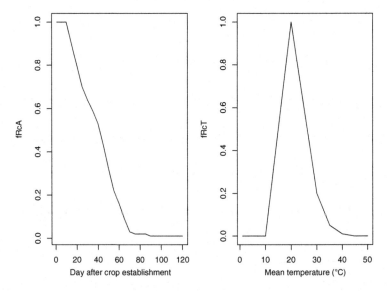

FIGURE A1.9 Functions for the Rc modifiers from Savary et al. (2012): fRcA (age effect; left) and fRcT (temperature effect, right).

TABLE A1.23 Partial Listing of the Output of EPIRICE Model

Day	DACE	H	L	II	P	TS	TOTDIS	Severity
180	15	2073	0	0	0	2073	0	0
181	16	2245	1	0	0	2246	1	0.004
...	...	...	...	...	...	...	...	...
206	41	11758	149	106	21	12035	277	2.125
207	42	12279	180	131	27	12616	337	2.465

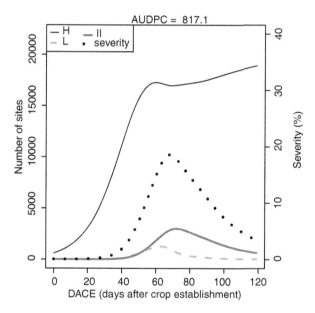

FIGURE A1.10 Dynamics of the state variables: number of healthy (H), latent (L), and infectious (II) sites and disease severity (fraction of leaf sites infected) using the EPIRICE model.

RH2M = 95, and RAIN = 0. The output starts on day sdate − 165 or day after crop establishment (DACE) = 0.

The dynamics of the number of healthy (H), latent (L), and infectious (II) sites and disease severity for this constant weather are shown in Figure A1.10.

A spatial representation of mean disease importance (area under disease progress curve, AUDPC) for South Asia (weather_SouthAsia data set) over the period from 1999 to 2008 is shown in Figure A1.11. To simplify, we use

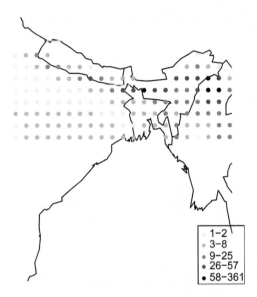

FIGURE A1.11   Map of simulated severity of rice leaf blast epidemics across north-east India, Nepal, Bangladesh, and Myanmar, using EPIRICE. Values represent AUDPC averaged over the period 1999–2008.

a single crop establishment date (sdate = 165, which corresponds to mid-June). In the original article by Savary et al. (2012), optimized crop establishment dates were computed from another crop model for each location from a crop growth model. One can see a strong variability in disease intensity (AUDPC values), corresponding to climatic variability among the different locations in this region.

REFERENCES

Jones, J.W., Graham, W.D., Wallach, D., Bostick, W.M., Koo, J., 2004. Estimating soil carbon levels using an ensemble Kalman filter. Transactions of the ASAE 47, 331–339.

Munier-Jolain, N.M., Chauvel, B., Gasquez, J., 2002. Long-term modelling of weed control strategies: analysis of threshold-based options for weed species with contrasted competitive abilities. Weed Research 42, 107–122.

Woli, Prem, 2010. Quantifying water deficit and its effect on crop yields using a generic drought index. Dissertation presented to the graduate school of the University of Florida in partial fulfillment of the requirements for the degree of doctor of philosophy. University of Florida.

Magarey, R.D., Sutton, T.B., Thayer, C.L., 2005. A simple generic infection model for foliar fungal plant pathogens. Phytopathology 95, 92–100.

EFSA, 2008. Scientific opinion of the Panel on Plant Health on a request from the European Commission on *Guignardia citricarpa* Kiely. The EFSA Journal Vol: 925, 1–108.

Magarey, R.D., Borchert, D.M., Fowler, G.L, Sutton, T.G., Colunga-Garcia, M., Simpson, J.A., 2007. NAPPFAST, an Internet System for the Weather-based Mapping of Plant Pathogens. Plant Disease 91, 336–445.

Darroch, B.A., Baker, R.J., 1990. Grain filling in three spring wheat genotypes: statistical analysis. Crop science 30, 525–529.

Savary, S., Nelson, A., Willocquet, L., Pangga, I., Aunario, J., 2012. Modeling and mapping potential epidemics of rice diseases globally. Crop Protection 34, 6–17.

Verhulst, P.F., 1839. Notice sur la loi que la population poursuit dans son accroissement. Correspondance mathématique et physique 1838 (no. 10), 113–121.

Wallach, D., 1980. An empirical mathematical model of a cotton crop subjected to damage. Field Crops Research 3, 7–25.

An Overview of the R Package ZeBook

1. INTRODUCTION

The ZeBook package is an R library containing material related to this book. It includes the code for many of the examples that are provided in this book (so that you can run them yourself), including the main models that are used as examples, and data sets for the exercises.

2. INSTALLATION

You can install the ZeBook package from the CRAN repository (http://cran .r-project.org/web/packages/ZeBook/), if you have an internet connection, by following the instructions:

Step 1: click on 'Packages' in the R menu bar and select 'Install package (s)....'.
Step 2: choose a location near you in the list of mirror sites for downloading which will appear.
Step 3: choose 'ZeBook' in the list of available packages.
Step 4: verify that there is a message in the console window indicating that installation was successful.

Another way to install the ZeBook package is to run the following instruction in the R console.

```
# change the repository argument and choose the nearest mirror
install.packages("ZeBook", repos = "http://cran.rstudio.com")
```

For more information on how to install a package, see Chapter 3.

3. FUNCTIONS AND DEMOS IN THE ZEBOOK PACKAGE

The package will continue to evolve over time, with the addition of new models and data sets useful for testing the methods presented in this book. The functions and data in version 0.5, June 2013, are shown in Table A2.1 (functions and data) and the demos in Table A2.2. The demos allow you

TABLE A2.1 Functions and Data Sets for Applying the Methods Embedded in the ZeBook Package (version 0.5, June 2013)

Functions for Models

magarey.define.param	Define values of the parameters for the Magarey model
magarey.model	The Magarey model
magarey.model2	The Magarey model, taking a vector of parameters as argument
magarey.simule	Wrapper function to run the Magarey model multiple times (for multiple sets of inputs)
watbal.model	WaterBalance model–calculate soil water over designated time period
watbal.update	WaterBalance model–calculate change in soil water for one day
watbal.model.arid	WaterBalance model–Variant with another order of calculation and ARID index
watbal.define.param	Define values of the parameters for the WaterBalance model
watbal.weather	Read weather data for the WaterBalance model (West of France Weather)
watbal.simobsdata	Soil water content measurements and associated simulations with WaterBalance model
maize.define.param	Define values of the parameters for the Maize model
maize.model	The basic Maize model
maize.model2	The basic Maize model for use with maize.simule
maize.RUEtemp	Calculate effect of temperature on RUE for Maize
maize.simule	Wrapper function to run Maize model for multiple sets of parameter values
maize.simule240	Wrapper function to run Maize model multiple times for multiple sets of parameter values and give Biomass at day 240
maize.multisy240	Wrapper function to run Maize model for multiple sets of input variables (site-year) and give Biomass at day 240
maize.simule_multisy240	Wrapper function to run Maize model for multiple sets of parameter values (virtual design) and multiple sets of input variables (site-year) and give Biomass at day 240
maize.weather	Read weather data for the Maize model
maize_cir.model	The Maize model with additional state variable CumInt
maize_cir_rue.model	The Maize model with temperature dependent RUE and CumInt

(Continued)

TABLE A2.1 (Continued)

maize_cir_rue_ear. model	The Maize model with temperature dependent RUE, CumInt, and ear growth
verhulst.model	The Verhulst (logistic) model—calculate daily values over designated time period
verhulst.update	The Verhulst (logistic) model—calculate change for one day
population.age.model	The PopulationAge model (Population Dynamics with Age Classes)
population.age. matrix.model	The PopulationAge model (Population Dynamics with Age Classes)—matrix form
population.age. model.ode	The PopulationAge model (Population Dynamics with Age Classes)—ode form
predator.prey.model	The PredatorPrey model (Predator-Prey Lotka-Volterra with logistic equation for prey)
exponential.model	The Exponential growth model of dynamic of population
exponential.model. bis	The Exponential growth model of dynamic of population—another form
exponential.model.ie	The Exponential growth model of dynamic of population—with improved Euler integration
carbonsoil.model	The CarbonSoil model—calculate daily values over designated time period
carbonsoil.update	The CarbonSoil model—calculate change in soil carbon for one year
weed.model	The Weed model—calculate daily values over designated time period
weed.update	The Weed model—calculate change for one year
weed.define.param	Define parameter values of the Weed model
weed.simule	Wrapper function to run the Weed model multiple times (for multiple sets of inputs)
epirice.model	The EPIRICE model (Disease model for rice)
epirice.define.param	Define values of the parameters for the EPIRICE model
epirice.multi.simule	Wrapper function to run EPIRICE multiple times (for multiple sets of inputs)
epirice.weather	Read weather data for EPIRICE (southern Asia weather)
cotton.model	The Cotton model (dynamic for numbers of Cotton fruiting points). See Chapter 6, Exercises section, for a full description
lactation.calf.model	The Lactation model
lactation.calf.model2	The Lactation model for use with lactation.calf.simule

(Continued)

TABLE A2.1 (Continued)

lactation.calf.simule	Wrapper function to run the Lactation model for multiple sets of parameter values
lactation.define.param	Define values of the parameters for the Lactation model
lactation.machine.model	The Lactation model with milking machine
lactation.machine.model2	The Lactation model for use with lactation.machine.simule
carcass.model	The Carcass (growth of beef cattle) model
carcass.EMI.model	The Carcass (growth of beef cattle) model with energy as input
carcass.EMI.model2	The Carcass model function for use with carcass.EMI.simule
carcass.EMI.simule	Wrapper function to the Carcass model for multiple sets of parameter values
carcass.EMI.multi	Wrapper function to run Carcass model on several animals with different conditions
carcass.define.param	Define values of the parameters for the Carcass model
carcass_data	Data of growth of beef cattle for Carcass model
seedweight.data	Wheat grain weight measurements after anthesis
Wheat_GPC	Grain Protein Contents in Wheat Grains
Sunflower Phomopsis	Phomopsis stem canker observations for Sunflower
WheatYieldGreece	National Wheat Yield evolution for Greece from FAO
Weather Databases	
weather_FranceWest	Weather series for western France from NASA POWER agroclimatology
weather_SouthAsia	Weather series for southern Asia from NASA POWER agroclimatology
weather_EuropeEU	Weather series for Europe EU from NASA POWER agroclimatology
Functions for Applying Methods	
param.runif	Generate a random plan as a data frame. Columns are parameters. Values have uniform distribution
param.rtriangle	Generate a random plan as a data frame. Columns are parameters. Values have triangle distribution
q.arg.fast.runif	Build the q.arg argument for the FAST function (sensitivity analysis)
AICf	Calculate AIC, Akaike's Information Criterion
goodness.of.fit	Calculate multiple goodness-of-fit criteria

TABLE A2.2 Demonstration R Scripts Embedded in the ZeBook Package (Version 0.5, June 2013)

Name of Demo	Description
ch03R_maize_parallel	ch03. R introduction. Improving performance with parallel code for maize.model
ch04Simulation.exponential	ch04. Simulation. Exponential model. Comparison of numerical integration with analytical solution. Influence of time step on error of integration. Improved Euler integration method
ch04Simulation.prey.predator	ch04. Simulation. Predator-Prey Lotka-Volterra model with deSolve integration library
ch04Simulation.population.age	ch04. Simulation. Population dynamics model with a representation of population with age classes
ch04Simulation.watbal	ch04. Simulation. WaterBalance model
ch04Simulation.maize	ch04. Simulation. Maize model
ch04Simulation.epirice	ch04. Simulation. EPIRICE model (may take up to 10 minutes)
ch05USA_Magarey	ch05. Uncertainty and Sensitivity analysis. Example on Magarey model
ch05USA_Weed	ch05. Uncertainty and Sensitivity analysis. Example of Monte Carlo, Morris, Fast, Sobol, and ANOVA for weed.model. Attention: run time can be long (up to 30 minutes)
ch05USA_Magarey	ch05. Uncertainty and Sensitivity analysis. Example with the magarey.model
ch05USA_Lactation	ch05. Sensitivity analysis. Example on Lactation model
ch05USA_Carcass	ch05. Sensitivity analysis. Example on Carcass model. Morris
ch07Bayes_Carbon	ch07. Bayes. Example with carbonsoil.model
ch08DLM_WheatYieldGreece	ch08. Data Assimilation. Example of Dynamic Linear Model (DLM) for analyzing yield time series
ch08DLM_CarbonSoil	ch08. Data Assimilation. Example of Dynamic Linear Model (DLM) with carbonsoil.model
ch08DLM_watbal	ch08. Data Assimilation. Example of Dynamic Linear Model (DLM) with watbal.model

(Continued)

TABLE A2.2 (Continued)

Name of Demo	Description
ch10maize1data	ch10. Case study, Exploration of data
ch10maize2evaluation	ch10. Case study, Maize model. Evaluation
ch10maize3uncertainty	ch10. Case study, Maize model. Uncertainty analysis (may take up to 20 minutes)
ch10maize4sensitivity	ch10. Case study, Maize model. Sensitivity analysis (may take up to 40 minutes)
ch10maize5aparameterOLS_gradient	ch10. Case study, Maize model. Parameter estimation with OLS and gradient method (may take up to 10 minutes)
ch10maize5bparameterOLS_simplex	ch10. Case study, Maize model. Parameter estimation with OLS simplex method (may take up to 120 minutes)
ch10maize5cparameterConcL	ch10. Case study, Maize model. Parameter estimation with concentrated likelihood—simplex method (may take up to 120 minutes)
ch10maize5dparameterConcL_log	ch10. Case study, Maize model. Parameter estimation with concentrated likelihood and a log transformation of variable—simplex method (may take up to 120 minutes)
ch10maize6aMSEP_OLS	ch10. Case study, Maize model. Estimation of MSEP by cross-validation after OLS parameter estimation (may take up to 120 minutes)
ch10maize6bMSEP_ConcL	ch10. Case study, Maize model. Estimation of MSEP by cross-validation after concentrated likelihood parameter estimation (may take up to 120 minutes)
ch10maize7scenario	ch10. Case study, Maize model. Use of final model for multi-site multi-year simulations (may take a few minutes)
ch10maizeB5aBayesEstimRUE	ch10. Case study, Maize model. Parameter estimation using a Bayesian approach, using MCMC—only RUE is estimated (may take up to a few hours)
ch10maizeB5bBayesEstimAllparam	ch10. Case study, Maize model. Parameter estimation using a Bayesian approach, using MCMC—all parameters (may take up to 4—5 days for 30,000 iterations and 2 chains)

(Continued)

TABLE A2.2 (Continued)

Name of Demo	Description
ch10maizeB7BayesUncertainty	ch10. Case study, Maize model. Use the posterior distribution for multi-site multi-year simulations. You must run ch10maizeB5bBayesEstimAllparam before this demo in the same session or adapt both scripts (may take up to a few hours)
ch10maizeB7Bayes_functionsMH	ch10. Case study, Maize model. Load specific functions for Bayesian approach. No calculation
ch10maizeS8DataAssimilation	ch10. Case study, Maize model. Data assimilation using DLM (may take a few minutes)

to run programs that were used to produce some of the examples in this book.

See index of help(ZeBook) for a complete and up-to-date listing. You can access a PDF version of the documentation on the CRAN repository (http://cran.r-project.org/web/packages/ZeBook/ZeBook.pdf).

4. HOW TO USE THE ZEBOOK PACKAGE

Once the package is installed, the current table of contents can be obtained.

```
# load package
library(ZeBook)
# show content of package
help(package = "ZeBook")
# show description of package
help(ZeBook)
# then click on Index
```

To use a specific function, check the documentation and run it.

```
# show documentation of a function described in the package
help(maize.model)
```

```
weather = maize.weather(working.year = 2010, working.site = 30,
weather_all = weather_EuropeEU)
maize.model(Tbase = 7, RUE = 1.85, K = 0.7, alpha = 0.00243, LAImax = 7,
TTM = 1200, TTL = 700, weather, sdate = 100, ldate = 250)
```

You can also play a full example embedded as a demo in the package.

```
# show demo in the ZeBook package
demo(package = "ZeBook")
# play a specific demo
demo(ch08DLM_CarbonSoil)
# find the where is a source of R code of a demo (to copy and edit it for example)
system.file("demo", "ch08DLM_CarbonSoil.r", package = "ZeBook")
```

You can also access the code of a demo from the index of the help page of the ZeBook package (help(ZeBook)) and then run it.

5. LIST OF PACKAGES NEEDED

Table A2.3 shows the additional R packages that are needed to run some of the examples or demonstration scripts included in the ZeBook package. You will need to install them and load them.

TABLE A2.3 Additional R packages Required to run Examples or Demonstration Scripts Included in the ZeBook Package

Package	Description	Used in Chapters
coda	Output analysis and diagnostics for MCMC (Markov Chain Monte Carlo simulations)	10
deSolve	General Solvers for Initial Value Problems of Ordinary Differential Equations (ODE), Partial Differential Equations (PDE), Differential Algebraic Equations (DAE), and Delay Differential Equations (DDE)	4
dlm	Bayesian and Likelihood Analysis of Dynamic Linear Models. Maximum likelihood, Kalman filtering and smoothing, and Bayesian analysis of normal linear State Space models, also known as Dynamic Linear Models	8
mnormt	The multivariate normal and t distributions	10
mvtnorm	Multivariate normal and t distributions	5
parallel	Support for parallel computation, including random-number generation	3

(Continued)

TABLE A2.3 (Continued)

Package	Description	Used in Chapters
pROC	Display and analyze ROC (receiver operating characteristic) curves	9
sensitivity	Sensitivity analysis	5
triangle	Provides the standard distribution functions for the triangle distribution	5,7
tseries	Time series analysis and computational finance	8
verification	Forecast verification utilities. This package contains utilities for verification of discrete, continuous, probabilistic forecasts, and forecasts expressed as parametric distributions	9

Index

Note: Page numbers followed by '*f*', '*t*' and '*b*' refer to figures, tables and boxes, respectively.

Printed in the United States
By Bookmasters